Water Cycle: Processes and Interactions

Water Cycle:
Processes and Interactions

Edited by Reggie Bing

SYRAWOOD
PUBLISHING HOUSE

New York

Published by Syrawood Publishing House,
750 Third Avenue, 9th Floor,
New York, NY 10017, USA
www.syrawoodpublishinghouse.com

Water Cycle: Processes and Interactions
Edited by Reggie Bing

International Standard Book Number: 978-1-64740-141-2 (Hardback)

Cataloging-in-Publication Data

Water cycle : processes and interactions / edited by Reggie Bing.
    p. cm.
Includes bibliographical references and index.
ISBN 978-1-64740-141-2
1. Hydrologic cycle. 2. Hydrology. 3. Water. 4. Cycles. I. Bing, Reggie.
GB848 .W38 2022
551.48--dc23

TABLE OF CONTENTS

Preface..VII

Chapter 1　New interpretation of the role of water balance in an extended Budyko hypothesis
　　　　　　in arid regions 1
　　　　　　C. Du, F. Sun, J. Yu, X. Liu and Y. Chen

Chapter 2　Future extreme precipitation intensities based on a historic event............................17
　　　　　　Iris Manola, Bart van den Hurk, Hans De Moel and Jeroen C. J. H. Aerts

Chapter 3　Daily Landsat-scale evapotranspiration estimation over a forested landscape in
　　　　　　North Carolina, USA, using multi-satellite data fusion 29
　　　　　　Yun Yang, Martha C. Anderson, Feng Gao, Christopher R. Hain,
　　　　　　Kathryn A. Semmens, William P. Kustas, Asko Noormets, Randolph H. Wynne,
　　　　　　Valerie A. Thomas and Ge Sun

Chapter 4　Evaporation from cultivated and semi-wild Sudanian Savanna in west Africa............50
　　　　　　Natalie C. Ceperley, Theophile Mande, Nick van de Giesen, Scott Tyler,
　　　　　　Hamma Yacouba and Marc B. Parlange

Chapter 5　Stochastic generation of multi-site daily precipitation focusing on extreme events............69
　　　　　　Guillaume Evin, Anne-Catherine Favre and Benoit Hingray

Chapter 6　Remapping annual precipitation in mountainous areas based on vegetation
　　　　　　patterns: a case study in the Nu River basin...87
　　　　　　Xing Zhou, Guang-Heng Ni, Chen Shen and Ting Sun

Chapter 7　Rainfall-runoff modelling using river-stage time series in the absence of reliable
　　　　　　discharge information: a case study in the semi-arid Mara River basin 103
　　　　　　Petra Hulsman, Thom A. Bogaard and Hubert H. G. Savenije

Chapter 8　Tree-, stand- and site-specific controls on landscape-scale patterns of transpiration............118
　　　　　　Sibylle Kathrin Hassler, Markus Weiler and Theresa Blume

Chapter 9　Precipitation pattern in the Western Himalayas revealed by four datasets 135
　　　　　　Hong Li, Jan Erik Haugen and Chong-Yu Xu

Chapter 10　Spatial and temporal variability of rainfall and their effects on hydrological
　　　　　　response in urban areas...149
　　　　　　Elena Cristiano, Marie-claire ten Veldhuis and Nick van de Giesen

Chapter 11 **The WACMOS-ET project – Evaluation of global terrestrial evaporation data sets** 169

D. G. Miralles, C. Jiménez, M. Jung, D. Michel, A. Ershadi, M. F. McCabe, M. Hirschi, B. Martens, A. J. Dolman, J. B. Fisher, Q. Mu, S. I. Seneviratne, E. F. Wood and D. Fernández-Prieto

Permissions

List of Contributors

Index

PREFACE

The water cycle is the continuous movement of water above, below and on the Earth's surface. The mass of water on Earth remains constant but it differs in its distribution among the reservoirs of ice, saline water, freshwater and atmospheric water. The variation in the amount of water present in various reservoirs depends on the climatic variables. Several physical processes such as evaporation, condensation, infiltration and precipitation move water from one reservoir to another. The water takes different forms such as solid, liquid and vapor while going through such processes. The water cycle is driven by the sun that heats water in oceans and seas and makes it evaporate as water vapor into the air. Sometimes ice and snow sublimate into water vapor. These water vapor move across the globe and form clouds, which collide and fall out on the upper atmospheric layers as precipitation. Most water comes back into the oceans, seas and on land as rain. This book outlines the processes and applications of water cycle in detail. It includes some of the vital pieces of work being conducted across the world, on various topics related to the water cycle. This book is a collective contribution of renowned group of international experts.

The information contained in this book is the result of intensive hard work done by researchers in this field. All due efforts have been made to make this book serve as a complete guiding source for students and researchers. The topics in this book have been comprehensively explained to help readers understand the growing trends in the field.

I would like to thank the entire group of writers who made sincere efforts in this book and my family who supported me in my efforts of working on this book. I take this opportunity to thank all those who have been a guiding force throughout my life.

Editor

New interpretation of the role of water balance in an extended Budyko hypothesis in arid regions

C. Du[1,2], F. Sun[1], J. Yu[1], X. Liu[1], and Y. Chen[3]

[1]Key Laboratory of Water Cycle and Related Land Surface Processes, Institute of Geographic Sciences and Natural Resources Research, Chinese Academy of Sciences, Beijing 100101, China

[2]University of Chinese Academy of Sciences, Beijing, 100049, China

[3]State Key Laboratory of Desert and Oasis Ecology, Xinjiang, Institute of Ecology and Geography, Chinese Academy of Sciences, Urumqi, 830011, China

Correspondence to: F. Sun (sunfb@igsnrr.ac.cn), J. Yu (yujj@igsnrr.ac.cn)

Abstract. The Budyko hypothesis (BH) is an effective approach to investigating long-term water balance at large basin scale under steady state. The assumption of steady state prevents applications of the BH to basins, which is unclosed, or with significant variations in root zone water storage, i.e., under unsteady state, such as in extremely arid regions. In this study, we choose the Heihe River basin (HRB) in China, an extremely arid inland basin, as the study area. We firstly use a calibrated and then validated monthly water balance model, i.e., the *abcd* model, to quantitatively determine annual and monthly variations of water balance for the sub-basins and the whole catchment of the HRB, and find that the roles of root zone water storage change and that of inflow from upper sub-basins in monthly water balance are significant. With the recognition of the inflow water from other regions and the root zone water storage change as additional possible water sources to evapotranspiration in unclosed basins, we further define the equivalent precipitation (P_e) to include local precipitation, inflow water and root zone water storage change as the water supply in the Budyko framework. With the newly defined water supply, the Budyko curve can successfully describe the relationship between the evapotranspiration ratio and the aridity index at both annual and monthly timescales, whilst it fails when only the local precipitation being considered. Adding to that, we develop a new Fu-type Budyko equation with two non-dimensional parameters (ω and λ) based on the deviation of Fu's equation. Over the annual timescale, the new Fu-type Budyko equation developed here has more or less identical performance to Fu's original equation for the sub-basins and the whole catchment. However, over the monthly timescale, due to large seasonality of root zone water storage and inflow water, the new Fu-type Budyko equation generally performs better than Fu's original equation. The new Fu-type Budyko equation (ω and λ) developed here enables one to apply the BH to interpret regional water balance over extremely dry environments under unsteady state (e.g., unclosed basins or sub-annual timescales).

1 Introduction

The Budyko hypothesis (hereafter BH) was postulated by a Russian climatologist, Mikhail Ivanovich Budyko, to analyze regional differences in long-term annual water and energy balance (Budyko, 1948). The BH's mean annual water balance is described by the evapotranspiration ratio and the climate aridity index. The BH becomes an effective approach to investigating the influence of climate change on mean annual runoff and evapotranspiration (Donohue et al., 2011; Xiong et al., 2014). There are various equations to describe the BH. Some empirical equations without parameters were proposed by Schreiber (1904), Ol'dekop (1911), Budyko (1948) and Pike (1964) (see Table 1). These equations explicitly include climate variations (radiation, precipitation, evapotranspiration and air temperature) and do not deal with recently recognized important catchment properties, such as characteristics of groundwater system, vadose zone properties, vegetation. Hence, attempts have been made to introduce physical parameters in these empirical equations (Mezentsev, 1955; Fu, 1981; Milly, 1993; Zhang et al., 2001; Yang et al., 2007, 2008). These physical parameters are a collection of myriad catchment characteristics (topography, vegetation, soil, and groundwater, etc.) and are therefore difficult to measure (Gerrits et al., 2009). These equations with a single parameter, however, provide the flexibility of using the BH over long-term timescales.

The BH assumes steady-state conditions. Firstly, the studied basin must be natural and closed, which means that the local precipitation is the only water source to the evapotranspiration. Recently, the BH has been widely used to investigate the interannual variability of precipitation partitioning (Gerrits et al., 2009), separation of runoff trends (H.-Y. Li et al., 2014; Xiong et al., 2015), evapotranspiration change (Savenije, 1997) and water storage change (Istanbulluoglu et al., 2012; Gao et al., 2014). These studies show that hydrological processes have been greatly affected by the climate change and intensive change of land cover owing to human activities. These human activities such as urbanization, withdrawing groundwater, hydraulic engineering, deforestation etc. are significantly changing natural hydrological cycle and breaking the original water balance to form a new balance under the new hydroclimatic conditions. For example, the transferring water becomes the new water source of the basin to evapotranspiration due to the implemented interbasin water transfer project (Bonacci and Andric, 2010). In dry regions, croplands expanded with irrigation, which increased water availability for evapotranspiration (Gordon et al., 2005). Land use/cover changes have also caused the change of runoff (J. Li et al., 2014). Nowadays, most of the inhabited basins have been developed or disturbed by large-scale human actives. Therefore, lots of basins were no longer closed or natural, and the relationship between annual evapotranspiration ratio and potential evapotranspiration ratio hardly meets the first condition of the BH, which presents a great challenge in applying the BH in unclosed basins.

Secondly, water storage change can be assumed to be negligible at the basin scale and at long-term timescale. However, over finer temporal scales, it becomes increasingly concerned of the importance of water storage in water balance in the Budyko framework. For example, Wang et al. (2009) found that the inter-annual water storage change should be considered due to the hysteresis response of the base flow to the inter-annual precipitation change in Nebraka Sand Hills. Zhang et al. (2008) considered the impacts of soil water and groundwater storage and developed a monthly water balance model based on the BH with application in 265 catchments in Australia. Yokoo et al. (2008) highlighted the importance of soil water storage change in determining both annual and seasonal water balances. Wang (2012) evaluated changes in inter-annual water storage at 12 watersheds in Illinois using the field observation of long-term groundwater and soil water and found that the impact of inter-annual water storage changes on the water supply in the BH need to be consid-

ered. Chen et al. (2013) defined the difference between rainfall and storage change as effective precipitation to develop a seasonal model for construction long-term evapotranspiration. Therefore, water storage change should be taken into account as the important part of the steady-state assumption of the BH (Zhang et al., 2008).

In summary, it has been more and more recognized that water systems are no longer natural to different extents (Sivapalan et al., 2011). Hence, it presents a great challenge to apply the BH to unsteady-state conditions (unclosed basins or intense water storage changes). The BH has been widely applied to mild arid basins with precipitation of 300–400 mm and aridity index of less than, for example, 5, such as over northern China (Yang et al., 2007), the southwestern regions of MOPEX catchments (Gentine et al., 2012; Carmona et al., 2014) and the west of Australia (Zhang et al., 2008). However, it is rare to apply the BH in extremely arid environments (say, the aridity index over 5), where water systems are typically unclosed with intense human interference and irrigation. For example, rivers in the arid region of northwestern China are typically from upper mountains with little human interference, and flow through middle regions with intensive irrigation and human interferences and finally into extremely dry desert plains. To investigate it in more detail, we choose the Heihe River basin (HRB), the second largest arid inland basin in northwestern China (mean annual aridity index = 10). Being an inland basin, the HRB consists of six sub-basins with different landscapes and climate conditions, where the upper mountainous basins are closed and natural with little human interference (long-term mean annual water storage change approaches zero), the middle basins are arid and intensively irrigated plain with strong human interference (mean annual evapotranspiration is higher than the local precipitation), and the lower basin is extremely dry Gobi desert plain without any runoff flowing out (evapotranspiration is mainly the local precipitation; mean annual evapotranspiration approaches mean annual precipitation). In this study, our aim is threefold. (1) We first test whether the BH is applicable to the unsteady-state condition in extremely arid basins. (2) If not, we in further improve the original BH by including observed water balance. (3) We finally extend the applicability of the BH at unclosed basin scale and annual or monthly timescales.

2 Theory and method

2.1 Annual and monthly water balance analysis

In the original BH, the basin is a natural hydrologic unit, and the only possible water source for evapotranspiration is the local precipitation. The annual or monthly water balance equation can be written as

$$P = \mathrm{ET} + Q_{\mathrm{out}} - Q_{\mathrm{in}} + \Delta S + \Delta G, \qquad (1)$$

Table 1. Different Budyko equations for the mean annual water-energy balance.

Number	Equation	Parameter	Reference
1	$\varepsilon = 1 - \exp(-\varphi)$	None	Schreiber (1904)
2	$\varepsilon = \varphi \tanh(1/\varphi)$	None	Ol'dekop (1911)
3	$\varepsilon = \{\varphi[1 - \exp(-\varphi)] \tanh(1/\varphi)\}^{0.5}$	None	Budyko (1958, 1974)
4	$\varepsilon = (1 + \varphi^{-2})^{-0.5}$	None	Pike (1964)
5	$\varepsilon = (1 + \varphi^{-\alpha})^{-1/\alpha}$	α – calibration factor	Mezentsev (1955); Chouldhury (1999); Yang et al. (2008)
6	$\varepsilon = \dfrac{1 + \omega\varphi}{1 + \omega\varphi + \varphi^{-1}}$	ω – coefficient of vegetation and water supply	Zhang et al. (2001)
7	$\varepsilon = \dfrac{\exp[\gamma(1 - 1/\varphi)] - 1}{\exp[\gamma(1 - 1/\varphi)] - \varphi^{-1}}$	γ – the ratio of the soil water storage capacity to precipitation	Milly (1993); Porporato et al. (2004)
8	$\varepsilon = 1 + \varphi - (1 + \varphi^{\omega})^{1/\omega}$	ω – a constant of integration	Fu (1981); Zhang et al. (2004); Yang et al. (2007)

Note: $\varepsilon = \mathrm{ET}/P$ evapotranspiration ratio (the ratio of mean annual evapotranspiration to mean annual precipitation); $\varphi = \mathrm{ET}_0/P$, aridity index (the ratio of mean annual potential evapotranspiration to mean annual precipitation).

where P is the annual or monthly precipitation (mm); ET is the sum of soil evaporation and vegetation transpiration (mm); Q_{out} is the outflow away from a basin (mm); Q_{in} is the channel inflow that is from the upper basin and/or inter-basin water transfer (mm); ΔS is the root zone water (namely, soil water) storage change, (mm); and ΔG is the groundwater storage change (mm).

Because of human interferences (land cover change, dams, irrigation and other withdrawals) to the hydrologic system worldwide, the water supply to evapotranspiration in a basin has changed. Local groundwater and root zone water and external water transfer also become new possible water sources. However, that new non-ignorable part of available water for evapotranspiration has yet been explicitly considered in the Budyko framework in an unclosed basin. More specifically, the inflow or/and inter-basin water transfer may affect the available water for evapotranspiration largely. By considering that, here we rearrange Eq. (1) as $P + Q_{\mathrm{in}} - \Delta S = \mathrm{ET} + Q_{\mathrm{out}} + \Delta G$ the available water for evapotranspiration in Eq. (1) as

$$P_{\mathrm{e}} = P + Q_{\mathrm{in}} - \Delta S, \tag{2}$$

where the total water supply to evapotranspiration in an unclosed basin is denoted as P_{e} and, for simplicity, P_{e} hereafter is defined as the equivalent precipitation of the BH at finer timescales. If ΔS is more than zero, it means the surplus water is stored in the vadose zone, which should be deducted from the water sources. If ΔS is less than zero, it means root zone water contributes to the evapotranspiration consumption. Note that the change of groundwater storage (ΔG) is the result of the exchange between groundwater and baseflow and is not directly interacted with evapotranspiration, so that ΔG is not included into the defined P_{e} in Eq. (2). It will be discussed in the results section.

2.2 Budyko hypothesis model at annual and monthly scales

As discussed above, in the original Budyko framework, the water supply to land evapotranspiration is mean annual precipitation, and the energy supply to land evapotranspiration is estimated by mean annual potential evapotranspiration. The general Budyko equation can be written as

$$\frac{\mathrm{ET}}{P} = F\left(\frac{\mathrm{ET}_0}{P}\right), \tag{3}$$

where $\frac{\mathrm{ET}}{P}$ is the evapotranspiration ratio; $\frac{\mathrm{ET}_0}{P}$ is the aridity index. F is the function to be determined. The general analytical solution to Eq. (3) over mean annual timescales is derived by Fu (1981) and is written as follows:

$$\mathrm{ET} = \mathrm{ET}_0 + P - (\mathrm{ET}_0^{\omega} + P^{\omega} + C)^{1/\omega}, \tag{4}$$

where ω is the parameter, which reflects the integrated effects of soil, vegetation and topography on separating the ET from the local precipitation (Sun, 2007). If the local precipitation is zero, evapotranspiration approaches zero due to no available water; C is zero constant. Note that another form of the BH is also given by Mezentsev (1955) (later, Choudhury, 1999, and Yang et al., 2008), which is, in fact, identical to Fu's equation (Zhou et al., 2015) with the parameters linearly related ($R^2 = 0.9997$) (Sun, 2007).

Water balance analysis in Sect. 2.1 concludes that the water supply in the BH under the unsteady-state condition is the equivalent precipitation instead of the local precipitation. So the annual (or monthly) evapotranspiration ratio is redefined as the ratio of annual (or monthly) evapotranspiration and equivalent precipitation, and the annual (or monthly) aridity index is redefined as the ratio of annual (or monthly) potential evapotranspiration and equivalent precipitation. They are

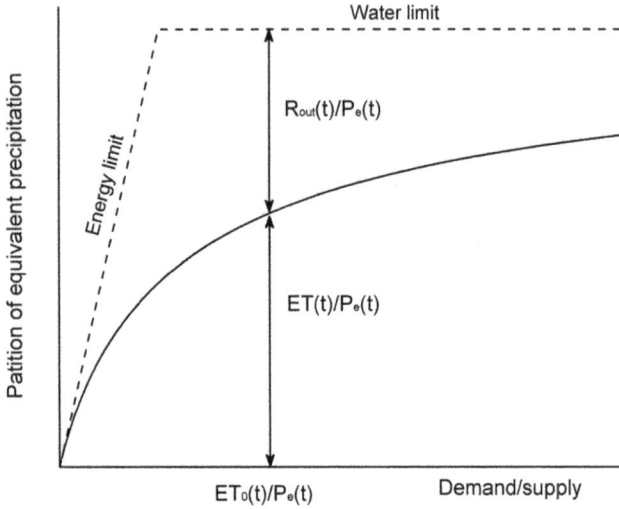

Figure 1. A schematic diagram of the BH under the unsteady-state condition.

described as follows:

$$\frac{ET}{P_e} = \frac{ET}{P + Q_{in} - \Delta S}, \tag{5}$$

$$\frac{ET_0}{P_e} = \frac{ET_0}{P + Q_{in} - \Delta S}. \tag{6}$$

If the equivalent precipitation can be evaporated by enough available energy ($ET_0 / P_e \to \infty$), then annual (or monthly) evapotranspiration may approach annual (or monthly) precipitation ($ET / P_e \to 1$). Such a condition is moisture-constrained. While, if the available energy to evaporate the annual (or monthly) precipitation is limited ($ET_0 / P_e \to 0$), the annual (or monthly) evapotranspiration may approach annual (or monthly) potential evapotranspiration ($ET / ET_0 \to 1$). Such condition is energy-constrained. Figure 1 describes partitioning of the equivalent precipitation into evapotranspiration, streamflow and groundwater storage change, which follows the BH. The Budyko equation under unsteady-state assumption can be written as

$$\frac{ET}{P_e} = F\left(\frac{ET_0}{P_e}\right). \tag{7}$$

Under the unsteady-state conditions for a region, when the local precipitation in the origin Fu's equation is zero, evapotranspiration may not be zero due to other water sources (e.g., inter-basin water transfer), so following the derivation of Fu (1981). Equation (4) can be rewritten as

$$\frac{ET}{P_e} = 1 + \frac{ET_0}{P_e} - \left[1 + \left(\frac{ET_0}{P_e}\right)^\omega + \lambda\right]^{1/\omega}, \tag{8}$$

where ω and λ are two fitting parameters and both non-dimensional. ω has been widely discussed and is greater than 1 (Fu, 1981; Yang et al., 2007). By meeting the constraints

formed by the BH, we can derive that $\lambda \geq -1$ (see the Appendix A). When $\lambda = 0$ (Fig. 2a), Eq. (8) is the same as Fu's equation in its original form (Fu, 1981; Zhang et al., 2004; Yang et al., 2007). For λ becomes positive, e.g., 1, the lower end of the Budyko curve adjusts to the right (Fig. 2b, c). And $\lambda = -1$ sets up the upper theoretical constraint of the Budyko curve (Fig. 2c, d). We speculated that λ may be related to rainfall intensity or hydraulic conductivity of soil.

2.3 A monthly water balance model: *abcd* model

Regional evapotranspiration and soil water cannot be measured directly and they are usually provided by monthly water balance models. Monthly water balance models were first developed in the 1940s. From that, many models have been developed in hydrological studies, such as the T model, $T\alpha$ model, P model, *abc* model and *abcd* model, that are often popular due to the relatively simple structure and fewer parameters (Fernandez et al., 2000).

Among these monthly models, the *abcd* model was proposed by Thomas (1981) has been widely applied to assess regional water resources due to its explicit model structure and only four parameters, of which two parameters pertain to runoff characteristics and the other two relate to groundwater sound physical meanings. Actually, the *abcd* model was originally developed and applied for monthly water balance instead of annual (Alley, 1984). Moreover, Savenije (1997) has verified that the *abcd* model to derive expressions for the evapotranspiration ratio has better agreement with observations than Budyko-type curves. Inputs to the *abcd* model are monthly precipitation and potential evapotranspiration. Outputs include monthly runoff (direct and indirect), soil water storage, groundwater storage and actual evapotranspiration. Therefore, this study employs the *abcd* model to provide monthly actual evapotranspiration and soil water storage.

The partitioning of monthly precipitation P_t in the model is as follows: runoff Q_t (direct and indirect), evapotranspiration ET_t, soil water storage S_t, and groundwater storage G_t. The partitioning is controlled by the magnitude of precipitation P_t, potential evapotranspiration ET_{0t}, and the initial storages in soil S_{t-1} and groundwater G_{t-1}. The following equation controls the partitioning:

$$Y_t(W_t) = \frac{W_t + b}{2a} - \sqrt{\left(\frac{W_t + b}{2a}\right)^2 - \frac{W_t b}{a}}, \tag{9}$$

where Y_t is the sum of monthly evapotranspiration and soil water storage at the end of the month, namely evapotranspiration opportunity. W_t is the sum of monthly precipitation and initial soil moisture, named as available water. The parameter a ($0 \sim 1$) means the propensity in a catchment for runoff to occur before the soil becomes saturated. The parameter b is the maximum value of Y_t. Wang and Tang (2014) demonstrated that Eq. (9) can be derived from the generalized proportionality hypothesis and is an equivalent Budyko-type equation. Available water partitioning between ET_t and

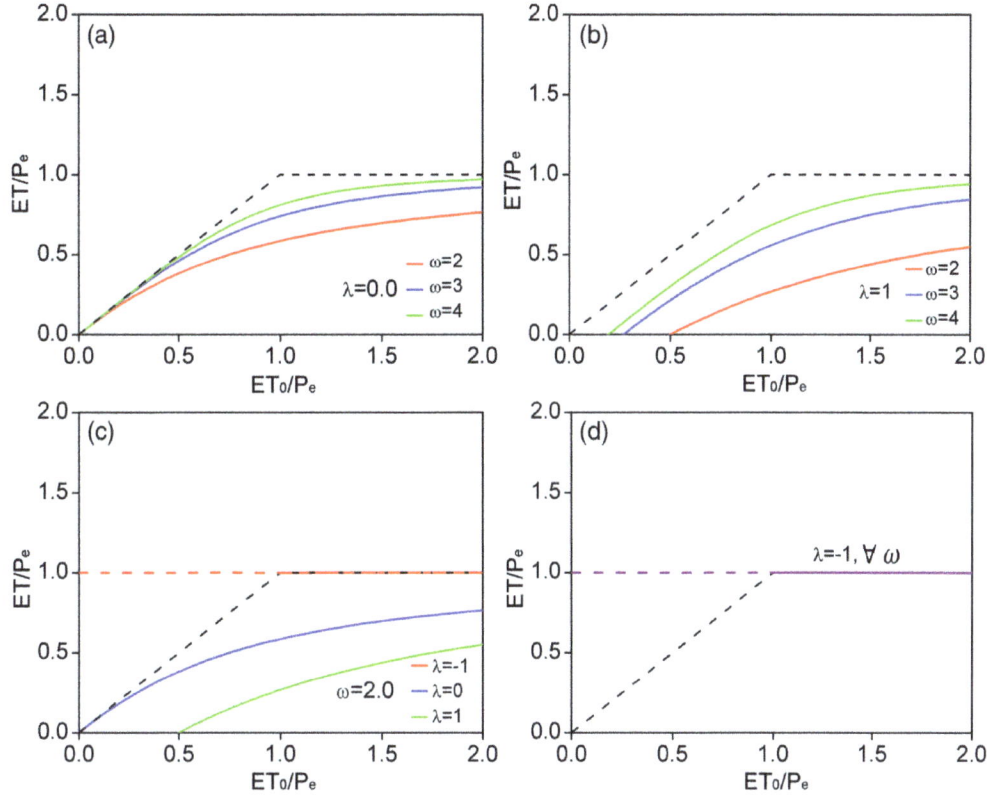

Figure 2. The Budyko curves in Eq. (8) with different combinations of parameters ω and λ.

S_t is controlled by the assumption that the loss rate of actual evaporation from soil water storage is proportional to the evapotranspiration capacity. So the soil water storage at the end of period t is written as

$$S_t = Y_t \exp(-ET_{0t}/b). \tag{10}$$

The actual evapotranspiration at the period t is the difference between evapotranspiration opportunity and soil water storage $(Y_t - S_t)$. The streamflow, including direct runoff and groundwater recharge, is determined by the difference between available water and evapotranspiration opportunity $(W_t - Y_t)$. The parameter c separates the direct runoff $(1 - c)(W_t - Y_t)$ and groundwater recharge $c(W_t - Y_t)$. Groundwater discharge dG_t as the base flow is determined by the parameter d and groundwater storage at the end of period t. The streamflow is sum of direct runoff and the base flow. For a given set of a, b, c and d and initial soil water storage and groundwater storage, the allocation of monthly precipitation can be computed one by one.

3 Study area and data

3.1 Study area

The HRB, originating from Qilian Mountains, is the second largest inland river basin in the arid area of the north-western China (Fig. 3). The drainage map and the basin border are extracted using a 90 m resolution digital elevation model (DEM) data from the Shuttle Radar Topography Mission (SRTM) website of NASA (http://srtm.csi.cgiar.org/SELECTION/inputCoord.asp) (basin length: 820 km; total area: 143 044 km^2; elevation: 870–5545 m). The HRB is in the middle of Eurasia and away from oceans, characterized with dry and windy climate, and very limited precipitation (mean annual precipitation: 126 mm yr^{-1}) but plentiful radiation (mean annual solar radiation: 1780 MJ m^{-2} yr^{-1}, ~ 660 mm yr^{-1} in the unit of evaporation).

The HRB is divided into six sub-basins according to basin characteristics, distributed along the eastern and western tributaries, shown in Fig. 3. Regions I and II are upper mountainous regions with the elevation of 3000–5500 m and belong to the cold and semiarid mountainous zone dominated by shrubs and trees with mean annual temperature of less than 2 °C and annual precipitation of 200–400 mm. And these two sub-basins are the water source area to the middle and lower reaches and have little interference of human activities. Regions III and V with annual precipitation of 100–250 mm are the main irrigation zone and residential area with more than 90 % of total population of the HRB. The two sub-basins are the main water-consuming regions and largely disturbed by human activities. Regions IV and VI located at lower reaches are extremely arid and the mean annual precipitation is less than 100 mm.

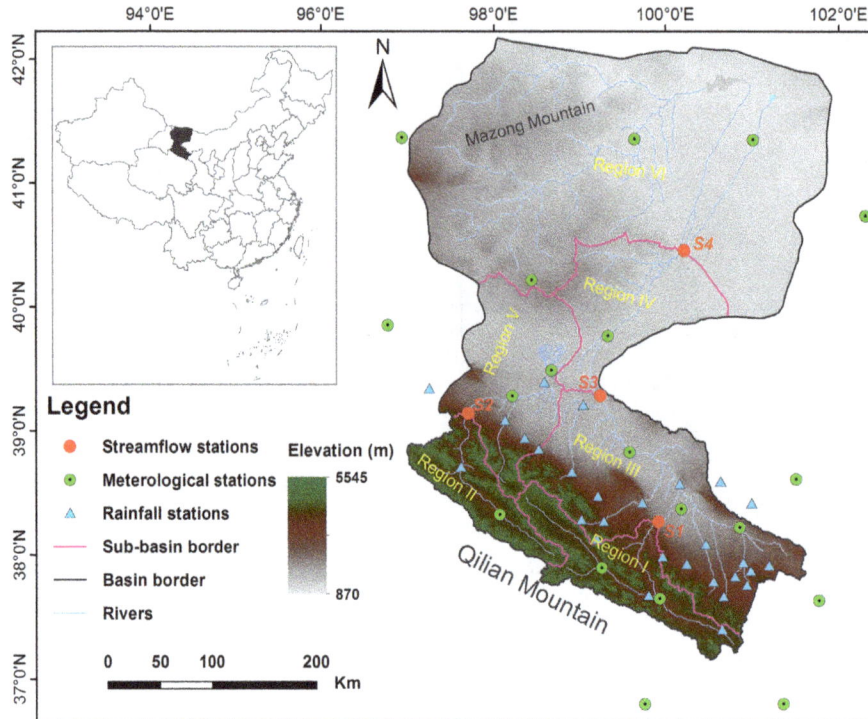

Figure 3. Location of study area and the distribution of hydrological stations and meteorological stations.

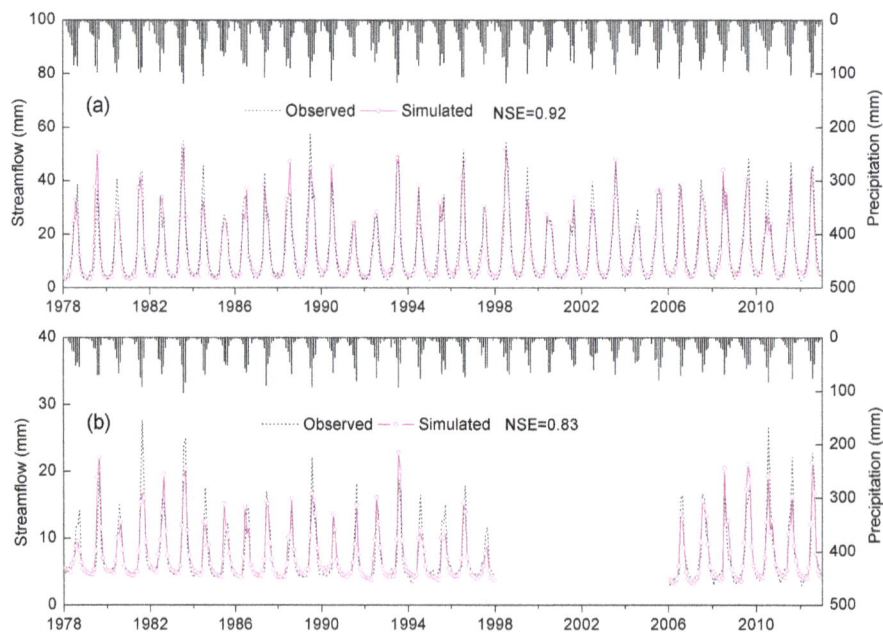

Figure 4. Time series of observed and simulated monthly streamflow using the *abcd* model in region I (**a**) and region II (**b**) during 1978–2012.

3.2 Data

The required data for Eq. (8) and the *abcd* model include monthly precipitation, potential evapotranspiration and runoff from those sub-basins in the HRB.

The daily precipitation data of all stations during 1978–2012 are obtained from the year book hydrology of China including 28 rainfall stations and the China Administration of Meteorology including 19 meteorological stations (Fig. 3). The monthly precipitation of each station is calculated by summing daily precipitation. The gridded data set with 1 km resolution across the whole basin is obtained by interpolation of the site data. The monthly precipitation of the six sub-basins is obtained by the extraction from the monthly precipitation in the whole basin. Daily meteorological data of 19 stations during 1978–2012 are also available. Daily potential evapotranspiration is estimated in each station using the FAO Penman–Monteith equation recommended by Allen et al. (1998). The monthly ET_0 at each station is the sum of the daily ET_0 and then interpolated to the whole basin. Finally, annual runoff, precipitation and potential evapotranspiration are obtained by summing monthly data.

The red points in Fig. 3 are the locations of hydrological stations. For the two upper streams, Gauge S1 controls region I and Gauge S2 controls region II. For the two middle streams, Gauges S1 and S3 control region III and Gauges S3 and S4 control region IV. For the two lower streams, regions V and VI without any runoff flowing out, Gauges S2 and S4 control their inflow, respectively (Fig. 3). Monthly runoff data are obtained from the year book of hydrology of China and are intended for calibrating the *abcd* model. The annual runoff is obtained by summing monthly runoff. The data time series for regions I and III are from 1978 to 2012. The same period is for regions II and V but with the period of 1998–2006 missing. The length of the data time series for regions IV and VI is from 1988 to 2012.

The natural runoff in regions III and IV were strongly disturbed by human activities and there is no runoff for regions V and VI and the whole basin. To validate the outputs of the *abcd* model for those regions, this study employs the evapotranspiration of remote sensing products from Heihe Plan Science Data Center (Wu et al., 2012) as a reference. The same data have been widely used as a reference for modeling evaluations and is supported by a State Key Research Program-Heihe Eco-hydrological Research Project of National Natural Science Foundation of China (Yan et al., 2014; Yao et al., 2014). The monthly evapotranspiration data sets (2000–2012) with 1 km spatial resolution over the HRB (http://westdc.westgis.ac.cn), are estimated by the ETWatch model based on multi-source remote sensing data (Wu et al., 2012).

4 Results

4.1 Calibration of the *abcd* model

In extremely dry basins like the HRB, the lack of observed hydro-climatic data presents great challenge. A monthly water balance model becomes an effective tool to estimate actual evapotranspiration, change in soil water storage and change in groundwater storage. This study employs the *abcd* water balance model due to its simple and sound physical structure tested and recommended by Alley (1984) and Fernandez et al. (2000). We calibrate and validate the *abcd*

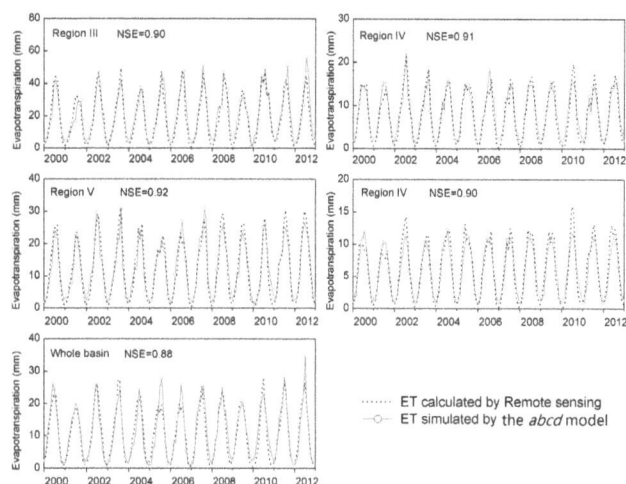

Figure 5. Comparison between ET simulated by the *abcd* model and ET calculated by remote sensing data for regions III–VI and the whole basin during 2000–2012. "WBM" denotes the *abcd* water balance model.

model using monthly time series of precipitation, potential evapotranspiration and runoff at each of the seven regions (the six sub-basins and the whole basin) and using the generalized pattern search optimization method. Nash–Sutcliffe efficiency (NSE) is used to assess the goodness of fit of the monthly water balance for the seven regions.

Figure 4 shows the results of the modeled streamflow at monthly timescales in regions I and II. Regions I and II are the water source area of the whole basin with little interference of human activities and both keep relatively natural steady state. The NSE for regions I and II is for 0.92 and 0.83, respectively. The results illustrate that the simulated monthly streamflow agrees well with the observation and other modeled components can be reasonable estimates, for instance, monthly actual evapotranspiration, soil water storage change and groundwater storage change in the two sub-basins.

The outputs from the *abcd* model being used include soil water storage and actual evapotranspiration. Only over the two upper sub-basins (regions I and II) is the streamflow used for the calibration and validation purpose.

Over the middle sub-basins (regions III, IV and also V), large areas of artificial oasis (cropland) is distributed and the streamflow water intensely disturbed by hydraulic engineering. Hence it becomes almost impossible to validate the *abcd* model by directly comparing the simulated and observed streamflow. Instead, we used the actual evapotranspiration by remote sensing to calibrate and validate the *abcd* model. For the new BH over regions III, IV and V, we use the observed Q_{in} from the upper sub-basin as the input. That is to be consistent with the remote sensing data, which are observed and hence human disturbed.

Table 2. The mean annual water balance of all regions.

Region	P (mm)	Q_{in} (mm)	ET (mm)	Q_{out} (mm)	ΔS (mm)	PWS(%)
I	351.9	–	165.3	169.3	0.0	0.0
II	220.7	–	143.9	85.2	0.1	0.0
III	223.6	66.1	253.2	37.5	−2.1	−0.9
IV	73.5	74.0	103.4	47.5	1.0	1.3
V	117.3	39.6	156.7	–	0.2	0.1
VI	66.8	7.9	74.7	–	0.0	0.0
Whole basin	125.8	–	125.5	–	0.2	0.2

"–" means no runoff; PWS represents the proportion of the root zone water storage change in the total precipitation.

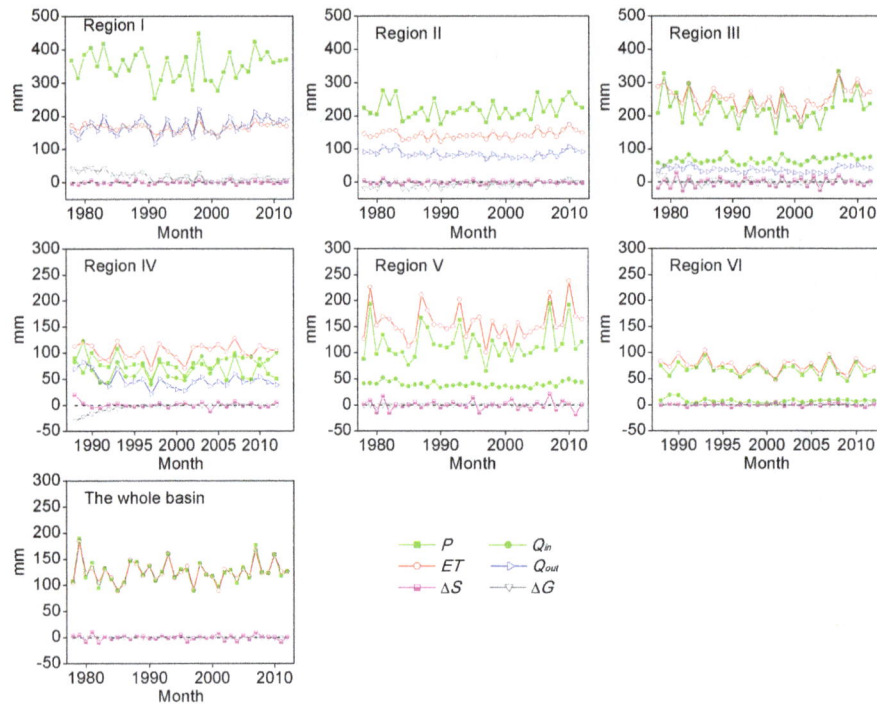

Figure 6. Variation of annual water balance for all the regions simulated using the *abcd* model.

4.2 Annual and monthly water balance analysis

To test the "steady-state" assumption of the Budyko framework, it is vital to examine whether changes in mean annual soil water storage in water balance approach zero. By using the monthly runoff, evapotranspiration, soil water and groundwater storage change from the *abcd* model and the observed monthly precipitation, the mean annual water balance of all regions are summarized in the Table 2. Regions I and II are located in mountainous area, where the mean annual soil water storage changes are almost zero with both 0.0 % of the corresponding precipitation. The mean annual soil water storage change in regions III and IV are relatively significant. For regions V and VI and the whole basin without any outflow, the mean annual soil water storage and groundwater storage changes both approach zero. In conclusion, the mean annual soil water storage changes for all regions are very small and can be ignored in mean annual water balance. These sub-basins and the whole basin keep natural basin characteristics and meet the second assumption of the BH that mean annual soil water storage can be ignored. However, no inflow only exists in regions I and II, which meets the first assumption of the BH that the local precipitation is the only potential water source to evapotranspiration. In other regions, water supply conditions have been changed by considerable inflow generally from upper sub-basins.

Because this study focuses on the application of the BH at the annual and monthly timescales, the annual and monthly water balance analysis is very critical to understanding the role of water storage and water source change in the BH. Figure 6 describes the variation of annual water balance for the six sub-basins and the whole basin. The most obvious in

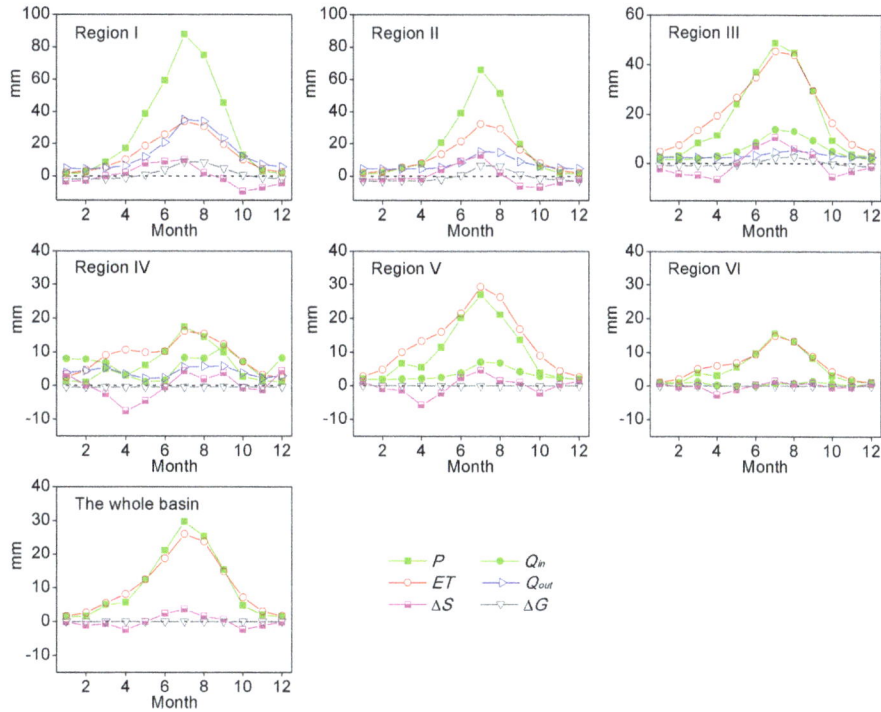

Figure 7. Variation of average monthly water balance for all regions using the *abcd* model.

Table 3. The fitting parameters of Fu's equation at annual scales.

Region	I	II	III	IV	V	VI	Whole basin
ω^*	1.34	1.45	2.05	1.42	20.28	13.05	17.60
ω^{**}	1.45	1.69	2.34	1.44	1.07	10.8	1.09
λ^{**}	0.25	0.67	0.62	0.08	−1	−1	−1

* means the calibrated values of ω in Fu's equation; Eq. (8) when $\lambda = 0$;
** means the calibrated values of ω and λ in Eq. (8).

Fig. 6 is that the proportion of soil water storage change in annual water balance is small compared with the annual precipitation. So the impact of soil water storage change on annual water balance is insignificant and can be also neglected. Moreover, annual evapotranspiration is higher than annual precipitation in regions III–VI and approaches annual precipitation over the whole basin. For water-limited regions, when inflow from other regions is available, the actual evapotranspiration increases with the increased water supply so that the actual evapotranspiration is more than the local precipitation. For the whole basin of the inland HRB, there is no water transferring with other basins, so the evapotranspiration almost approaches the precipitation at the annual timescale due to little variations in the soil water storage changes. In conclusion, the facts that soil water storage change in all basins can be ignored in annual water balance meet the second assumption of the BH, and the results that the annual water balance in regions III–VI and the whole basin have been disturbed do not meet the first assumption of the BH. Therefore,

except for regions I and II, the original BH cannot be directly used for those sub-basins and the whole basin.

Different from the annual timescale, the impacts of monthly changes of soil water storage and groundwater storage behave differently (Fig. 7). The variations of monthly groundwater storage change for all regions are similar to those of runoff (Fig. 7). For those regions with no runoff (regions V and VI and the whole basin), the modeled groundwater storage change is almost zero. This means that the groundwater storage can hardly contribute to the evapotranspiration while the variation of soil water storage is tightly coupled with the evapotranspiration (Fig. 7). For regions I and II and during the winter season, the evapotranspiration is more than the precipitation; the extra water source required by the evapotranspiration is from root zone water storage. After the summer season, the precipitation sharply decreases, but the evapotranspiration slowly decreases by consuming the root zone water storage recharged during the summer season. For regions III–VI, the water supply is more complicated

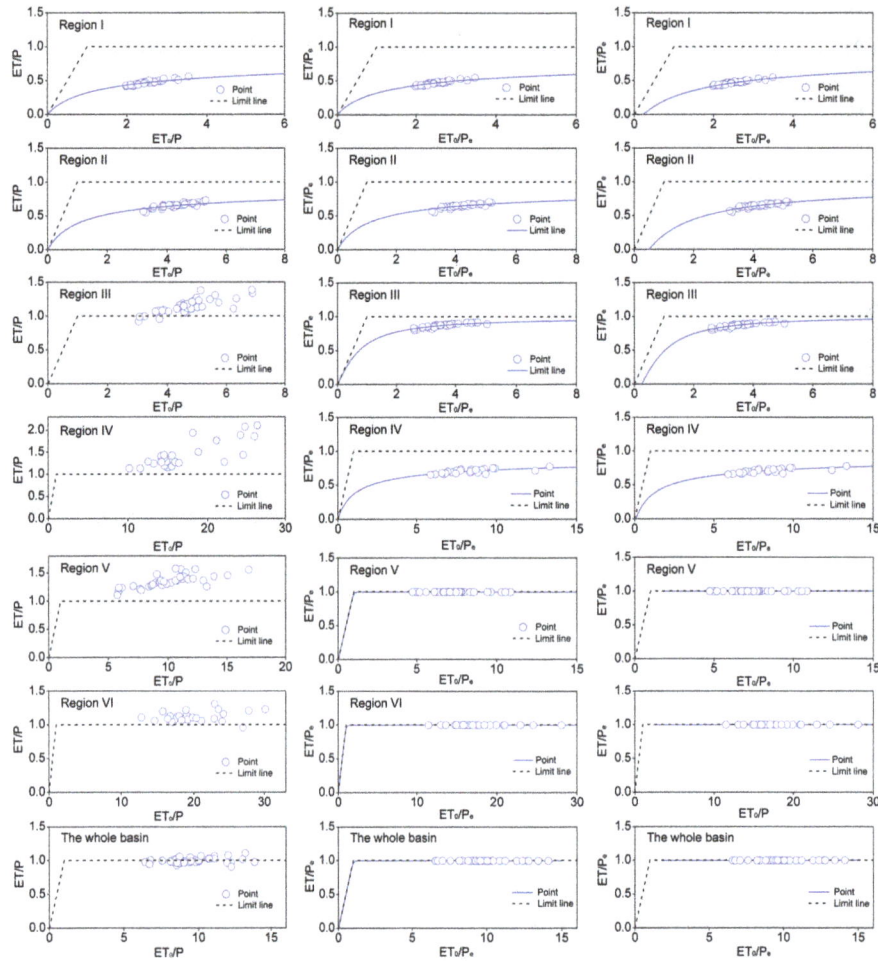

Figure 8. Comparison of the original Budyko curves (left panel) and the new Fu-type Budyko curves (middle panel, with $\lambda = 0$) and the new Fu-type Budyko curves (right panel, with $\lambda > 0$) for regions I–VI and the whole basin at the annual timescale.

by the interference of monthly inflow water, and the monthly variations of root zone water storage. As shown in Fig. 7, it can be concluded that both the soil water storage change and inflow water have obvious effect on the monthly water balance, whilst the impact of monthly groundwater storage change is negligible.

In summary, due to the complications of the water transfer and soil water storage change, the two assumption conditions for applying the original BH are difficult to meet for the sub-basins and the whole HRB on the monthly timescales, which in turn requires new treatments in the BH as further investigated in the following sections.

4.3 The annual Budyko curve analysis

Figure 8 (left panel) plots the original Budyko curves for the six sub-basins and the whole basin. For regions I and II, the points of annual evapotranspiration ratio and aridity index fall in the domain of water and energy limit boundary and they can be well fitted by Fu's equation. The relationship be-

tween water and energy in regions I and II can be described by the original BH as expected in the section above. However, the points of evapotranspiration ratio and aridity index for other regions exceed the water limit boundary. And the results show the relationship of water and energy in regions III–VI, and the whole basin is inconsistent with the original BH. After using the equivalent precipitation instead of the local precipitation, the new Fu-type Budyko curves (Eq. (8) with $\lambda = 0$) for all regions are shown in Fig. 8 (middle panel). Compared with the original Budyko curve, the new curves for regions I and II did not behave differently, because the two basins are natural and closed. The obvious change between the improved and original Budyko curves are for regions III and IV. For the whole basin and regions V and VI, the new curves fall on the upper limit of $ET / P_e = 1$ due to no runoff flowing out. These improved Budyko curves can be fitted using Fu's equation and the parameters are listed in Table 3. Interestingly for the annual timescale, the fitted performances of Fu's equation and Eq. (8) are almost identical. Therefore, the new Fu-type Budyko curves (Eq. 8) with fitted

Table 4. The fitting parameters of the improved Budyko equation at the monthly scales.

Region	Parameter	May–Aug	Apr and Sep	Mar and Oct	Feb and Nov	Jan and Dec
I	ω^*	1.40	1.39	1.35	1.28	1.22
	ω^{**}	1.50	1.51	1.48	1.40	1.33
	λ^{**}	0.16	0.24	0.31	0.32	0.28
II	ω^*	1.54	1.53	1.47	1.37	1.29
	ω^{**}	1.70	1.72	1.67	1.57	1.48
	λ^{**}	0.34	0.54	0.63	0.66	0.60

Region	Parameter	Apr–Sep	Mar and Oct	Feb and Nov	Jan and Dec	
III	ω^*	2.20	2.05	1.86	1.71	
	ω^{**}	2.31	2.15	1.97	1.90	
	λ^{**}	0.18	0.19	0.22	0.39	
IV	ω^*	1.51	1.42	1.33	1.25	
	ω^{**}	1.75	1.59	1.51	1.41	
	λ^{**}	0.92	0.53	0.56	0.41	

Region	Parameter	May–Aug	Apr and Sep	Mar and Oct	Feb and Nov	Jan and Dec
V	ω^*	35.5	29.8	28.0	22.5	23.7
	ω^{**}	1.02	1.03	1.03	1.04	1.04
	λ^{**}	−1	−1	−1	−1	−1
VI	ω^*	17.3	18.9	15.5	12.7	13.1
	ω^{**}	1.02	1.02	1.03	1.03	1.04
	λ^{**}	−1	−1	−1	−1	−1
The whole basin	ω^*	28.5	22.4	18.8	16.3	15.7
	ω^{**}	1.02	1.02	1.04	1.04	1.03
	λ^{**}	−1	−1	−1	−1	−1

* Means the calibrated values of ω in Fu's equation; Eq. (8) when $\lambda = 0$;
** means the calibrated values of ω and λ in Eq. (8).

values of λ (right panel, Fig. 8) do not show much difference from those curves with λ set zero.

In summary, if a basin (sub-basin) is closed, the original BH can be applicable at the annual timescale. However under unsteady state, the new Fu-type BH, instead of the original BH, is more applicable to describe the annual water balance.

4.4 The monthly Budyko curves analysis

Again as expected based on the monthly water balance analysis, the points of monthly evapotranspiration ratio and aridity index exceed the water limit boundary for all the basins (Fig. 9, left panel). The value of evapotranspiration ratio can be up to 40, which means that the local precipitation in original water balance is well below the actual water supply to the evapotranspiration. The new Fu-type Budyko curves at the monthly timescale are shown in Fig. 9 in the middle panel (Eq. 8 with setting $\lambda = 0$) and in the right panel (Eq. 8 with calibrated λ). It is remarkable that the points of monthly

evapotranspiration ratio and aridity index distribute regularly in the Budyko framework (in Fig. 9, middle panel and right panel). The improved Budyko curves with calibrated λ perform better than Fu's original equation (i.e., $\lambda = 0$) by 5–10 % in terms of NSE. The fitting parameter λ introduced in this study (Eq. 8) can add further improvement to the BH, in spite of obviously deserving further investigations.

The fitted values of the parameters in the Budyko curves for regions I to VI are listed in the Table 4. These curves and the parameters have significantly seasonal characteristics. For example, the Budyko curves in regions I and II can be divided to five groups (Fig. 9). The values of the integrated parameter ω in Eq. (8) gradually decrease from the summer months to winter months. The absolute values of parameters λ gradually increase, which illustrates that the points in summer months are more centralized than those in winter months. Moreover, in regions V and VI and the whole basin, all the equivalent precipitation is consumed by evapo-

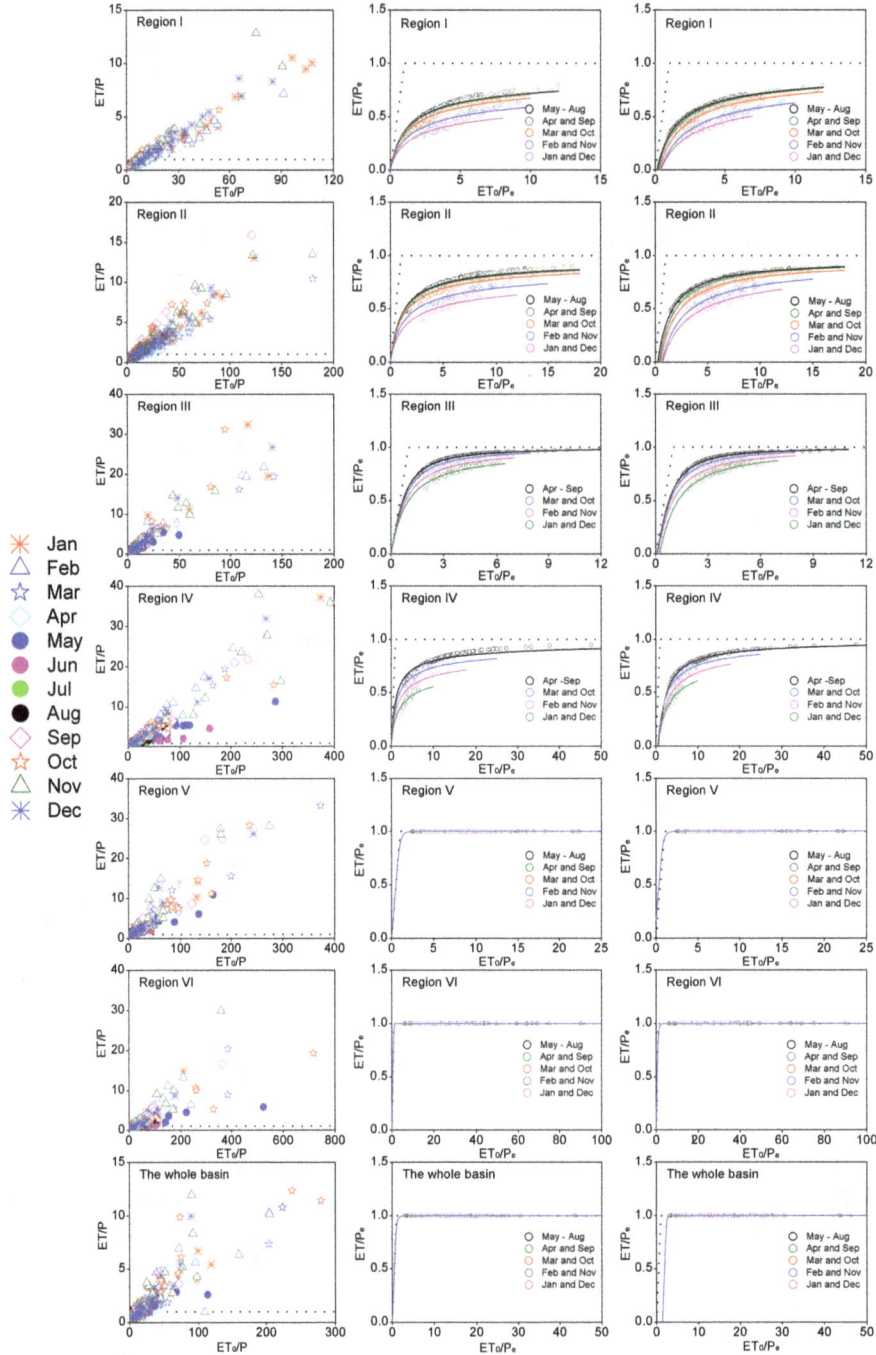

Figure 9. Comparison of the original Budyko curves (left panel) and the new Fu-type Budyko curves (middle panel, with $\lambda = 0$) and the new Fu-type Budyko curves (right panel, with $\lambda > 0$) for regions I–VI and the whole basin at the monthly timescale.

transpiration, and therefore the ratio of evapotranspiration to the equivalent precipitation is almost 1.

4.5 Storage change and inflow water impact on the BH

In this study, we intended to extend the BH to the annual and sub-annual time scales by explicitly considering the root zone water storage and new water source from other regions.

To further investigate it, we choose regions I and III as typical cases in Fig. 10. In region I, as there is no inflow into the region, we can separate the impact of soil water storage and groundwater storage on the BH (Fig. 10a). With subtle differences, the impacts of changes in root zone water storage and groundwater storage on water balance can be almost ignored at annual scales. Region III is another extreme case

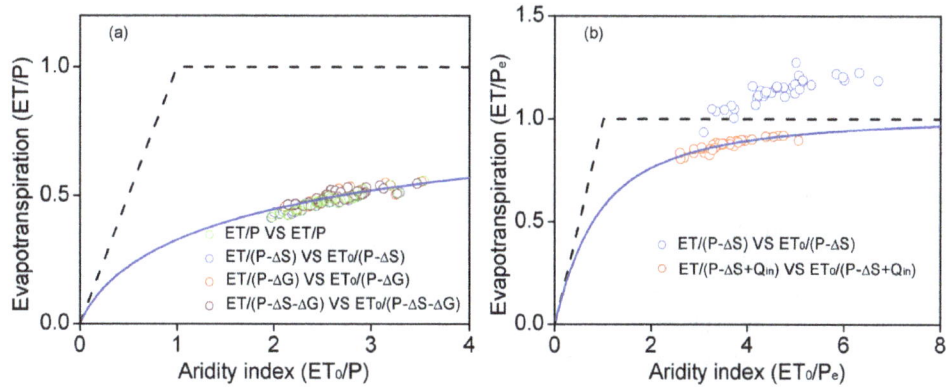

Figure 10. Different presentations of the annual water balance for (**a**) region I and (**b**) region III.

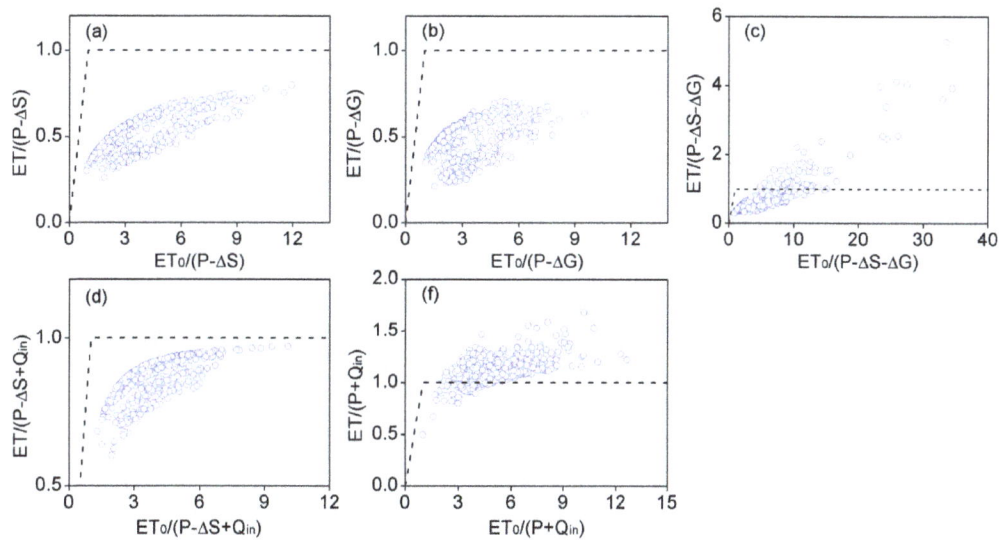

Figure 11. Five presentations of monthly water balance for region III considering different combinations in the water supply to evapotranspiration.

where only if the role of the inflow water is considered, can the BH perform well under unsteady state (Fig. 10b).

In Fig. 11, we further adopted the approach presented by Chen et al. (2013) to examine the impacts of soil water storage, groundwater storage and inflow water on monthly water balance. We test different combinations in monthly water balance in region III, a midstream sub-basin of the HRB (Fig. 11a–c), and found that when the equivalent precipitation includes the root zone water storage change the BH performs well at the monthly scale. However, the inclusion of the groundwater storage change into the equivalent precipitation does not improve as much (Fig. 11b, c). By examining the impact of monthly inflow water on the BH in region III (Fig. 11d, f), we find that inflow water at the monthly scale has as much impact as that at annual scale. The results presented above highlight the fact that the water supply cannot be the local precipitation only, but should have included root zone water storage change and inflow water.

5 Conclusions

The Budyko hypothesis (BH) is a useful approach to depicting and understanding the long-term mean water balance at a large basin scale under a steady-state condition. However, river systems worldwide have in fact been disturbed by human activities to different extents. That is important for extremely arid environments (say, the aridity index over 5), especially in China, where water systems are typically unclosed with intense human inference and irrigation. That presents a great challenge if one is applying the BH to those regions under unsteady state, e.g., unclosed or significant variation in soil water storage, or those timescales finer than a year.

To investigate it, we choose an extremely arid inland basin, the Heihe River basin in China, as the study area, which is divided into six sub-basins based on catchment hydrologic characteristics. We first calibrate and validate a widely used monthly water balance model, i.e., the *abcd* model.

For the two upper sub-basins, the simulated monthly water balance is compared against monthly streamflow from hydrological gauges, and for the other sub-basins and the whole catchment, the simulated evapotranspiration is compared with widely used remote sensing ET products in the HRB. The *abcd* model can successfully simulate the monthly water balance and capture the inter-annual variations (NSE over 0.85). Based on that, we find that the role of root zone water storage change in monthly water balance is significant but almost negligible over timescales longer than a year. And the impact of inflow water from upper sub-basins is also significant and does not rely on the timescale. We conclude that the upstream basins in the HRB are almost closed basins, which meet the two steady-state conditions of the BH, and other sub-basins become an unclosed basin due to the impact of the inflow water and human interference.

With the recognition that the inflow water from other regions and the water storage change are both new possible water sources to evapotranspiration in unclosed basins, we define the equivalent precipitation (P_e) including the local precipitation, inflow water and water storage change as the water supply, instead of just the local precipitation, in the Budyko framework. (The evapotranspiration ratio and the aridity index are also redefined using the equivalent precipitation.) In addition to the new definition of the water supply, we develop a new Fu-type Budyko equation with two non-dimensional parameters (ω and λ) based on the deviation by Professor Baopu Fu, i.e., Fu's equation, to consider the effect of the change of root zone water storage and the inflow water on the water and energy constraints. Over the annual timescale, the new Fu-type Budyko equation developed here has a more or less identical performance to Fu's equation for the sub-basins and the whole catchment. However, for the monthly timescale, the new Fu-type Budyko equation performs better than Fu's original equation when the ratio of evapotranspiration to equivalent precipitation less than 1, and performs the same when the evapotranspiration ratio is very close to 1. The new Fu-type Budyko equation (ω and λ) developed in this study enables one to apply the BH to interpret regional water balance over extremely dry environments under unsteady state (e.g., unclosed basins or sub-annual timescales).

Appendix A: Derivation of the constraints of λ in new Fu-type Budyko equation

For an unclosed basin or region, the water supply to evapotranspiration is defined as equivalent precipitation ($P_e = P + Q_{in} - \Delta S$). Evapotranspiration ratio: $\varepsilon = ET / P_e$ and aridity index: $\phi = ET_0 / P_e$. The Budyko equation is written the same as Eq. (8).

$$\varepsilon = 1 + \phi - (1 + \phi^\omega + \lambda)^{1/\omega} \tag{A1}$$

According to the constrained boundary of the BH, (1) the evapotranspiration ratio is less than or equal to the aridity index, namely $\varepsilon \leq \phi$, and (2) the evapotranspiration ratio is no more than 1; i.e., $\varepsilon \leq 1$.

With $\varepsilon \leq \phi$, we can have

$$1 + \phi - (1 + \phi^\omega + \lambda)^{1/\omega} \leq \phi. \tag{A2}$$

Therefore,

$$\phi^\omega + \lambda \geq 0, \tag{A3}$$

where $\phi \geq 0$ and $\omega > 1$.

For the other constraint, $\varepsilon \leq 1$ we can derive

$$1 + \lambda \geq 0. \tag{A4}$$

Acknowledgements. This research was supported by the National Natural Science Foundation of China (41271049), the Chinese Academy of Sciences (CAS) Pioneer Hundred Talents Program, and an open research fund for the State Key Laboratory of Desert and Oasis Ecology, Xinjiang, Institute of Ecology and Geography, Chinese Academy of Sciences. The authors would like to thank Hubert H. G. Savenige and Wang Ping for their suggestions and support in this research. We are particularly grateful to the two reviewers Wang Dingbao and Gao Hongkai for their efforts and helpful comments.

Edited by: H. Li

References

Allen, R. G., Pereira, L. S., Raes, D., and Smith, M.: Crop evapotranspiration - Guidelines for computing crop water requirements, FAO Irrigation and drainage paper 56, 24, 55–56, 1998.

Alley, W. M.: On the Treatment of Evapotranspiration, Soil Moisture Accounting, and Aquifer Recharge in Monthly Water Balance Models, Water Resour. Res., 20, 1137–1149, doi:10.1029/WR020i008p01137, 1984.

Bonacci, O. and Andric, I.: Impact of an inter-basin water transfer and reservoir operation on a karst open streamflow hydrological regime: an example from the Dinaric karst (Croatia), Hydrol. Process., 24, 3852–3863, doi:10.1002/hyp.7817, 2010.

Budyko, M. I.: Evaporation Under Natural Conditions, Gedrometeoizdat, St. Petersburg, Russia, 29 pp., 1948.

Budyko, M. I.: The Heat Balance of the Earth's Surface, Department of Commerce, Weather Bureau, 144–155, 1958.

Budyko, M. I.: Climate and Life: Volume 18 International geophysics series, Academic Press, 1–508, 1974.

Carmona, A. M., Sivapalan, M., Yaeger, M. A., and Poveda, G.: Regional patterns of interannual variability of catchment water balances across the continental US: A Budyko framework, Water Resour Res, 50, 9177–9193, doi:10.1002/2014wr016013, 2014.

Chen, X., Alimohammadi, N., and Wang, D.: Modeling interannual variability of seasonal evaporation and storage change based on the extended Budyko framework, Water Resour. Res., 49, 6067–6078, doi:10.1002/wrcr.20493, 2013.

Choudhury, B. J.: Evaluation of an empirical equation for annual evaporation using field observations and results from a biophysical model, J. Hydrol., 216, 99–110, doi:10.1016/s0022-1694(98)00293-5, 1999.

Donohue, R. J., Roderick, M. L., and McVicar, T. R.: Assessing the differences in sensitivities of runoff to changes in climatic conditions across a large basin, J. Hydrol., 406, 234–244, doi:10.1016/j.jhydrol.2011.07.003, 2011.

Fernandez, W., Vogel, R. M., and Sankarasubramanian, A.: Regional calibration of a watershed model, Hydrol. Sci. J., 45, 689–707, doi:10.1080/02626660009492371, 2000.

Fu, B. P.: On the Calculation of the Evaporation from Land Surface, Sci. Atmos. Sin., 5, 23–31, 1981.

Gao, H., Hrachowitz, M., Schymanski, S. J., Fenicia, F., Sriwongsitanon, N., and Savenije, H. H. G.: Climate controls how ecosystems size the root zone storage capacity at catchment scale, Geophys. Res. Lett., 41, 7916–7923, doi:10.1002/2014GL061668, 2014.

Gentine, P., D'Odorico, P., Lintner, B. R., Sivandran, G., and Salvucci, G.: Interdependence of climate, soil, and vegetation as constrained by the Budyko curve, Geophys. Res. Lett., 39, L19404, doi:10.1029/2012gl053492, 2012.

Gerrits, A. M. J., Savenije, H. H. G., Veling, E. J. M., and Pfister, L.: Analytical derivation of the Budyko curve based on rainfall characteristics and a simple evaporation model, Water Resour Res, 45, W04403, doi:10.1029/2008wr007308, 2009.

Gordon, L. J., Steffen, W., Jonsson, B. F., Folke, C., Falkenmark, M., and Johannessen, A.: Human modification of global water vapor flows from the land surface, P. Natl. Acad. Sci. USA, 102, 7612–7617, doi:10.1073/pnas.0500208102, 2005.

Istanbulluoglu, E., Wang, T., Wright, O. M., and Lenters, J. D.: Interpretation of hydrologic trends from a water balance perspective: The role of groundwater storage in the Budyko hypothesis, Water Resour. Res., 48, W00H16, doi:10.1029/2010wr010100, 2012.

Li, H.-Y., Sivapalan, M., Tian, F., and Harman, C.: Functional approach to exploring climatic and landscape controls of runoff generation: 1. Behavioral constraints on runoff volume, Water Resour. Res., 50, 9300–9322, doi:10.1002/2014WR016307, 2014.

Li, J., Tan, S., Chen, F., and Feng, P.: Quantitatively analyze the impact of land use/land cover change on annual runoff decrease, Nat. Hazards, 74, 1191–1207, doi:10.1007/s11069-014-1237-x, 2014.

Mezentsev, V. S.: More on the calculation of average total evaporation, Meteorol. Gidrol., 5, 24–26, 1955.

Milly, P. C. D.: An Analytic Solution of The stochatic Storage Problem Applicable to Soil Water, Water Resour. Res., 29, 3755–3758, doi:10.1029/93wr01934, 1993.

Ol'dekop, E. M.: On evaporation from the surface of river basins, Trans. Meteorol. Obs., 4, p. 200, 1911.

Pike, J. G.: The Estimation of Annual Run-off from Meteorological Data in a Traopical Climate, J. Hydrol., 2, 116–123, 1964.

Porporato, A., Daly, E., and Rodriguez-Iturbe, I.: Soil water balance and ecosystem response to climate change, Am. Nat., 164, 625–632, doi:10.1086/424970, 2004.

Savenije, H. H. G.: Determination of evaporation from a catchment water balance at a monthly time scale, Hydrol. Earth Syst. Sci., 1, 93–100, doi:10.5194/hess-1-93-1997, 1997.

Schreiber, P.: Ueber die Beziehungen zwischen dem Niederschlag und der Wasserfvhrung der flvsse in Mitteleuropa, Z Meteorol, 21, 442–452, 1904.

Sivapalan, M., Yaeger, M. A., Harman, C. J., Xu, X., and Troch, P. A.: Functional model of water balance variability at the catchment scale: 1. Evidence of hydrologic similarity and space-time symmetry, Water Resour Res, 47, W02522, doi:10.1029/2010wr009568, 2011.

Sun, F.: Study on Watershed Evapotranspiraiton based on the Budyko Hypothesis, Doctor of Engineering, Tsinghua University, 147 pp., 2007.

Thomas, H. A.: Improved methods for national water assessment, Water Resources Council, Washington, D. C., contract: WR15249270, 1–59, 1981.

Wang, D.: Evaluating interannual water storage changes at watersheds in Illinois based on long-term soil moisture and groundwater level data, Water Resour. Res., 48, WO35032, doi:10.1029/2011WR010759, 2012.

Wang, D. and Tang, Y.: A one-parameter Budyko model for water balance captures emergent behavior in darwinian hydrologic models, Geophys. Res. Lett., 41, 4569–4577, doi:10.1002/2014GL060509, 2014.

Wang, T., Istanbulluoglu, E., Lenters, J., and Scott, D.: On the role of groundwater and soil texture in the regional water balance: An investigation of the Nebraska Sand Hills, USA, Water Resour. Res., 45, W10413, doi:10.1029/2009wr007733, 2009.

Wu, B., Yan, N., Xiong, J., Bastiaanssen, W. G. M., Zhu, W., and Stein, A.: Validation of ETWatch using field measurements at diverse landscapes: A case study in Hai Basin of China, J. Hydrol., 436, 67–80, doi:10.1016/j.jhydrol.2012.02.043, 2012.

Xiong, L. H., Yu, K. X., and Gottschalk, L.: Estimation of the distribution of annual runoff from climatic variables using copulas, Water Resour. Res., 50, 7134–7152, doi:10.1002/2013wr015159, 2014.

Yan, H., Zhan, J., Liu, B., and Yuan, Y.: Model Estimation of Water Use Efficiency for Soil Conservation in the Lower Heihe River Basin, Northwest China during 2000–2008, Sustainability, 6, 6250–6266, doi:10.3390/su6096250, 2014.

Yang, D., Sun, F., Liu, Z., Cong, Z., Ni, G., and Lei, Z.: Analyzing spatial and temporal variability of annual water-energy balance in nonhumid regions of China using the Budyko hypothesis, Water Resour. Res., 43, W04426, doi:10.1029/2006wr005224, 2007.

Yang, H., Yang, D., Lei, Z., and Sun, F.: New analytical derivation of the mean annual water-energy balance equation, Water Resour. Res., 44, W03410, doi:10.1029/2007wr006135, 2008.

Yao, Y., Liang, S., Xie, X., Cheng, J., Jia, K., Li, Y., and Liu, R.: Estimation of the terrestrial water budget over northern China by merging multiple datasets, J. Hydrol., 519, 50–68, doi:10.1016/j.jhydrol.2014.06.046, 2014.

Yokoo, Y., Sivapalan, M., and Oki, T.: Investigating the roles of climate seasonality and landscape characteristics on mean annual and monthly water balances, J. Hydrol., 357, 255–269, doi:10.1016/j.jhydrol.2008.05.010, 2008.

Zhang, L., Dawes, W. R., and Walker, G. R.: Response of mean annual evapotranspiration to vegetation changes at catchment scale, Water Resour. Res., 37, 701–708, doi:10.1029/2000wr900325, 2001.

Zhang, L., Hickel, K., Dawes, W. R., Chiew, F. H. S., Western, A. W., and Briggs, P. R.: A rational function approach for estimating mean annual evapotranspiration, Water Resour. Res., 40, W02502, doi:10.1029/2003wr002710, 2004.

Zhang, L., Potter, N., Hickel, K., Zhang, Y., and Shao, Q.: Water balance modeling over variable time scales based on the Budyko framework – Model development and testing, J. Hydrol., 360, 117–131, doi:10.1016/j.jhydrol.2008.07.021, 2008.

Zhou, S., Yu, B., Huang, Y., and Wang, G.: The complementary relationship and generation of the Budyko functions, Geophys. Res. Lett., 42, 1781–1790, doi:10.1002/2015gl063511, 2015.

2

Future extreme precipitation intensities based on a historic event

Iris Manola[1], Bart van den Hurk[2,3], Hans De Moel[2], and Jeroen C. J. H. Aerts[2]

[1]Meteorology and Air Quality, Department of Environmental Sciences, Wageningen University, Wageningen, the Netherlands

[2]Institute for Environmental Studies, Vrije Universiteit (VU), Amsterdam, the Netherlands

[3]The Royal Netherlands Meteorological Institute (KNMI), De Bilt, the Netherlands

Correspondence: Bart van den Hurk (bart.van.den.hurk@knmi.nl)

Abstract. In a warmer climate, it is expected that precipitation intensities will increase, and form a considerable risk of high-impact precipitation extremes. This study applies three methods to transform a historic extreme precipitation event in the Netherlands to a similar event in a future warmer climate, thus compiling a "future weather" scenario. The first method uses an observation-based non-linear relation between the hourly-observed summer precipitation and the antecedent dew-point temperature (the P_i–T_d relation). The second method simulates the same event by using the convective-permitting numerical weather model (NWP) model HARMONIE, for both present-day and future warmer conditions. The third method is similar to the first method, but applies a simple linear delta transformation to the historic data by using indicators from The Royal Netherlands Meteorological Institute (KNMI)'14 climate scenarios. A comparison of the three methods shows comparable intensity changes, ranging from below the Clausius–Clapeyron (CC) scaling to a 3 times CC increase per degree of warming. In the NWP model, the position of the events is somewhat different; due to small wind and convection changes, the intensity changes somewhat differ with time, but the total spatial area covered by heavy precipitation does not change with the temperature increase. The P_i–T_d method is simple and time efficient compared to numerical models. The outcome can be used directly for hydrological and climatological studies and for impact analysis, such as flood-risk assessments.

1 Introduction

It is expected that climate change will increase the frequency and intensity of extreme precipitation events (e.g., Stocker et al., 2014; Pachauri et al., 2014). Different types of flooding may result from extreme precipitation, while the antecedent soil conditions also play a role on stream discharge levels (Ivancic and Shaw, 2015; Wasko and Sharma, 2017). In urban environments, extreme precipitation may lead to local-scale inundations, causing damage to houses and infrastructure within a time frame of several hours. On a larger river-basin scale, extreme rainfall over a period of days to several weeks may lead to river or flash floods, which may cause fatalities and can be catastrophic for the economy (e.g., Koks et al., 2015) and ecosystems (e.g., Knapp et al., 2008).

For the management of these risks, it is important to understand how the risk of extreme precipitation will change under future weather conditions. Current knowledge of climate change and possible future climate scenarios are developed within the Intergovernmental Panel on Climate Change (IPCC; Pachauri et al., 2014). For regional and national applications, tailored climate change scenarios have been developed, such as those for the Netherlands (Van den Hurk et al., 2014, henceforth "KNMI'14"). An important element for the successful application of climate change scenarios within a local to regional context is that they are tailored towards the needs of policy makers who use them in order to assess the effectiveness of adaptation strategies in reducing the risk of adverse effects, such as from flooding. Therefore, users of regional climate scenarios are increasingly involved in tailoring climate change information, in order to ensure that climate-scenario information is comprehensible and applicable to policy making (Van den Hurk et al., 2014).

In flood-risk management there is a need for climate scenarios that provide information on how extreme weather events may look like in the future (Aerts et al., 2014; Ward et al., 2014). The preferable way to obtain such information is to perform numerical (climate) model simulations that are sufficiently long to resolve the climate change trend (e.g., > 30 years) and which have a sufficiently high resolution to adequately resolve important dynamical and thermodynamic interactions, such as convective processes. Currently, such long and precise model simulations are lacking (Bürger et al., 2014) due to computational and data-storage constraints. Therefore, (a combination of) climate modeling and statistical corrections are usually employed, using shorter time series and providing projections of future climate, such as the official climate change scenarios for the Netherlands (KNMI'14; Van den Hurk et al., 2014). For example, a common approach to dealing with climate change in flood-risk studies can be described as a "delta-change" technique. In such a statistical approach, results usually from regional and global climate models are used to derive the (seasonal) change in precipitation characteristics, such as the wet-day frequency and the median or extreme precipitation. This change factor is subsequently applied to an observed time series or individual event in order to generate (extreme) rainfall under a changed climate (Lenderink et al., 2007; Fowler et al., 2007; Van Pelt et al., 2012; Räty et al., 2014). Another approach that is used to study precipitation extremes is to improve the low spatial and temporal resolution of long model simulations by means of statistical and dynamical downscaling techniques (Maraun et al., 2010). Such simulated time series can also be improved by using bias-correction methods that are derived from present-day simulations (Teutschbein et al., 2012; Bakker et al., 2014). Nevertheless, bias continues to exist and the uncertainties remain quite high.

Recently, a novel "future weather" concept has been proposed in order to provide high-resolution information on relevant characteristics of specific future extreme events, such as the duration and intensity of heavy rainfall (Hazeleger et al., 2015). According to this concept, historically observed events are used as a reference and modified with the use of numerical weather-prediction models, so that the outcome shows how the same event would occur in a future warmer climate. By applying a future situation to past events that are known to flood-risk managers, it is much easier for them to interpret the impact of such hypothetical future conditions. Hazeleger et al., (2015) used a high-resolution global-atmospheric model to simulate a future extreme weather event, by imposing future boundary conditions on historic numerical weather prediction (NWP) simulations. Lenderink

and Attema (2015) developed future scenarios of local precipitation events by perturbing the temperature and humidity boundary conditions of simulations in the regional models RACMO2 and HARMONIE, in order to mimic a 2 °C warmer world. A similar method is the "pseudo-global-warming" method, which involves the simulation of observed events modifying the meteorological forcing by a climate change difference (Schär et al., 1996; Michaelis et al., 2017). For example, Trapp and Hoogewind (2016) applied climate change differences from CMIP5 (Coupled Model Intercomparison Project Phase 5) simulations on the high-resolution Weather Research and Forecasting (WRF) model to reveal how typically observed extreme tornadoes might be realized under conditions of the late 21st century.

Among the four methods described above (delta change, downscaling techniques, bias correction, and future weather), the delta change and the future weather are employed in this paper. The main aim is to compare a "future weather" simulation with two alternative "delta-change" scaling methods, of which one is developed in this study. All three methods are applied to the same case study of extreme precipitation that took place in the Netherlands on 28 July 2014. The future weather method uses the outcome of the high-resolution numerical weather prediction model HARMONIE (Seity et al., 2011). This model was forced with boundary conditions representing both the historic event and future conditions, in order to obtain information on how the event would behave in the future. The first scaling method follows a non-linear delta transformation (Lenderink and Van Meijgaard, 2008; Lenderink and Attema, 2015), based on the observational behavior of precipitation intensity (P_i) as a function of the dew-point temperature (T_d) (henceforth "P_i–T_d relation"). The transformation was superimposed directly on the historical data assuming a future warmer world. The second scaling-method is a simplistic linear delta-change technique, which takes results from the KNMI'14 scenarios (Van den Hurk et al., 2014) in order to develop the future event.

This paper is organized as follows: Sect. 2 begins with a flowchart summary of the steps that are followed, and subsequently illustrates and discusses the three methods for projecting the future event. In Sect. 3, the observed Dutch case study is presented and then simulated in the HARMONIE model. Section 4 presents the future event for each method, firstly individually and secondly by a quantitative and qualitative comparison. The final section summarizes the research and concludes this paper.

2 Methods

The steps that were followed in this paper are summarized in Fig. 1. Overall, three methods were used to transform an observed event (28 July 2014, in the Netherlands) into a future event, assuming a warmer climate. The first scaling method, which is the P_i–T_d non-linear delta transformation, is based on summertime hourly radar precipitation data and hourly-

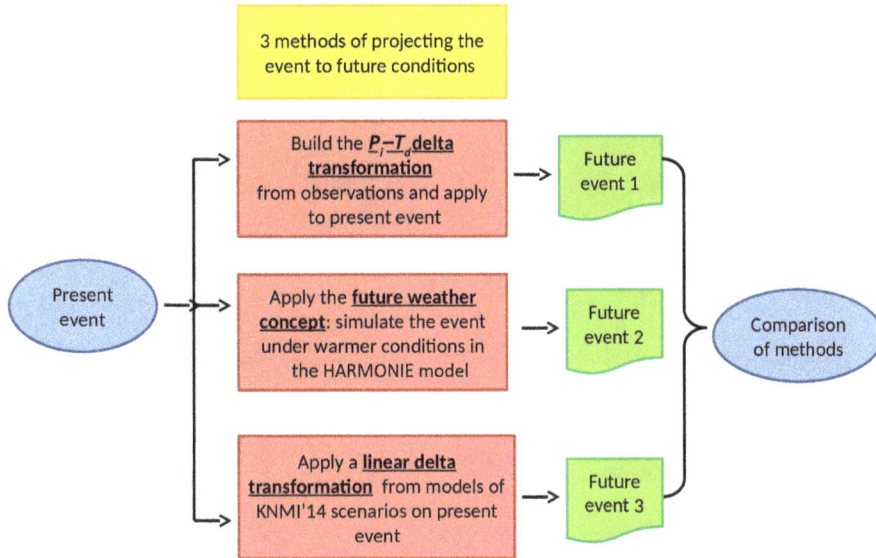

Figure 1. In order to produce a future precipitation event, three methodologies were followed, each departing from the same historic event. Future event 1: using observed precipitation intensity–dew-point temperature relations in order to create change factors. Future event 2: using the weather model HARMONIE with perturbed initial conditions from ECMWF's ENS. Future event 3: using delta-change factors retrieved from climate model simulations.

observed dew-point temperature for the years 2008–2015. In order to create the future precipitation event, the P_i-T_d transformation was applied to the precipitation data assuming a 2 °C warmer T_d. In the future weather method, the historic event was simulated using an ensemble of seven runs from the weather model HARMONIE, using both the historic boundary and the initial conditions from the Ensemble Prediction System (ENS) of ECMWF (Molteni et al., 1996; Leutbecher et al., 2008) for that particular day. Subsequently, the relevant future ensemble was simulated by perturbing the initial and boundary conditions to represent a unified increase of 2 °C, while maintaining the relative humidity (RH) constant. This assumption is based on long-term projections for the Netherlands, where the RH shows either no change or a small decrease (Lenderink et al., 2011 and KNMI'14). In the linear delta-change method (or delta transformation method) a factor was used to perturb the event, again assuming a warming of 2 °C.

The three methodologies were statistically compared and evaluated. As the HARMONIE model has shown to sufficiently simulate observed events (Attema et al., 2014; Koutroulis et al., 2015) and as it is the method that involves the highest level of physical sophistication, its outcome is used here as a benchmark for the evaluation of the P_i-T_d method and the linear-delta method. The comparison between the P_i-T_d method and the future simulation provides information on how the explicitly modeled interactions affect the results, compared to the statistical methods. Comparing the P_i-T_d method and the linear-transformation points, reveals the added value of enhancing the sophistication of the statistical scaling approach.

2.1 The Dutch case study

The Netherlands is a low-lying country that is shaped by the river deltas of the Meuse and Rhine rivers. It is vulnerable to flooding from storms surging from the North Sea, as well as river flooding. In addition, extreme precipitation events inundate urban areas and agricultural fields, frequently leading to considerable damage. Observations show that the temperature in the Netherlands rose by 1.8 °C since the beginning of the 20th century, clearly exceeding the global average (KNMI'14 scenarios; Van den Hurk et al., 2014). This has led to an increase in atmospheric moisture, a 25 % increase in the annual mean precipitation, and an increase of 12 % °C^{-1} in the hourly intensity of the most extreme showers of the 20th century. In the KNMI'14 scenarios, the temperature is projected to rise another 1–2.3 degrees until 2050, leading to more frequent and intense extreme precipitation events.

The extreme precipitation event that is analyzed here took place on the 28 July 2014 and resulted in blocked highways, the disruption of air transportation, and flooded buildings and public facilities. An analysis of the 325 Dutch rain-monitoring stations shows that an event of such intensity has a 5- to 15-year return period (Van Oldenborgh and Lenderink, 2014). It consisted of scattered, strong convective cells that started in the early morning in the west and southwest of the country, and reached the central-eastern region in the afternoon. The daily accumulated-precipitation intensity reached 140 mm locally (Fig. 2a). The small scale of the convective events underlines the need for high-resolution convection-permitting modeling. As the most severe damages are usually reported over urban areas (Ward et al.,

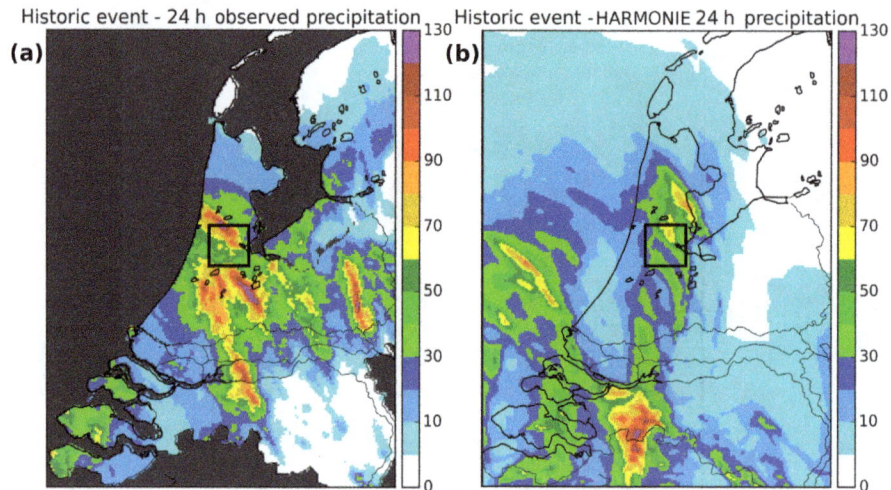

Figure 2. The daily accumulated precipitation in $mm\,d^{-1}$ over the Netherlands for 28 July 2014. **(a)** From radar observations (only with available data from the Netherlands' inland regions) and **(b)** As simulated by a representative member from the HARMONIE ensemble run. The black box shows the area of the city of Amsterdam.

2013), this analysis mainly focuses on the period between 08:00 and 09:00 LT (local time), the time of day in which the most precipitation fell over the city of Amsterdam.

2.2 The historic event in the HARMONIE model

In order to simulate the historic and future events, two ensemble simulations were carried out with the high-resolution weather forecasting model HARMONIE (Seity et al., 2011, cycle 40): one ensemble under present climate conditions and one under future climate conditions (2 °C warmer). HARMONIE uses non-hydrostatic convection-permitting dynamics and AROME physics with a horizontal resolution of $1\,km \times 1\,km$, 60 levels in the vertical direction, and a time step of 1 min. The output is given every hour. The initial and boundary conditions are taken from the ECMWF's ENS ensemble runs, are updated every hour, and have a $0.28° \times 0.28°$ grid size ($\sim 32\,km \times 20\,km$). The ENS is built to predict the probability distribution of forecast states, taking into account the random analysis error and model uncertainties. In order to select the best-fitted initial and boundary conditions for the simulation of the present event in HARMONIE, the ENS ensemble of 51 members was run for the day of the event. From the outcome, seven runs that performed closest to the radar observations were selected. These runs initiated the HARMONIE ensemble at 12:00 LT on 27 July 2014 and ran for 36 h, rendering an hourly output of the simulated historic event. This starting time was selected as the precipitable pattern was closer to that of the radar observations, compared to the runs initiated at 00:00 LT on 28 July.

The outcome of the present ensemble simulation under the initial conditions for 28 July shows that the HARMONIE captures sufficiently well the convective nature, the approx-

imate size of the cells, and the maximum intensity of precipitation, as well as the duration and the approximate time evolution of the event in all of its seven members. However, the location of the reported events was not very accurate (Fig. 2b). Clear discrepancies can be found in the position and number of convective cells between the simulation and the observations, and between the individual ensemble members (not shown here). The relatively low predictability of the exact position of the cells is due to the unstable, chaotic character of the specific event and to the imperfection of the model's initial and boundary conditions.

2.3 Scaling method 1: the P_i–T_d from observations

In this section, the methodology for expressing the precipitation intensity as a function of the dew-point temperature is discussed and compared to CC (Clausius–Clapeyron) scaling. The method was applied to the historic event using a perturbed input temperature in order to depict the expected intensity changes for a warmer climate.

2.3.1 The P_i–T_d relation

An important thermodynamic expression for the formation of precipitation in the atmosphere is the CC relation, according to which the maximum holding capacity of water vapor in an air mass increases by approximately $7\,\% \,°C^{-1}$ of warming (Trenberth et al., 2003). When the intensity of heavy precipitation is limited by the local availability of atmospheric moisture and is not sensitive to the atmospheric dynamic advection processes, it can be expected that the precipitation intensity increases at the same rate. However, both observations and model simulations show deviations from the CC scaling (Haerter and Berg, 2009; Bürger et al., 2014), as the dynamics and feedbacks between atmospheric processes also play an important role in the formation of precipitation.

For example, the relation between extreme precipitation intensity and temperature has been found to reach 2 times that of the CC scaling, i.e., up to $14\,\%\,°C^{-1}$ of warming (Lenderink and Van Meijgaard, 2008; Sugiyama et al., 2010; Panthou et al., 2014; Attema et al., 2014; Allan, 2011; Berg et al., 2013). This scaling relation shows some large spatial inhomogeneity (Wasko et al., 2016), with the strong scaling found mainly in the mid- and high latitudes, while in the tropics extreme precipitation intensities are even found to exhibit a decrease with increasing dew-point temperatures (Utsumi et al., 2011). The exceedance of the CC scaling for extreme precipitation is suggested to be related to the large- and small-scale dynamics of the atmosphere, and to the vertical stability (Loriaux et al., 2013; Lenderink and Attema 2015). Other studies indicate that statistical factors account for temperature-related changes in precipitation types, with an increasing contribution of convective warmer rain as temperature rises (Haerter and Berg, 2009). Other processes that potentially play a role are the increase in convective available potential energy (CAPE) with temperature and the positive feedback that is induced by the release of latent heat energy during the condensation of water vapor, thereby enhancing convection (Panthou et al., 2014). In general, the relation between precipitation intensity and temperature varies with region, season, duration, and type of precipitation, and is different for low and high temperatures, ranging from below CC to super CC. The scaling can be expressed using either absolute temperature (T) or dew-point temperature (T_d) as a reference. Preference is given to T_d, as this quantity contains explicit information on both temperature and the near-surface humidity (Lenderink et al., 2011).

In addition, large-scale circulation, vertical stability, cloud microphysics, moist adiabatic lapse rate, soil-water scarcity, and other factors play a role. The CC or below CC rates are mainly followed by long, synoptic, colder rain, while the super CC is mainly found in short-lived, warmer convective rain (Panthou et al., 2014; Lenderink et al., 2011; Mishra et al., 2012; Singleton and Toumi, 2013).

The precipitation data used to build the P_i–T_d relation is hourly data from the gauge-adjusted Dutch Doppler weather-radar data set (Overeem et al., 2011). In this data set, the pixel area is approximately $0.9\,km \times 0.9\,km$ and is available for 8 years (2008–2015). The radar operates on the C-band and measures precipitation depths based on composites of reflectivities from two Dutch radar stations: De Bilt and Den Helder. The hourly dew-point temperature was derived from 37 KNMI weather stations in the Netherlands, for the same period as the precipitation data. The sample size of the observed data for the temperature range between $7\,°C$ and $21\,°C$ can be considered large enough to eliminate statistical artifacts that may occur, since it contains $97\,\%$ of the 8 years of hourly summer data.

One advantage of the radar's high resolution is that small-scale convective precipitation ($\sim 1\,km$) is resolved explicitly. Following Lenderink et al. (2011), only precipitating areas were taken into account ($> 0.1\,mm\,h^{-1}$). Rainfall intensity data were first classified into 15 non-uniform percentile classes, ranging from the 25th to the 99.9th percentile and placed in bins of $2\,°C$ T_d width overlapping with steps of $1\,°C$. The sensitivity to the temperature bin width was tested by comparing a 1 and $0.5\,°C$ bin width, and was found to be insignificant. In order to match the precipitation data to antecedent air-mass properties that are characteristic for the formation of the precipitation events, T_d was measured 4 hours prior to the precipitation time. This time shift also avoids the contamination of the temperature and RH records by the changes that the precipitation process imposes, such as temperature drops due to descending colder, dry air from convective downdrafts or to heat release from the evaporation of precipitation (Lenderink et al., 2011; Bao et al., 2017).

The P_i–T_d scaling was calculated separately for low to very high percentiles and is illustrated in Fig. 3. For low to medium temperatures, there is little change in the precipitation intensity with the temperature for all percentiles, while for higher temperatures and percentiles below the medium, a monotonic increase with T_d of about $5\,\%\,°C^{-1}$ is shown. This behavior is usually attributed to large-scale precipitation and passing synoptic systems. For warmer temperatures between about 15–20 °C, precipitation intensity increases rapidly with temperature. For medium percentiles the intensity increases at a rate of over 2 times CC ($14\,\%\,°C^{-1}$ of warming) and rising up to $21\,\%$ for higher percentiles (a 3 times CC rate). This rate levels off at very high percentiles and at dew-point temperatures above $21\,°C$, possibly due to a limitation of the moisture supply to sustain the high precipitation intensities (Hardwick-Jones et al., 2010), clouds reaching the tropopause, or statistical artifacts (Wasko et al., 2015). The extreme 3 times CC rate is attributed to short-lasting, warm, convective precipitation events. To confirm this, a comparison to winter conditions was made where the larger synoptic systems are dominant. Also, values were computed from daily averaged data, to filter out the effect of short-duration convective events. The rate is almost uniformly CC during the winter, while for daily summer data the rate is below CC for small percentiles and above CC for larger percentiles (not shown). This rapid increase in P_i with T_d is also visible in Fig. 1 of Loriaux et al. (2013), where a 3 times CC rate in the P_i–T_d relation is illustrated at the hourly and sub-hourly precipitation, for the 90th percentile of a temperature band, similar to the one that is discussed here. For very high percentiles the rate decreases to 2 times CC.

2.3.2 The P_i–T_d as delta transformation

A multi-decadal observational analysis in the Netherlands shows that the trend in extreme precipitation can be ex-

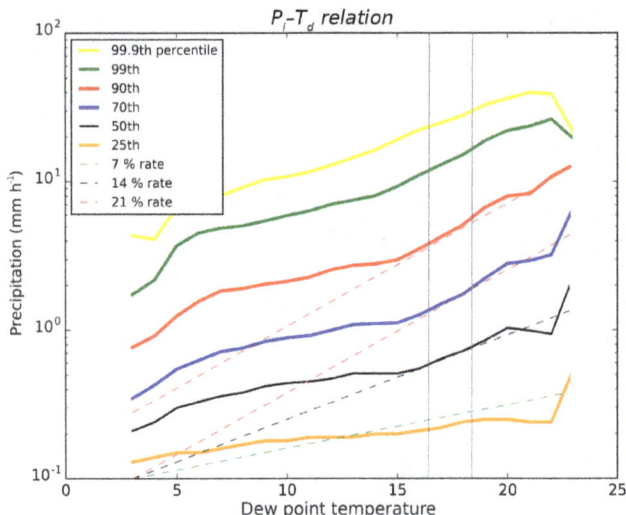

Figure 3. Precipitation intensity as a function of the dew-point temperature from hourly observations for June–August 2008–2015. The solid lines indicate the different percentiles of precipitation intensity. The dotted lines indicate the 7, 14, and 21 % °C^{-1} scaling. The two grey vertical lines indicate the mean temperature on 28 July 2014 in the Netherlands and a 2 °C warmer future event.

plained by changes in dew-point temperatures (Lenderink et al., 2011; Lenderink and Attema 2015). In the same study, a similar long-term trend between T and T_d indicates an almost constant RH with time, which implies that changes in T scale with changes in T_d. The KNMI'14 scenarios also project no change or a small decrease in the future RH, depending on the scenario.

In the P_i–T_d transformation, the dynamics of the atmosphere and the RH are assumed to be unchanged. Starting from the historic event, the dew-point temperature and precipitation intensity per grid point are calculated and attributed to a point in the P_i–T_d graph (Fig. 3), for the corresponding P_i percentile. The increase in the precipitation intensity, ΔP_i, is found by moving the initial point along the isolines in Fig. 3 by 2 °C. The procedure is repeated for each grid point individually. This method only examines the possible changes in the intensity of precipitation in already precipitable areas and does not allow changes in the spatial scale of the event. The mean dew-point temperature of the event of 28 July 2014 is 17.3 °C with a spatial variance of 1.8 °C. In addition to the application to the observed records, the P_i–T_d method is applied to the seven members of the simulated historic event shown in Fig. 4a below (to be discussed later).

2.4 Simulating the future weather

For the future event, the HARMONIE ensemble was run again, with the temperature of the initial and boundary conditions being increased by 2 °C at all levels and time steps. The RH was kept constant in order to ensure that the provided moisture remained adequate. Due to the constant RH,

the temperature change approximately scales with the T_d change, resulting in a roughly equal change of 2 °C in T_d. Attema et al. (2014) show that the simplification of the homogeneous increase in temperature and RH do not result in significant differences compared to non-homogeneous changes to the temperature and humidity profiles that were derived from a long climate change simulation.

2.5 Scaling method 2: linear delta transformation from climate models

A common approach to account for climate change effects in hydrological assessments is known as the delta change approach (Andreasson et al., 2004). The change signal between a control (current climate) situation and a future climate condition is used to adjust an observed climate record (such as temperature and precipitation). This adjusted series is subsequently used as input for the hydrological assessment (such as flood simulation). This approach is widely used (e.g., see Hay et al., 2000; Andreasson et al., 2004; and references therein), as it is relatively easy to use and requires only a couple of change factors that can directly be retrieved from either global climate model (GCM) runs or climate scenarios (such as the KNMI'14). Such change factors can differ in terms of complexity, ranging from a single change factor for all values to separate change factors for different months, seasons, and percentiles. In some cases, specific statistical tools have been developed that adjust observed time series by using various parameters that are related to climate change (such as amount of wet days, change in mean, change in extreme; Bakker and Bessembinder, 2012), as used in Te Linde et al. (2010).

In our case, the linear delta approach is applied with the KNMI'14 scenarios (Van den Hurk et al., 2014), which are based on the global climate scenarios from the latest IPCC report (Stocker et al., 2014), but tailored to the area of the Netherlands. Four KNMI'14 climate change scenarios were developed for 2050 and 2085. We selected a scenario in which T_d is expected to rise by 2 °C by 2050. Furthermore, the mean temperature in the selected scenario is expected to increase by 2.3 °C and mean summertime RH is expected to decrease by 2.5 %. According to this scenario, the maximum hourly intensity of the precipitation per year will increase by a maximum of 25 %. In order to up-scale the intensity of the historic event with this linear-delta factor, the entire range of the historic precipitation is increased by 25 % (assuming a steady increase with temperature, an increase of 11.8 % °C^{-1} of T_d warming for a total rise of 2 °C, leads to a $1.118^2 = 1.25$ or 25 % of increase in the intensities).

3 Results

Figure 4a shows the historic event as simulated by a representative ensemble member in HARMONIE at 09:00 LT. The relevant future event for the same member as simulated by the HARMONIE model is shown in Fig. 4b, as resulted by the P_i–T_d method in Fig. 4c and by the linear delta transformation in Fig. 4d. It is shown that the maximum P_i is clearly increased in all three methods.

As the P_i–T_d and linear delta methods only modify precipitable areas, the future spatial pattern remains unchanged compared to the historic simulated event. Conversely, the simulated future event differs in both intensity and precipitable pattern. The main body of the precipitable area is shifted towards the northeast in this member, mainly due to changes in horizontal winds. The variability between the different members primarily results from alterations in the horizontal winds and the convection, due to changes in the surface temperatures, which may shift or change the structure of the clouds. As the event evolves in time, the dynamic heat fluxes and the rapid drying of the soil induce temperature deviations that reach $\pm 4\,°C$ locally, thereby influencing the convection and the horizontal winds.

One interesting outcome in the simulated future weather method is that, despite the temperature increase and the moisture supply, the overall size of the future precipitable domain in all members remains relatively similar to the historic event. A possible explanation could be that, due to the stronger updrafts (caused by extensive warming, and resulting in increased convection and P_i), stronger downdrafts might be imposed at the outskirts of the clouds, thereby preventing them from expanding further. This may also explain the low or negative scaling that is observable in the low percentiles: as the P_i increases faster within the same domain and reaches higher maxima in the future event, there are smaller chances of finding light precipitation.

The box plots of Fig. 5 depict the intensity increase in the three methods compared to the simulated historic event for all seven members and for various precipitation percentiles at 09:00 LT and supplementary at 14:00 LT, when the event evolves towards its decaying phase. In the P_i–T_d method, following the observed scaling of Fig. 3, the lower percentiles (25th) increase with a rate of ΔP_i around the CC rate $(7\,\%\,°C^{-1})$. The medium percentiles (50th) increase between 2 times CC and over 3 times CC, and the high percentiles increase from 2 times CC up to 3 times CC. The rate of increase decreases slightly for the very high percentiles, reaching a maximum rate of 2 times CC. There are no considerable differences between the intensity increase at 09:00 LT and 14:00 LT, while some variance is observable between the different members, due to slightly different initial conditions of P_i and T_d across the ensemble. In the linear delta method the increase is a constant $11.8\,\%\,°C^{-1}$ with no variance between the members. The overall duration of the event in both P_i–T_d and linear delta methods remains unchanged compared to the historic event.

On the other hand, the simulated future weather method in HARMONIE in Fig. 5 shows deviations in the response of the model in the morning and in the afternoon. The main P_i increase takes place during the first hours of the event,

while the rate of increase later reduces, possibly due to the reduced moisture supply that results from the extensive precedent rain. In more detail, the very high percentiles in the morning increase at a rate that lies between 2 times CC and 3 times CC, the high percentiles even exceed 3 times CC and the medium percentiles cover the range of both the high and the very high percentiles. The ΔP_i for the lower percentiles varies considerably between the different ensemble members, ranging from a negative ΔP_i to a 3 times CC rate. In the afternoon, the overall rate of increase is substantially decreased, with an average intensity increase of CC or lower, while some negative values appear in all percentiles.

Overall, the total increase in the precipitable water for the entire event duration for a $2\,°C$ of warming in the P_i–T_d method is 36 %, which is about $17\,\%\,°C^{-1}$, the total increase in the future weather method is 27 % (or $13\,\%\,°C^{-1}$) and the total increase in the linear delta transformation is 25 % (or $11.8\,\%\,°C^{-1}$).

4 Discussion

All three methods analyzed in this study show an overall increase in the precipitation, together with temperature. Some discrepancies occur in the changes of intensities, duration, and the percentile distribution of the future precipitation, as well as in the spatial patterns, the position, and the number and size of the precipitable cells. A summary of the main results is found in Table 1.

The fitted lognormal distributions for the frequency of occurrence of the different precipitation intensities (Fig. 6) show strong similarities between the three methods and a clear distinction between present and future. The entire spectrum of the future events is shifted towards higher intensities. The chances of moderate precipitation are reduced and there is a distinct increase in the frequency of occurrence for $P_i > 15\,mm\,h^{-1}$. For example, the probability of the occurrence of intensities higher than $20\,mm\,h^{-1}$ is increased by over 35 %. A Kolmogorov–Smirnov test was performed to compare the goodness of fit for various distributions (the beta, gamma, Pareto, and lognormal) to conclude that the best fitting distribution for the current data is the lognormal.

Unlike in the P_i–T_d and the linear method, the future model simulations show a non-uniform change in P_i with time and space. In the model, the most intense precipitation increase takes place during the first hours, while the rate of increase later drops, possibly due to a drying of the atmosphere resulting from the exceedance of the water that precipitates in the early hours. HARMONIE tends to simulate stronger increases in the very high and low precipitation intensities in the first hours of the event, while the P_i–T_d method follows a structured and more constant increase that depends only on the P_i and T_d of the historic event at every hour (Fig. 5). The total amount of precipitable water that falls in the future P_i–T_d event is slightly larger than in the HAR-

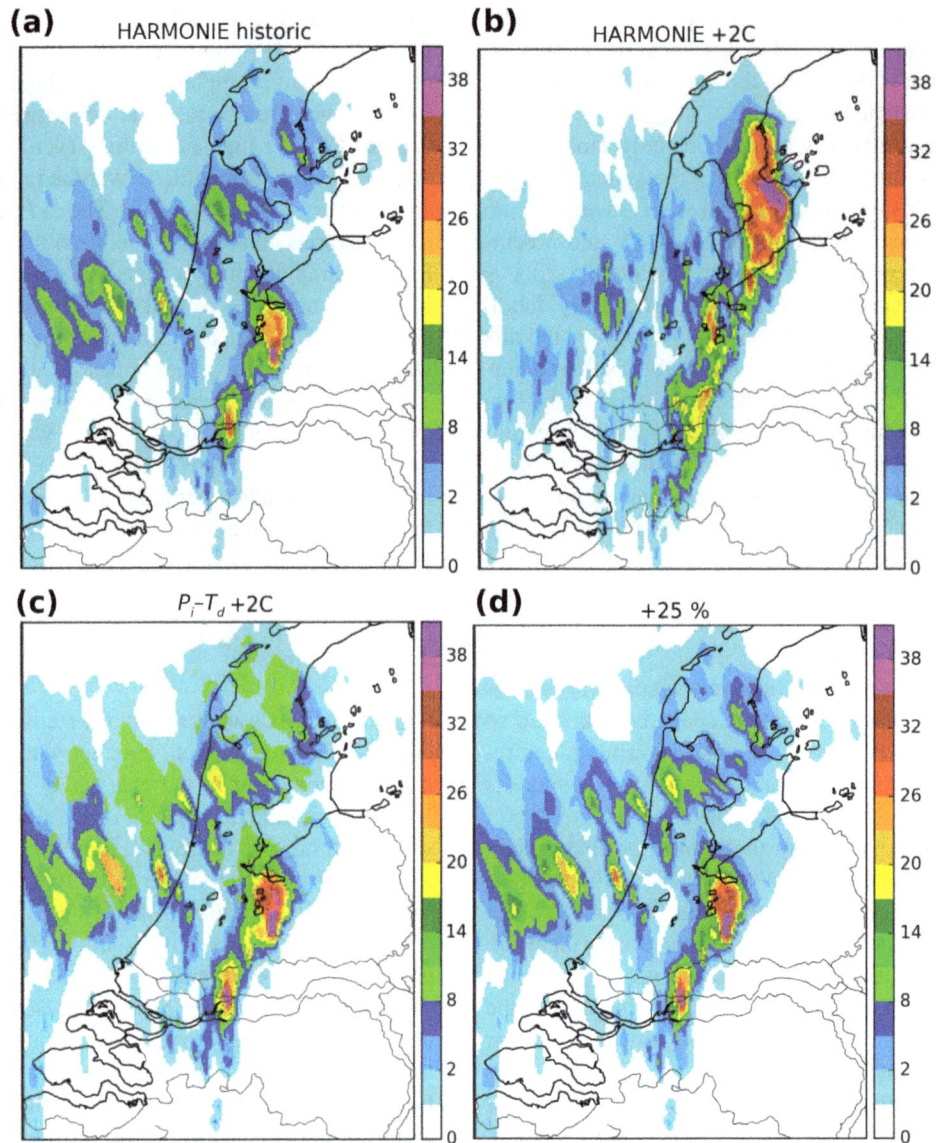

Figure 4. Hourly precipitation intensities at 09:00 LT (mm h^{-1}). **(a)** HARMONIE simulation under historic initial and boundary conditions and **(b)** simulated under future conditions. **(c)** Application of the P_i–T_d method and in **(d)** the uniform scaling of $+11.8\,\%\,°C^{-1}$.

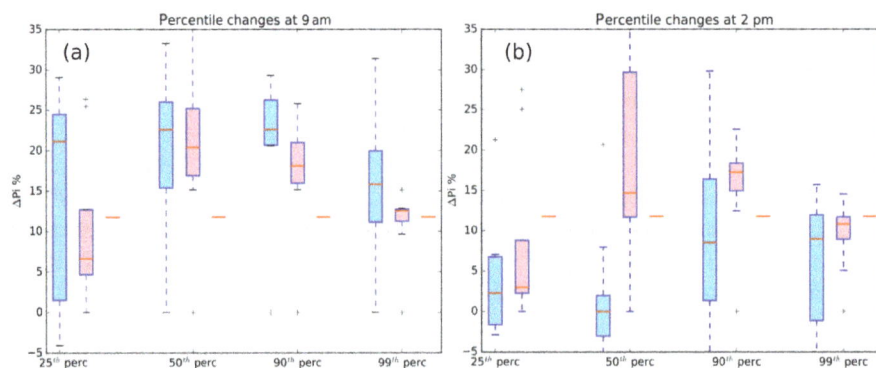

Figure 5. The change in the hourly precipitation per degree of warming compared to the simulated historic ensemble for the different percentiles and for the three methods over all precipitable points at 09:00 LT **(a)** and 14:00 LT **(b)**. The blue box plots represent the ensemble of the future model run and the pink box plots represent the outcome of the P_i–T_d method, starting from the ensemble of the historic simulation. Each ensemble contains seven runs. The single red lines indicate the relevant linear transformation of $+11.8\,\%\,°C^{-1}$.

Table 1. Summary of the findings from all ensemble members and at all hours of the event. The numbers show the percentages of change in comparison to the historic ensemble for the three methods and the total precipitation that fell, as well as for the changes in the average intensity, the very extreme percentiles and the precipitable surface area. The noted percentages are per degree of warming, assuming a linear increase.

	Future P_i–T_d	Future HARMONIE	Future +25 %
Total P	17 %°C^{-1}	13 %	11.8 %
Average P	17 %	11 %	11.8 %
99.9th P	12 %	10.5 %	11.8 %
Precipitable area	0	2 ± 3 %	0

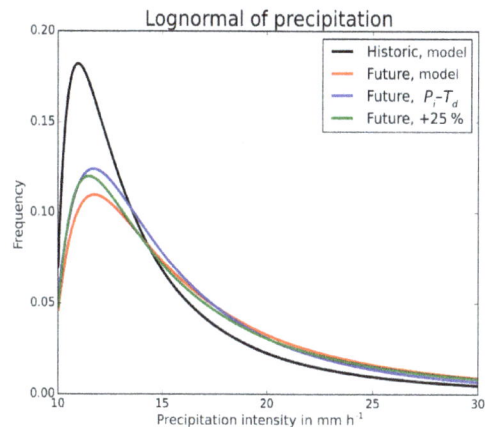

Figure 6. The fitted lognormal distribution of precipitation intensities for the present and future events, for all members and methods at 09:00 LT.

MONIE future event, due to the model's reduction of the P_i increase in the late hours of the event. The linear method, on the other hand, results in an overall underestimation of the total precipitable amount of water, as it underestimates the P_i increase for the moderate and high percentiles. The duration of the event in the model does not change in the future simulation, in agreement with Chan et al. (2016), in which future simulations with a convective-permitting model were made to show a clear intensification of sub-hourly rainfall, but no change in rainfall duration.

An intrinsic discrepancy between the model and the delta methods is the ability to shift or build new convective cells, due to the advection of moisture as a result of changes in wind and temperature patterns, which lead to changes in the precipitable spatial patterns. Nevertheless, in the model, the total precipitable coverage remains practically unchanged with temperature change, as is also assumed in the two statistical methods. This case study finding might contradict the recent observational study of Lochbihler et al. (2017), where Dutch radar precipitation data were used to conclude that on average the precipitable cells increase with increasing temperature and precipitation intensity, especially at higher dew point temperatures. On the other hand, Wasko et al. (2016) found evidence that precipitation intensity in Australia increases with temperature, while the storm's spatial extent decreases, as a redistribution of moisture towards the center takes place at the cost of the outer region of the precipitable area. The model study of Guinard et al. (2015) supports that the changes in precipitable structures with temperature are sensitive to the climatic region and the season.

It is of interest to investigate whether the different characteristics over sea and land (specifically the more unified temperatures over sea, the possibility of additional moisture provision, and the differences in wind characteristics) could induce deviations in the behavior of the future event's individual development over sea and over land. However, this experimental setting does not allow for such an analysis, as the spatial domain is rather limited. Changes in the horizon-

tal winds may therefore shift the clouds from over land to over sea or vice versa, thereby obfuscating the analysis.

Overall, the P_i–T_d method appears to render reliable results when qualitatively compared to the model and linear-transformation methods, while it is also faster, less expensive, and less complicated. The P_i–T_d relation has to be derived explicitly for different locations and different seasons, and is recommended to be used only within the range of well-documented dew-point temperatures for a specific area (e.g., $T_d > 7\,°C$ and $T_d < 21\,°C$ for the Netherlands in the summer).

5 Conclusions

New methods are emerging to project future extreme precipitation as it develops under climate change, grounded in existing events. For water managers, such future weather approaches have the advantage that they take a known extreme event as the basis and simulate its characteristics in a future warmer climate. However, such an approach requires high-resolution modeling and can be computationally demanding. In this paper, we compare two novel methods for a historic event in the Netherlands and one existing scenario method for projecting future extreme precipitation events starting from historic events, which can be used for climate research and impact studies.

The first method is a non-linear P_i–T_d relation and is used here as a delta transformation in order to project how a historic extreme precipitation event would intensify under future warmer conditions. We show that the hourly summer precipitation from radar observations with the dew-point temperature (the P_i–T_d relation) for moderate to warm days can increase by up to 21 %°C^{-1} of warming: a relation that is 3 times higher than the theoretical CC relation. The rate of change depends on the initial precipitation intensity, whereby low percentiles increase at a rate below CC, the medium and the very high percentiles (99.9th) at 2 times CC, and the moderately high and high percentiles at 3 times CC (90th). In the second method, the future extreme event is simulated in the HARMONIE model, alternating the historic initial conditions to represent a warmer atmosphere. Finally, the third method applies a linear delta transformation over the simulated historic event. The linear delta arises from the KNMI'14 scenarios, according to which all precipitation percentiles experience an increase of $11.8\,\%\,°C^{-1}$ in their intensities.

The comparison between the three future weather methods shows a comparable increase in the precipitation intensities, which range from below CC to a 3 times CC rate of change per degree of warming, depending on the initial percentiles. Some divergence is found in the distribution of the intensity changes, the time evolution of the event, and the position of the precipitable cells, due to the intrinsic discrepancies between the methods.

While the P_i-T_d method focuses primarily on the contribution of the thermodynamics and statistics in order to conclude the behavior of the precipitation with temperature, the future weather method in HARMONIE explores both the atmospheric dynamics and the thermodynamics, as well as on their interactions. Each run can evolve differently with time, while resolving the complicated atmospheric dynamics may increase the noise in the outcome.

A noteworthy discrepancy is that, in the HARMONIE model, the intensity changes are not uniform with time, as the main P_i increase takes place during the first hours of the event, while the rate of increase later reduces, possibly due to an exhaustion of atmospheric moisture resulting from the extensive precedent rain. Overall, the total increase in the precipitable amount of water increased by about $13\,\%\,°C^{-1}$ in the model method, $17\,\%\,°C^{-1}$ in the P_i-T_d method, and $11.8\,\%\,°C^{-1}$ in the linear method. Due to small wind and convection changes in the model, the clouds' position and patterns are displaced. Nevertheless, in the model, total-spatial precipitable coverage remains practically unchanged with temperature change, as is also assumed in the statistical methods.

The P_i-T_d method also has limitations, as it focuses on the precipitation-intensity changes, while it does not answer questions on spatial distribution and time evolution. Different precipitation types may also show different behavior with the temperature increase. For example, observations in Molnar et al. (2015) have shown that the intensity increase with temperature in convective events is higher than that of the synoptic storms. It should also be stated that none of the three methods include information on changes in the return period of events, or changes in the synoptic state of the atmosphere. For example, it is suggested that in the future rate of precipitation, intensities with temperature may decline over the UK, due to the more frequent occurrence of anticyclonic systems (Chan et al., 2016), indicating that there is a possibility for

some change in the future P_i-T_d scaling, depending on the region.

The P_i-T_d method projects precipitation events at different temperatures and is simple to use, requires little time, and is computationally and resource efficient, while it continues to offer rather robust results compared to a relevant non-hydrostatic model simulation. In all cases, the variance in the results with the P_i-T_d method is smaller than with the model method, allowing for a more straightforward and deterministic analysis if the outcome is to be used for impact studies. This method is suggested for use within well-documented temperature ranges deriving from observations in order to avoid statistical artifacts in the P_i-T_d scaling. Therefore it is not recommended to be used for very high (or very low) temperatures.

The outcome of the P_i-T_d future event can be used in several applications, such as impact and risk analyses to assess the economic and environmental damages of a future extreme event over an urban (or rural, industrial) areas, supporting policy makers to evaluate adequate adaptation measures against future disasters. It can also be used in several regional hydrological or larger spatial-scale climatological studies.

Competing interests. The authors declare that they have no conflict of interest.

Acknowledgements. This research has been funded by Amsterdam Water Science, the Amsterdam Academic Alliance (AAA), and the NWO Vici grant 453-13-006. We would like to thank KNMI for their support to offer access to the HARMONIE model.

Edited by: Carlo De Michele

References

Aerts, C. J. H. J., Botzen, W. J. W., Emanuel, K., Lin, N., de Moel, H., and Michel-Kerjan, E. O.: Evaluating flood resilience strategies for coastal megacities, Science, 344, 473–475, 2014.

Andreasson, J., Bergstrom, S., Carlsson, B., Graham, L. P., and Lindstrom, G.: Hydrological change – climate change impact simulations for Sweden, AMBIO: a journal for the human environment, 33, 228–234, https://doi.org/10.1579/0044-7447-33.4.228, 2004.

Attema, J., Loriaux, J. M., and Lenderink, G.: Extreme precipitation response to climate perturbations in an atmospheric mesoscale model, Environ. Res. Lett., 9, 014003, https://doi.org/10.1088/1748-9326/9/1/014003, 2014.

Bakker, A., Bessembinder, J., de Wit, A., Van den Hurk, B. J. J. M., and Hoek, S. B.: Exploring the efficiency of bias corrections of regional climate model output for the assessment of future crop yields in Europe, Reg. Environ. Change, 14, 865–877, 2014.

Bakker, A. M. R. and Bessembinder, J. E.: Time series transformation tool: description fo the program to generate time series consistent with the KNMI'06 climate scenarios, Royal Netherlands Meteorological Institute, Technical Report TR-326, 2012.

Bao, J., Sherwood, S. C., Alexander, L. V., and Evans, J. P.: Future increases in extreme precipitation exceed observed scaling rates, Nat. Clim. Change, 7, 128–132, https://doi.org/10.1038/nclimate3201, 2017.

Berg, P., Moseley, C. and Haerter, J. O.: Strong increase in convective precipitation in response to higher temperatures, Nat. Geosci., 6, 181–185, 2013.

Bürger, G., Heistermann, M., and Bronstert, A.: Towards Subdaily Rainfall Disaggregation via Clausius–Clapeyron, J. Hydrometeorol., 15, 1303–1311, 2014.

Chan, S. C., Kendon, E. J., Roberts, N. M., Fowler, H. J., and Blenkinsop, S.: Downturn in scaling of UK extreme rainfall with temperature for future hottest days, Nat. Geosci., 9, 24–28, 2016.

Fowler, H. J., Blenkinsop, S., and Tebaldi, C.: Linking climate change modelling to impacts studies: recent advances in downscaling techniques for hydrological modelling, Int. J. Climatol., 27, 1547–1578, https://doi.org/10.1002/joc.1556, 2007.

Guinard, K., Mailhot, A., and Caya, D.: Projected changes in characteristics of precipitation spatial structures over North America, Int. J. Climatol., 35, 596–612, https://doi.org/10.1002/joc.4006, 2015.

Haerter, J. O. and Berg, P.: Unexpected rise in extreme precipitation caused by a shift in rain type?, Nat. Geosci., 2, 372–373, 2009.

Hardwick Jones, R., Westra, S., and Sharma, A.: Observed relationships between extreme sub-daily precipitation, surface temperature, and relative humidity, Geophys. Res. Lett., 37, L22805, https://doi.org/10.1029/2010GL045081, 2010.

Hay, L. E., Wilby, R. L. and Leavesley, G. H.: A comparison of delta change and downscaled GCM scenarios for three mountainous basins in the United States, J. Am. Water Resour. As., 36, 387–397, 2000.

Hazeleger, W., Van den Hurk, B. J. J. M., Min, E., Van Oldenborgh, G. J., Petersen, A. C., Stainforth, D. A., Vasileiadou, E., and Smith, L. A.: Tales of future weather, Nat. Clim. Change, 5, 107–113, 2015.

Ivancic, T. J. and Shaw, S. B.: Examining why trends in very heavy precipitation should not be mistaken for trends in very high river discharge, Clim. Change, 133, 681–693, https://doi.org/10.1007/s10584-015-1476-1, 2015.

Knapp, A. K., Beier, C., Briske, D. D., Classen, A. T., Luo, Y., Reichstein, M., Smith, M. D., Smith, S. D., Bell, J. E., Fay, P. A., and Heisler, J. L.: Consequences of more extreme precipitation regimes for terrestrial ecosystems, Bioscience, 58, 811–821, 2008.

Koks, E. E., Bočkarjova, M., Moel, H., and Aerts,, J. C. J. H.: Integrated direct and indirect flood risk modeling: development and sensitivity analysis, Risk Anal., 35, 882–900, 2015.

Koutroulis, A. G., Grillakis, M. G., Tsanis I. K., and Jacob, D.: Exploring the ability of current climate information to facilitate local climate services for the water sector, Earth Perspec., 2, 1–19, 2015.

Lenderink, G. and Van Meijgaard, E.: Increase in hourly precipitation extremes beyond expectations from temperature changes, Nat. Geosci., 1, 511–514, 2008.

Lenderink, G. and Attema, J.: A simple scaling approach to produce climate scenarios of local precipitation extremes for the Netherlands, Environ. Res. Lett., 10, 085001, https://doi.org/10.1088/1748-9326/10/8/085001, 2015.

Lenderink, G., Buishand, A., and van Deursen, W.: Estimates of future discharges of the river Rhine using two scenario methodologies: direct versus delta approach, Hydrol. Earth Syst. Sci., 11, 1145–1159, https://doi.org/10.5194/hess-11-1145-2007, 2007.

Lenderink, G., Mok, H. Y., Lee, T. C., and van Oldenborgh, G. J.: Scaling and trends of hourly precipitation extremes in two different climate zones – Hong Kong and the Netherlands, Hydrol. Earth Syst. Sci., 15, 3033–3041, https://doi.org/10.5194/hess-15-3033-2011, 2011.

Leutbecher, M. and Palmer, T. N.: Ensemble forecasting, J. Comput. Phys., 227, 3515–3539, 2008.

Lochbihler, K., Lenderink, G., and Siebesma, A. P.: The spatial extent of rainfall events and its relation to precipitation scaling, Geophys. Res. Lett., 44, 8629–8636, https://doi.org/10.1002/2017GL074857, 2017.

Loriaux, J. M., Lenderink, G., De Roode S. R., and Siebesma, A. P.: Understanding convective extreme precipitation scaling using observations and an entraining plume model, J. Atmos. Sci., 70, 3641–3655, 2013.

Maraun, D., Wetterhall, F., Ireson, A. M., Chandler, R. E., Kendon, E. J., Widmann, M., Brienen, S., Rust, H. W., Sauter, T., Themeßl, M., and Venema, V. K. C.: Precipitation downscaling under climate change: Recent developments to bridge the gap between dynamical models and the end user, Rev. Geophys., 48, RG3003, https://doi.org/10.1029/2009RG000314, 2010.

Michaelis, A. C., Willison, J., Lackmann, G. M., and Robinson, W. A.: Changes in winter North Atlantic extratropical cyclones in high-resolution regional pseudo-global warming simulations, J. Climate, 30, 6905–6925, 2017.

Mishra, V., Wallace, J. M., and Lettenmaier, D. P.: Relationship between hourly extreme precipitation and local air temperature in the United States, Geophys. Res. Lett., 39, L16403, https://doi.org/10.1029/2012GL052790, 2012.

Molnar, P., Fatichi, S., Gaál, L., Szolgay, J., and Burlando, P.: Storm type effects on super Clausius-Clapeyron scaling of intense rainstorm properties with air temperature, Hydrol. Earth Syst. Sci., 19, 1753–1766, https://doi.org/10.5194/hess-19-1753-2015, 2015.

Molteni, F., Buizza, R., Palmer, T. N., and Petroliagis, T.: The ECMWF Ensemble Prediction System: methodology and validation, Q. J. Roy. Meteor. Soc., 122, 73–119, 1996.

Overeem A., Leijnse, H., and Uijlenhoet, R.: Measuring urban rainfall using microwave links from commercial cellular communication networks, Water Resour. Res., 47, W12505, https://doi.org/10.1029/2010WR010350, 2011.

Pachauri, R. K., Allen, M. R., Barros, V. R., Broome, J., Cramer, W., Christ, R., Church, J. A., Clarke, L., Dahe, Q., Dasgupta, P., and Dubash, N. K.: Climate Change 2014: Synthesis Report, Contribution of Working Groups I, II and III to the Fifth Assessment Report of the Intergovernmental Panel on Climate Change, p. 151, 2014.

Panthou, G., Mailhot, A., Laurence, E., and Talbot, G.: Relationship between surface temperature and extreme rainfalls: A multi-time-scale and event-based analysis, J. Hydrometeorol., 15, 1999–2011, 2014.

Räty, O., Räisänen, J., and Ylhäisi, J. S.: Evaluation of delta change and bias correction methods for future daily precipitation: intermodel cross-validation using ENSEMBLES simulations, Clim. Dynam., 42, 2287–2303, 2014.

Schär, C., Frei, C., Lüthi, D., and Davies, H. C.: Surrogate climate-change scenarios for regional climate models, Geophys. Res. Lett., 23, 669–672, 1996.

Seity, Y., Brousseau, P., Malardel, S., Hello, G., Bénard, P., Bouttier, F., Lac, C., and Masson, V.: The AROME-France convective-scale operational model, Mon. Weather Rev., 139, 976–991, 2011.

Singleton A. and Toumi, R.: Super-Clausius–Clapeyron scaling of rainfall in a model squall line, Q. J. Roy. Meteor. Soc., 139, 334–339, 2013.

Stocker, T. F., Qin, D., Plattner, G. K., Tignor, M., Allen, S. K., Boschung, J., Nauels, A., Xia, Y., Bex, B., and Midgley, B. M.: IPCC, 2013: climate change 2013: the physical science basis. Contribution of working group I to the fifth assessment report of the intergovernmental panel on climate change, Cambridge University Press, 2014.

Sugiyama, M., Shiogama, H., and Emori, S.: Precipitation extreme changes exceeding moisture content increases in MIROC and IPCC climate models, P. Natl. Acad. Sci., 107, 571–575, 2010.

Te Linde, A. H., Aerts, J. C. J. H., Bakker, A. M. R., and Kwadijk, J. C. J.: Simulating low probability peak discharges for the Rhine basin using resampled climate modeling data, Water Resour. Res., 46, W03512, https://doi.org/10.1029/2009WR007707, 2010.

Teutschbein, C. and Seibert, J.: Bias correction of regional climate model simulations for hydrological climate-change impact studies: Review and evaluation of different methods, J. Hydrol., 456, 12–29, 2012.

Trapp, R. J. and Hoogewind, K. A.: The realization of extreme tornadic storm events under future anthropogenic climate change, J. Climate, 29, 5251–5265, 2016.

Trenberth, K. E., Dai, A., Rasmussen, R. M., and Parsons, D. B.: The changing character of precipitation, B. Am. Meteorol. Soc., 84, 1205–1217, 2003.

Utsumi, N., Seto, S., Kanae, S., Maeda, E. E., and Oki, T.: Does higher surface temperature intensify extreme precipitation?, Geophys. Res. Lett., 38, L16708, https://doi.org/10.1029/2011GL048426, 2011.

Van den Hurk, B., Siegmund, P., and Tank, A. K. (Eds.): KNMI'14: Climate Change Scenarios for the 21st Century-a Netherlands Perspective, Royal Netherlands Meteorological Institute (KNMI), 2014.

Van Oldenborgh, G. J. and Lenderink, G.: Een eerste blik op de buien van maandag 28 juli 2014, Meteorologica, 3, 28–29, 2014.

van Pelt, S. C., Beersma, J. J., Buishand, T. A., van den Hurk, B. J. J. M., and Kabat, P.: Future changes in extreme precipitation in the Rhine basin based on global and regional climate model simulations, Hydrol. Earth Syst. Sci., 16, 4517–4530, https://doi.org/10.5194/hess-16-4517-2012, 2012.

Ward, P. J., van Pelt, S. C., de Keizer, O., Aerts, J. C. J. H., Beersma, J. J., Van den Hurk, B. J. J. M., and Linde, A. H.: Including climate change projections in probabilistic flood risk assessment, J. Flood Risk Manage., 7, 141–151, 2014.

Ward, P. J., Jongman, B., Weiland, F. S., Bouwman, A., van Beek, R., Bierkens, M. F. P., Ligtvoet, W., and Winsemius, H. C.: Assessing flood risk at the global scale: model setup, results, and sensitivity, Environ. Res. Lett., 8, 044019, https://doi.org/10.1088/1748-9326/8/4/044019, 2013.

Wasko, C., Sharma, A., and Johnson, F.: Does storm duration modulate the extreme precipitation-temperature scaling relationship?, Geophys. Res. Lett., 42, 8783–8790, 2015.

Wasko, C., Sharma, A., and Westra, S.: Reduced spatial extent of extreme storms at higher temperatures, Geophys. Res. Lett., 43, 4026–4032, 2016.

Wasko, C. and Sharma, A.: Global assessment of flood and storm extremes with increased temperatures, Sci. Rep., 7, 7945, https://doi.org/10.1038/s41598-017-08481-1, 2017.

3

Daily Landsat-scale evapotranspiration estimation over a forested landscape in North Carolina, USA, using multi-satellite data fusion

Yun Yang[1], Martha C. Anderson[1], Feng Gao[1], Christopher R. Hain[2], Kathryn A. Semmens[3], William P. Kustas[1], Asko Noormets[4], Randolph H. Wynne[5], Valerie A. Thomas[5], and Ge Sun[6]

[1]USDA ARS, Hydrology and Remote Sensing Laboratory, Beltsville, MD, USA

[2]Marshall Space Flight Center, Earth Science Branch, Huntsville, AL, USA

[3]Nurture Nature Center, Easton, PA, USA

[4]Department of Forestry and Environmental Resources, North Carolina State University, Raleigh, NC, USA

[5]Department of Forest Resources and Environmental Conservation, Virginia Polytechnic Institute and State University, Blacksburg, VA, USA

[6]Eastern Forest Environmental Threat Assessment Center, Southern Research Station, USDA Forest Service, Raleigh, NC, USA

Correspondence to: Yun Yang (yun.yang@ars.usda.gov)

Abstract. As a primary flux in the global water cycle, evapotranspiration (ET) connects hydrologic and biological processes and is directly affected by water and land management, land use change and climate variability. Satellite remote sensing provides an effective means for diagnosing ET patterns over heterogeneous landscapes; however, limitations on the spatial and temporal resolution of satellite data, combined with the effects of cloud contamination, constrain the amount of detail that a single satellite can provide. In this study, we describe an application of a multi-sensor ET data fusion system over a mixed forested/agricultural landscape in North Carolina, USA, during the growing season of 2013. The fusion system ingests ET estimates from the Two-Source Energy Balance Model (TSEB) applied to thermal infrared remote sensing retrievals of land surface temperature from multiple satellite platforms: hourly geostationary satellite data at 4 km resolution, daily 1 km imagery from the Moderate Resolution Imaging Spectroradiometer (MODIS) and biweekly Landsat thermal data sharpened to 30 m. These multiple ET data streams are combined using the Spatial and Temporal Adaptive Reflectance Fusion Model (STARFM) to estimate daily ET at 30 m resolution to investigate seasonal water use behavior at the level of individual forest stands and land cover patches. A new method, also exploiting the STARFM algorithm, is used to fill gaps in the Landsat ET retrievals due to cloud cover and/or the scan-line corrector (SLC) failure on Landsat 7. The retrieved daily ET time series agree well with observations at two AmeriFlux eddy covariance flux tower sites in a managed pine plantation within the modeling domain: US-NC2 located in a mid-rotation (20-year-old) loblolly pine stand and US-NC3 located in a recently clear-cut and replanted field site. Root mean square errors (RMSEs) for NC2 and NC3 were 0.99 and 1.02 mm day^{-1}, respectively, with mean absolute errors of approximately 29 % at the daily time step, 12 % at the monthly time step and 0.7 % over the full study period at the two flux tower sites. Analyses of water use patterns over the plantation indicate increasing seasonal ET with stand age for young to mid-rotation stands up to 20 years, but little dependence on age for older stands. An accounting of consumptive water use by major land cover classes representative of the modeling domain is presented, as well as relative partitioning of ET between evaporation (E) and transpiration (T) components obtained with the TSEB. The study provides new insights about the effects of management and land use change on water yield over forested landscapes.

1 Introduction

Evapotranspiration (ET) is a major component of the water balance and connects hydrologic and biological processes (Hanson et al., 2004; Wilson et al., 2001). ET varies with climate, vegetation type and phenological stage and is directly affected by land management strategies and climate change (Pereira et al., 2002). ET is also a key variable in most ecohydrological models and ecosystem service assessments (Abramopoulos et al., 1988; Kannan et al., 2007; Olioso et al., 1999; Tague and Band, 2004; Sun et al., 2011a, b). In spite of the importance of ET, routine estimation of ET at high spatial (plot level) and temporal (daily) resolution has not yet been achieved with acceptable accuracy over landscape and regional scales (Wang and Dickinson, 2012).

Current forest ET estimation methods span a range of spatial scales: from individual plants, to tower footprints, to watershed scales (Fang et al., 2015). These methods include in situ measurement, simulation using hydrologic and land surface models which are normally driven by weather data and estimation from satellite remote sensing data. Techniques for measuring ET include weighing lysimeters (Wullschleger et al., 1998), sap flow (Klein et al., 2014; Smith and Allen, 1996) and plant chambers (Cienciala and Lindroth, 1995), soil water budgets (Cuenca et al., 1997), eddy covariance (EC; Baldocchi et al., 2001) and catchment water balance (Pan et al., 2012). While EC is a widely used observation method and provides an important data source to many research fields (Baldocchi et al., 2001), it measures turbulent fluxes over a relatively small footprint area (10^2–10^4 m^2), which is determined by the microclimate conditions around the flux tower and the instrument height. Catchment water balance is also a frequently used method, calculating ET from long-term precipitation and streamflow observations with the assumption that the soil water storage change is negligible (Domec et al., 2012; Wilson et al., 2001). All these observation methods have their inherent advantages and limitations, especially when considering both temporal and spatial resolution issues.

Another group of forest ET estimation methods is empirically based, establishing a relationship between ET with other parameters: for example, precipitation, reference ET and vegetation indices (leaf area index – LAI, normalized difference vegetation index – NDVI and enhanced vegetation index – EVI) (Johnson and Trout, 2012; Mutiibwa and Irmak, 2013; Nemani and Running, 1988; Sun et al., 2011a; Zhang et al., 2004). Many studies have applied process-based ecohydrological models to estimate ET (Chen and Dudhia, 2001; Tague and Band, 2004; Tian et al., 2010). These models usually estimate ET from potential ET, which is then downregulated based on weather data and soil and vegetation characteristics. However, with the focus on predicting runoff and the soil water profile, studies using hydrologic models generally do not evaluate the performance of ET simulation. To simplify the physical processes, many models as-

sume the plant growth rate is static. This assumption can result in errors in simulating ET dynamics, especially over shorter time periods (seasonally, monthly, weekly or daily) (Méndez-Barroso et al., 2014; Tian et al., 2010). Often physical process-based models involve hundreds of input variables/parameters, many of which are not easily measured or known in a spatially distributed manner at watershed and regional scales. Although models can be calibrated using local or watershed-scale observations, there is the often-mentioned problem of equifinality, where different sets of parameters during calibration give the same simulation results due to the inherent complexity of the system (Von Bertalanffy, 1968; Beven and Freer, 2001).

Mapping ET using satellite remote sensing data has been widely applied since the 1980s due to growing interest in the spatial dynamics of water use at the watershed and regional scales (Kalma et al., 2008). Of particular interest in the water resource community are surface energy balance methods based on remotely sensed land-surface temperature (LST) retrieved from thermal infrared (TIR) imagery, which provides proxy information regarding the surface moisture status (Hain et al., 2011; Anderson et al., 2012a). LST captures signals of crop stress and variable soil evaporation that are often missed by crop coefficient remote sensing techniques, which are based on empirical regressions with reflectance-based vegetation indices. Furthermore, diagnostic estimates of ET from the surface energy balance provide an independent estimate of landscape water use that is a valuable benchmark for comparison with estimates based on water balance or hydrologic modeling (Hain et al., 2015; Yilmaz et al., 2014). Finally, the range in spatial resolution and coverage of existing TIR data sources enables mapping of ET from the plot or field scale (< 100 m resolution) up to continental or global coverage at 1–5 km resolution.

The Atmosphere–Land Exchange Inverse Model (ALEXI; Anderson et al., 1997, 2007a, b) and associated flux disaggregation algorithm (DisALEXI; Anderson et al., 2004; Norman et al., 2003) are examples of a multiscale energy balance modeling approach that can utilize LST data from multiple satellite platforms with TIR sensing capabilities. The regional ALEXI model uses time-differential measurements of morning LST rise from geostationary satellites to estimate daily flux patterns at 3–10 km resolution and continental scales. Using higher-resolution LST information from polar-orbiting systems, DisALEXI enables downscaling of ALEXI fluxes to finer scales, better resolving land use and moisture patterns over the landscape and better approximating the spatial scale of ground-based flux observations. Landsat data (30–120 m) can be used to retrieve ET at the field scale, which is particularly useful for water management applications. However, due to the lengthy revisit interval (8 to 16 days), further lengthened by cloud contamination, the number of useful Landsat scenes that can be acquired during a growing season is limited. The Moderate Resolution Imaging Spectroradiometer (MODIS) has a shorter revisit inter-

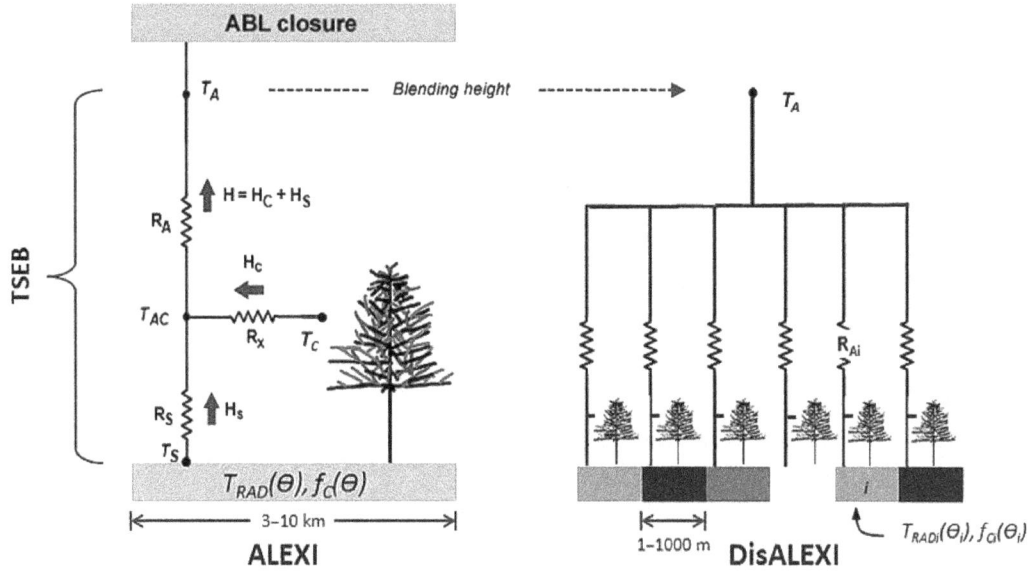

Figure 1. Schematic diagram of the ALEXI and DisALEXI modeling schemes. The left panel shows TSEB as employed to partition the income radiometric temperature ($T_{RAD}(\theta)$, θ is view angle) into canopy (subscript "c") and soil (subscript "s") components based on vegetation coverage ($f(\theta)$). Sensible heat (H) is regulated by the aerodynamic resistance (R_a), bulk leaf boundary layer resistance (R_x) and soil surface boundary layer resistance (R_s). ALEXI combines the TSEB and ABL models to estimate air temperature (T_A) at the blending height. The right panel represents the disaggregation of ALEXI output to finer scales based on LST and $f(\theta)$ information from Landsat and MODIS.

val (approximately daily) but is too coarse (1 km in the TIR bands) for field-scale ET estimation. Cammalleri et al. (2013) proposed a data fusion method to combine ET estimates derived from geostationary, MODIS and Landsat TIR data, attempting to exploit the spatiotemporal advantages of each class of satellite to map daily ET at a sub-field scale. This ET fusion approach has been successfully applied over rainfed and irrigated corn, soybean and cotton fields (Cammalleri et al., 2013, 2014), as well as irrigated vineyards (Semmens et al., 2015). The work described here constitutes the first application to forest land cover types, representing a substantially different roughness and physiological regime than that of shorter crops. This presents a modeling challenge in terms of accurately defining turbulent exchange coefficients, as well as describing radiation transport through the canopy.

In this paper, ALEXI and DisALEXI are applied over a commercially managed loblolly pine (*Pinus taeda*) plantation, representing a range in stand age, to estimate daily field-scale ET using the data fusion methodology. Retrieved 30 m ET time series are evaluated at two flux tower sites, sited in mature and recently clear-cut pine stands. The primary science objectives are to (1) study the accuracy of ALEXI and DisALEXI ET retrievals over forested sites, (2) evaluate the models' ability to capture the dynamics of fluxes over the contrasting canopy structures in both pine and the adjacent vegetation and (3) investigate the utility of daily field-scale ET retrievals for water resource management in forested systems. Additionally, we present a novel methodological ad-

vancement, based on data fusion, for filling gaps in Landsat-based ET retrievals due to partial cloud cover as well as the scan-line corrector (SLC) failure in Landsat 7. This technique facilitates more complete use of the existing Landsat archive for investigating water use dynamics at the landscape scale.

2 Methods

2.1 Thermal-based multiscale ET retrieval

The regional ALEXI and the associated DisALEXI models are based on the the Two-Source Energy Balance (TSEB) land-surface representation of Norman et al. (1995), with further refinements by Kustas and Norman (1999, 2000). The combined modeling system is described schematically in Fig. 1. Rather than treating the land surface as a homogeneous surface, the TSEB partitions modeled surface fluxes and observed directional radiometric surface temperature between soil and vegetation components:

$$T_{RAD}(\emptyset)^4 = f(\emptyset)T_c^4 + [1 - f(\emptyset)]T_s^4, \qquad (1)$$

where $\emptyset$ is the thermal view angle, $f(\emptyset)$ is the fractional vegetation cover apparent at the thermal view angle, T_{RAD} is the directional radiometric temperature, T_c is the canopy temperature and T_s is the soil temperature (K). In remote sensing applications, $f(\emptyset)$ can be estimated from retrievals of LAI using the Beer–Lambert law. The surface energy balance for the canopy, soil and the combined system is represented in Eq. (2):

$$R_{n,s} = H_s + \lambda E_s + G \tag{2a}$$

$$R_{n,,c} = H_c + \lambda E_c \tag{2b}$$

$$R_n = H + \lambda E + G, \tag{2c}$$

where the subscripts "c" and "s" represent fluxes from the canopy and soil components, and R_n is net radiation, λE is latent heat flux, H is sensible heat and G is the soil heat flux (all in units of $W\,m^{-2}$). Component surface temperatures in Eq. (1) are used to constrain R_n, H and G; canopy transpiration (λE_c) is initially estimated with a modified Priestley–Taylor approach under the unstressed conditions assumption, and then iteratively downregulated if T_c indicates canopy stress, ruled by the assumption that condensation under daytime clear-sky conditions is unlikely, while soil evaporation (λE_s) is computed as a residual to the soil energy budget. Further information regarding the TSEB model formulation is provided by Kustas and Anderson (2009).

Roughness length (Z_m) impacts the aerodynamic resistance (R_a), which is the resistance to heat transport across the layer between the nominal momentum exchange surface within the canopy and the height of the air temperature boundary condition (Z_T (m)). The aerodynamic resistance (R_a) can be expressed as Eq. (3) (Brutsaert, 1982):

$$R_a = \frac{\left[\ln\left(\frac{Z_T-d}{Z_m}\right) - \psi_h\right]\left[\ln\left(\frac{Z_u-d}{Z_m}\right) - \psi_m\right]}{k^2 u}, \tag{3}$$

where k is the von Karman constant (0.4); u ($m\,s^{-1}$) is the wind speed measured at height Z_u (m); d (m) is the displacement height; Z_m (m) is the roughness length, which can be estimated from the nominal canopy height (h_c(m)), $Z_m \approx h_c/8$ (Shaw and Pereira, 1982); and ψ_h and ψ_m are the stability corrections for heat and momentum transport, respectively. Additional boundary layer resistances linking the bulk canopy and the soil surface to the in-canopy momentum exchange node (R_X and R_S, respectively; see Fig. 1) are defined as in the series TSEB model formulation described in Norman et al. (1995).

The regional-scale ALEXI model applies the TSEB in time-differential mode using measurements of morning LST rise obtained from geostationary platforms (Anderson et al., 1997, 2007a). Energy closure over this morning period is obtained by coupling the TSEB with a simple model of atmospheric boundary layer (ABL) development (Fig. 1). In this study, instantaneous morning fluxes from ALEXI have been upscaled to daily total latent heat flux by conserving the ratio of λE to solar radiation, following the recommendations of Cammalleri et al. (2014). Daily latent heat flux (in energy units of $MJ\,m^{-2}\,day^{-1}$) is converted to ET (in mass units of $mm\,day^{-1}$) by dividing by the latent heat of vaporization ($\lambda = 2.45\,MJ\,kg^{-1}$).

This time-differential approach used in ALEXI reduces model sensitivity to errors in LST retrieval due to atmospheric and surface emissivity effects, but it does constrain ET estimates to the relatively coarse spatial scales typical of geostationary satellites. To estimate ET at the finer scales required for many management applications, the ALEXI fluxes can be spatially disaggregated using the DisALEXI approach (Anderson et al., 2004; Norman et al., 2003). DisALEXI uses, as an initial estimate, air temperature estimates diagnosed by ALEXI at a nominal blending height at the interface between the TSEB and ABL submodels, along with high spatial resolution images of surface temperature data and vegetation cover fraction from polar-orbiting or airborne systems, to run the TSEB at sub-pixel scales over each ALEXI pixel area. The TSEB fluxes are reaggregated and compared with the ALEXI pixel flux, and the air temperature boundary condition is iteratively modified until the fluxes are consistent at the ALEXI pixel scale. More details on the ALEXI/DisALEXI multiscale modeling system can be found in Anderson et al. (2004, 2011, 2012b).

2.2 Data processing and fusion system

In this study, ET retrievals generated with DisALEXI using TIR data from MODIS (near daily, at 1 km resolution) and Landsat (periodic, sharpened to 30 m resolution) have been fused to produce daily Landsat-scale ET time series. The major components of the processing stream are described in greater detail below, including a data-mining sharpener (DMS; Gao et al., 2012b) tool that is used to improve the spatial resolution of the LST inputs to DisALEXI, the Spatial and Temporal Adaptive Reflectance Fusion Model (STARFM; Gao et al., 2006) and a gap-filling procedure that is applied to ALEXI and MODIS and Landsat-DisALEXI retrievals prior to disaggregation and fusion. The gap-filling and fusion processes are schematically represented in Fig. 2.

2.2.1 Data-mining sharpener (DMS)

In both the Landsat and MODIS imaging systems, the TIR sensors have significantly lower spatial resolution than the shortwave instruments on the same platform. For Landsat, TIR resolution varies from 60 m (Landsat 7) to 100 m (Landsat 8) to 120 m (Landsat 5), while the shortwave images are processed to 30 m. For MODIS, TIR resolution is 1 km while the shortwave resolution is 250 m. Particularly for Landsat, there is a benefit to mapping ET at 30 m rather than the native TIR resolution, as boundaries in land cover and moisture variability are much better defined.

The DMS sharpening tool implemented within the ET fusion package enables this higher-resolution mapping. The DMS technique creates regression trees between TIR band brightness temperatures and shortwave spectral reflectances both globally across the full scene and within a localized moving window (Gao et al., 2012b). The original TIR data

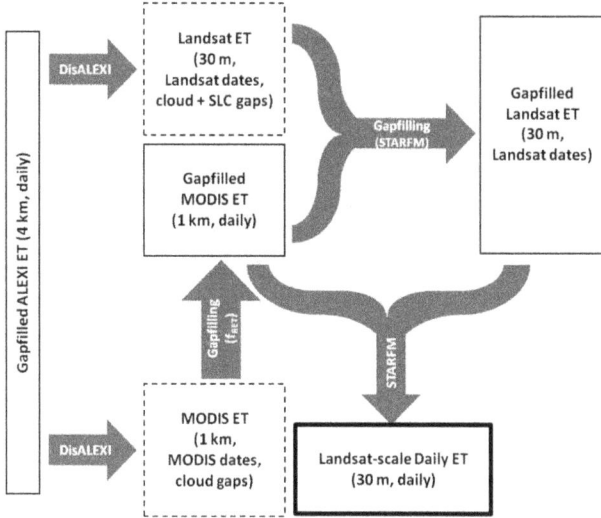

Figure 2. Flowchart describing the Landsat gap-filling and data fusion method. The arrows represent the methods applied. The boxes represent the datasets with different spatial and temporal characteristics created during the process. The dashed boxes indicate ET products with partially filled scenes (due to clouds or SLC gaps); solid boxes identify gap-filled scenes and the thick box highlights the final gap-filled 30 m daily product.

are sharpened from their native spatial resolution to finer resolution with DMS, with the choice of using all of the available shortwave bands or a subset of these bands or even a single vegetation index as input to the regression tree. The sharpened results from the global and local models are combined based on a weighting factor calculated from the residuals of the two sharpened results.

2.2.2 Spatial and Temporal Adaptive Reflectance Fusion Model (STARFM)

The STARFM algorithm fuses spatial information from Landsat imagery with temporal information from the coarser but more frequently collected MODIS imagery to produce daily estimates at Landsat-like scale. STARFM was originally designed for application to surface reflectance images, but has demonstrated utility in fusing higher-order satellite products as well, as long as there is sufficient consistency between the Landsat and MODIS retrievals.

First, ET data from both MODIS and Landsat retrievals are extracted onto a common 30 m grid. A moving searching window method is then used in STARFM to estimate values at the center pixel of the moving window:

$$L\left(x_{p/2}y_{p/2}t_0\right) = \sum_{i=1}^{p}\sum_{j=1}^{p}\sum_{k=1}^{n} W_{ijk} \times \left(M\left(x_i, y_j, t_0\right)\right.$$
$$\left. + L\left(x_i, y_j, t_k\right) - M\left(x_i, y_j, t_k\right)\right), \qquad (4)$$

where p is the size of the moving window and $(x_{p/2}, y_{p/2})$ is the center pixel of the moving window that needs to be es-

timated at time t_0. (x_i, y_j) is the pixel location, $M(x_i, y_j, t_0)$ is the MODIS pixel value at time t_0, $M(x_i, y_j, t_k)$ is the MODIS pixel value at time t_k and $L(x_i, y_j, t_k)$ is the Landsat pixel value at time t_k. W_{ijk} is the weighting factor that determines how much each pixel in the moving window contributes to the estimation of the center pixel value. In the ET fusion system, STARFM uses the weighting function derived from Landsat ET and MODIS ET retrieved on the same date and MODIS ET on the prediction date to get Landsat-like ET estimations on all prediction dates between Landsat overpasses.

2.2.3 ET gap-filling methods

Spatiotemporal gaps in TIR-based ET retrievals occur for a variety of reasons, including cloud cover, frequency of sensor overpass, limitations imposed to avoid distortions in LST data acquired at large off-nadir view angles and other sensor issues. Prior to disaggregation and fusion, the input ET fields must be gap filled, both spatially and temporally, to the extent possible to ensure relatively gap-free output time series.

Due to the high temporal frequency of data acquisition from both geostationary and MODIS systems, the ALEXI and DisALEXI–MODIS retrievals can be reasonably gap filled and interpolated to daily time steps in all but the cloudiest of circumstances. Time intervals between clear-sky Landsat acquisitions are too lengthy in general, motivating the need for data fusion to fill temporal gaps. Spatial gaps in Landsat ET retrievals have been filled using a method based on STARFM, as described below.

ALEXI and MODIS-DisALEXI

Gaps in the daily ET maps from ALEXI and MODIS-DisALEXI were filled using the method described by Anderson et al. (2012b). Daily reference ET is first calculated using the Food and Agriculture Organization (FAO) Penman–Monteith formulation for a grass reference site (Allen et al., 1998). The ratio of actual-to-reference ET (f_{RET}) is computed and then smoothed and gap filled at each pixel using a Savitzky–Golay filter. Gap-filled daily ET is recovered by multiplying this f_{RET} series by daily reference ET.

Landsat-DisALEXI

To ensure optimal spatial coverage in the fused 30 m daily time series, the Landsat-based ET retrievals on Landsat overpass dates must also be gap filled to the extent possible. Gaps in Landsat ET result from cloud cover, or in the case of Landsat 7, missing pixels due to the SLC failure that has been occurring since May 2003, resulting in striped gaps in all but the center of each scene. In the case of Landsat, the time interval between usable overpasses may be too long to justifiably use the f_{RET} approach applied to the ALEXI and MODIS time series. Therefore, an alternate method has been developed to fill cloud gaps/stripes to create filled scenes for ingestion into STARFM.

The method involves running STARFM for the partly cloudy or striped prediction date using Landsat-retrieved ET from surrounding clear dates. The cloud/stripe-impacted areas in the Landsat retrieval are then filled as a weighted function of the STARFM estimated Landsat-like ET and the Landsat-retrieved ET. This weighting is implemented to reduce impacts of bias that may exist between the STARFM estimate and the actual retrieval in the area of the gap, which could otherwise result in a notably patchy fill. The weighting function is computed within a moving window, predicting ET at the center pixel. The weighting value of each pixel in the moving window is calculated based on land cover type, spatial distance to the center predicting pixel and pixel value and is then normalized to a 0–1 value. Pixels that have the same land cover type as the prediction pixel, are nearby and have a similar value are assigned a higher weighting score. The resulting filled value is computed as

$$\text{Filled Value} = A_{\text{L}} - A_{\text{S}} + \sum_{i}^{p} \sum_{j}^{p} \left(W_{i,j} \times S_{i,j} \right), \qquad (5)$$

where A_{L} is the average of pixels in the moving window in Landsat-retrieved ET, A_{S} is the average of pixels in the moving window in STARFM fused ET on the same day as the Landsat-retrieved ET, i and j is the pixel location in the moving window, p is the moving window size, W is the weighting score and S is the STARFM value.

The searching distance is predefined based on the heterogeneity of the study area. A larger searching distance normally requires a longer computing time and can result in more random noise. A searching distance that is too small might not be able to provide a sufficient number of similar surrounding pixels to predict the value of the center pixel. As described above, pixels that are far away from the center pixel have lower weighting than pixels that are close to the center pixel. When the gap area is large and contiguous (more than 80 % of the moving window), there are not enough good pixels that can provide useful information for the gap fill. In this case, the gaps are left unfilled.

Previous Landsat gap-filling techniques have focused on filling spectral reflectance fields. Chen et al. (2011) applied a similar weighting function in a moving window to fill the Landsat 7 SLC-off images using an appropriate thematic mapper (TM) image or SLC-on enhanced thematic mapper plus (ETM+) image. Roy et al. (2008) used both MODIS BRDF/Albedo products and Landsat observations to predict Landsat reflectance with a semi-physical fusion approach. In contrast, the methods described here are a novel application of data fusion to filling SLC-off gaps in ET retrievals.

The cloud mask used in this study is the Fmask (function of mask) data from the Level 2 surface reflectance product distributed by EROS (Earth Resources Observation and Science) center. Fmask uses Landsat top-of-atmosphere reflectance and brightness temperature as inputs to produce cloud, cloud shadow, water and snow mask for Landsat images (Zhu and Woodcock, 2012). Cloud physical properties are first used to identify potential cloud pixels and clear-sky pixels and then normalized temperature, spectral variability and brightness probability functions are combined to estimate cloudy area. The cloud shadow area is derived from the darkening effect of the cloud shadows in the near-infrared band, view angle of the satellite sensor and the solar illumination angle. In this study, we flagged pixels with Fmask class 2 (cloud_shadow) and 4 (cloud) in the cloud mask file as cloud impacted.

3   Experimental site and datasets

3.1   Study area

The study area (Fig. 3) is located over the Parker Tract in the lower coastal plain of North Carolina. The Parker Tract consists of loblolly pine plantations of different ages and native hardwood forests (Noormets et al., 2010). The plantations are commercially managed for timber production by the Weyerhaeuser Company. The study area is flat, about 3 m a.s.l. and has been ditched (fourth-order ditches at 100 m spacing) to manage the water table and improve tree productivity (Domec et al., 2012). The soil is Belhaven series Histosol, with a 50–85 cm organic layer over coarse glacial outwash sand (Sun et al., 2010). The study area is classified as outer coastal plain mixed forest province (Bailey, 1995). The long-term (1945–2008) monthly temperature ranges between 26.6 °C in July and 6.4 °C in January, with an annual mean temperature of 15.5 °C. The long-term annual precipitation is around 1320 ± 211 mm, relatively evenly distributed throughout the year.

Evaluation of the DisALEXI ET estimates was performed at two AmeriFlux tower sites in this area: US-NC2 (35–48° N, 76–40° W) and US-NC3 (35–48° N, 76–39° W). US-NC2 is a mid-rotation plantation stand with 90 ha area, which was established after clear-cutting a previous rotation of loblolly pine, replanted with 2-year-old seedlings at 1.5 m by 4.5 m spacing in 1992. The stand has been fertilized twice – at establishment and in 2010, following a thinning in 2009. The tree density during the study period in 2013 was 171 trees per hectare with a standing biomass of 42.6 t C ha^{-1} in the overstory and 6.5 t C ha^{-1} in the understory. The understory was composed of red maple, greenbrier and volunteer loblolly pine. US-NC3 was established in 2013 in a stand that was clear-cut in 2012, located approximately 1.5 km from US-NC2. US-NC3 was replanted with seedling loblolly pines after the clear cut. Trees in the US-NC2 site were 22 years old in 2013, were 19.0 m tall and had a mean LAI of 3.77 m^2 m^{-2}, whereas NC3 was freshly planted with 2-year-old seedlings, which were 0.2 m tall and had no overstory leaf area. The mostly herbaceous understory contained $85 \pm 52\,\mathrm{g\,C\,m^{-2}}$ at NC3.

Both NC flux towers are equipped with similar instrumentation, and biophysical data are collected routinely. These measurements are described below. This study focuses on data collected during the 2013 growing season, starting after the launch of Landsat 8 on 11 February 2013 and continuing until 8 November 2013.

3.2 Micrometeorological and land management data

At both NC2 and NC3, energy fluxes were measured using an open-path eddy covariance system, which includes a CSAT3 three-dimensional sonic anemometer (Campbell Scientific instrument – CSI, Logan, UT, USA[1]), a CR5000 data logger (CSI), an infrared gas analyzer (IRGA, Model LI-7500, LI-COR, Lincoln, NE, USA) and a relative humidity and air temperature sensor (model HMP-45C; Vaisala Oyj, Helsinki, Finland) (Sun et al., 2010). Soil heat flux was measured at NC2 with three heat flux plates (model HFT3, CSI, Logan, UT, USA) at the depth of 2 cm. The soil heat flux plates were placed in three contrasting microsites – one in a row of trees, in relative shade, another between rows in a mostly open environment and one about halfway in between. Measurements of G at the NC3 site are not available for 2013 due to an instrument failure. Net radiation was measured with four-component net radiometers (Kipp & Zonen CNR-1, Delft, the Netherlands) at each of the two towers. Precipitation was measured by two tipping bucket types of rain gages (TE-525, CSI; Onset data-logging rain gage, Onset Computer Corporation, USA).

Flux observations at 30 min time steps were quality checked, as judged by atmospheric stability and flux stationarity (Noormets et al., 2008). The 30 min data were then gap filled using the monthly regression between observed and potential ET models created from good quality observed data. The energy imbalance problem was checked and the average closure ratio of the 30 min dataset at the NC2 site was 0.88 during daytime when net radiation was larger than 0. Since there were no soil heat flux observations at the NC3 site, there was no closure information and the observed latent heat was used to compare with the simulated data. The 30 min energy fluxes during the daytime were summed up to get daily energy fluxes for validation.

Stand age maps and tree planting history for the study area were obtained from the Weyerhaeuser Company. The stand age ranges from 1 to 89 years, with most stands under 30 years of age. Since this information is proprietary, the stand age maps cannot be displayed; however, these data

were used statistically to assess the relationships between water use and stand age. A 60 m buffer inside the edge of each field was applied to exclude the pixels mixed with roads or other fields. All the other pixels were used to assess the relationships between water use and stand age.

3.3 ALEXI/DisALEXI model inputs

The ET estimation process involves fusion of data from three major geostationary and polar-orbiting satellite systems: GOES, MODIS and Landsat. In addition, each ET retrieval pulled meteorological inputs (air temperature, wind speed, vapor pressure, atmospheric pressure and insolation) from a common gridded dataset, generated at hourly time steps and relatively coarse spatial resolution (32 km) as part of the North American Regional Reanalysis (NARR).

LST data from the GOES imager instruments were used to run ALEXI over the continental US for 2013 at 4 km resolution (Anderson et al., 2007a). In addition, MODIS (4-day) LAI products (MCD15A3) were aggregated from 1 to 4 km and interpolated to daily scale using the smoothing algorithm developed by Gao et al. (2008). LAI is used to estimate $f(\emptyset)$ for Eq. (1), to compute radiation transmission to the soil surface and to assign land cover class-dependent vegetation heights for roughness parameterization (see Anderson et al., 2007a).

MODIS products used in the MODIS disaggregation include instantaneous swath LST (MOD11_L2; Wan et al., 2004), geolocation data (MOD03), NDVI (MOD13A2; Huete et al., 2002), LAI (MCD15A3; Myneni et al., 2002), albedo (MCD43GF; Schaaf et al., 2010; Sun et al., 2017) and land cover (MCD12Q1; Friedl et al., 2002). The LST swath data, at 1 km spatial resolution, were converted to geographic coordinates using an interactive data language (IDL)-based MODIS reprojection tool. The NDVI product (1 km spatial resolution) is produced at 16-day intervals, LAI (1 km) at 4 days and albedo (1 km) at 8 days. All data were quality checked using a data quality filter. The MODIS NDVI, LAI and albedo data were bilinearly interpolated to estimate daily values. MODIS LST was sharpened using NDVI to reduce the off-nadir pixel smearing effect.

Landsat 8 thermal infrared and shortwave surface reflectance data from 2013 used to run DisALEXI were obtained from USGS. Eight relatively cloud-free ($> 75\,\%$) Landsat scenes (path 14 and row 35) and clear conditions over the tower site were available during the 2013 growing season, including one Landsat 7 scene and seven from Landsat 8 (Table 1). Landsat-scale LAI was retrieved from Landsat shortwave surface reflectance data using MODIS LAI products as reference (Gao et al., 2012a). LST was sharpened to 30 m using the blue, green, red, near infrared, SWIR1 and SWIR2 bands (refer to Sect. 2.2.1 for more details about Landsat LST sharpening).

Land cover type was used in both the Landsat and MODIS disaggregation to set pixel-based vegetation parameters in-

---

[1] The use of trade, firm, or corporation names in this article is for the information and convenience of the reader. Such use does not constitute an official endorsement or approval by the United States Department of Agriculture or the Agricultural Research Service of any product or service to the exclusion of others that may be suitable.

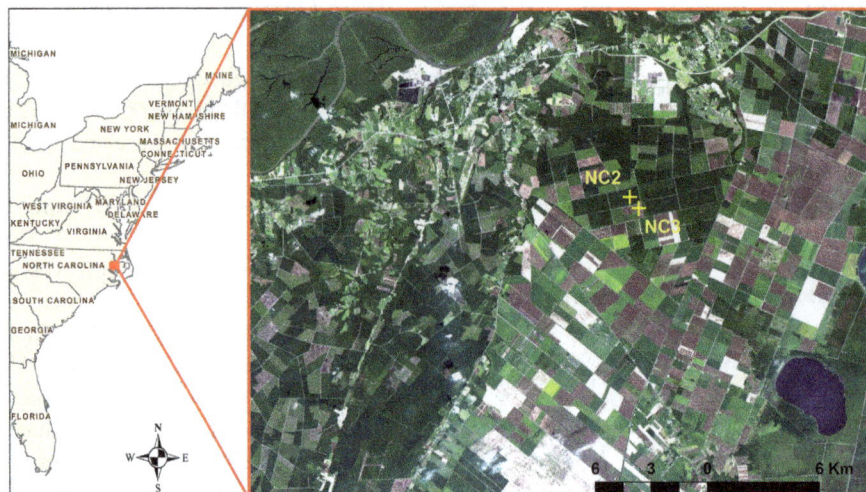

Figure 3. A Landsat 8 true color image (5 September 2013) showing the North Carolina study area. The yellow crosses indicate the location of the NC2 and NC3 flux towers.

Table 1. Landsat overpass dates used in the study.

Sensor	Landsat 7	Landsat 8	Landsat 8	Landsat 8	Landsat 8	Landsat 8	Landsat 8	Landsat 8
DOY	96	104	136	152	200	248	312	328
% Cloudiness	0.1	0	16.4	5.1	23.8	0.3	0.1	0.1
% SLC gap	37	n/a	n/a	n/a	n/a	n/a	n/a	n/a

n/a means not applicable.

cluding seasonal maximum and minimum vegetation height (used in the surface roughness formulations), leaf size and leaf absorptivity in the visible, near-infrared (NIR) and TIR bands following Cammalleri et al. (2013). For DisALEXI using Landsat, the 30 m national land cover dataset (NLCD) for 2006 (Fry et al., 2011; Wickham et al., 2013) was used. For the MODIS disaggregation, the NLCD was resampled to 1 km resolution using the dominant class in each pixel.

4   Results

4.1   Performance of the Landsat gap-filling algorithm

Examples of results from the Landsat gap-filling method are shown in Figs. 4 and 5. For DOY 96 (Fig. 4), L7 SLC stripes and also a few cloudy areas were filled by combining the direct Landsat retrieval (left panel) with the STARFM ET prediction for DOY 96 generated using a Landsat–MODIS image pair from DOY 104. The cloudy areas in DOY 200 (Fig. 5) were filled using a Landsat–MODIS pair from DOY 152. In each case, the size of the overlapped moving window was 420 m by 420 m. This means that contiguous gaps larger than the window were not filled since there were not enough candidate pixels to create the statistical relationship needed between the direct Landsat retrieval and

STARFM ET. The white rectangular box located in the northeast area of both figures contains no data because more than 40 % of Landsat pixels within the 4 km ALEXI ET pixel were affected by a large water body.

In each of these cases, the spatial patterns within missing regions in the direct retrievals appear reasonably reconstructed with the ET gap-filling method, with no obviously patchy artifacts in the gap-filled ET images. Even linear structures were restored: for example, roads and field boundaries. The gap-filling method works even better when the small or linear objects are very different from surrounding pixels. The choice of moving window size can affect the gap-filling results and may require scene-based adjustment, with the need to balance the risk of inappropriate candidate pixel selection if the window is too large, or a lack of candidate pixels if the window is too small.

Figure 6 shows a synthetic study used to assess the accuracy of the gap-filling procedure. Here, a direct Landsat ET retrieval for DOY 152 was artificially masked using SLC stripes from DOY 96. The final panel shows the gap-filled image. Comparing the original values with the gap-filled values yields an R^2 of 0.89 and a mean absolute error (MAE) of $-0.01\ \mathrm{mm\,day^{-1}}$, with the average of original values as $5.81\ \mathrm{mm\,day^{-1}}$ and the average of gap-filled values as $5.80\ \mathrm{mm\,day^{-1}}$.

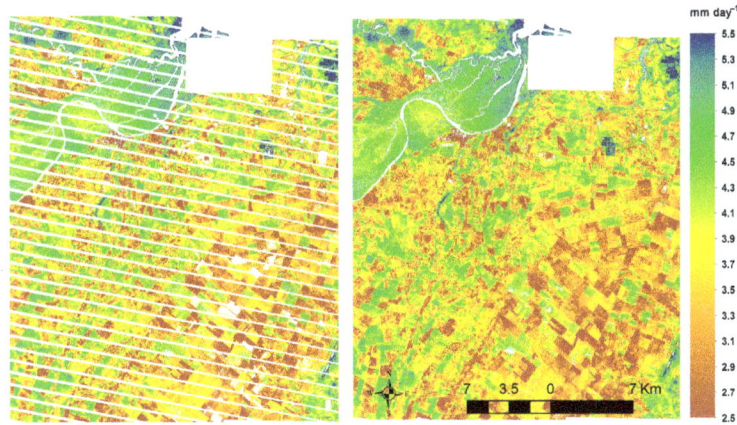

Figure 4. Example of gap-filling SLC-off stripes in a Landsat 7 ET image for DOY 96, 2013. The left image is Landsat-retrieved ET with stripes, while the right image has been gap filled using the method described in Sect. 2.2.3.

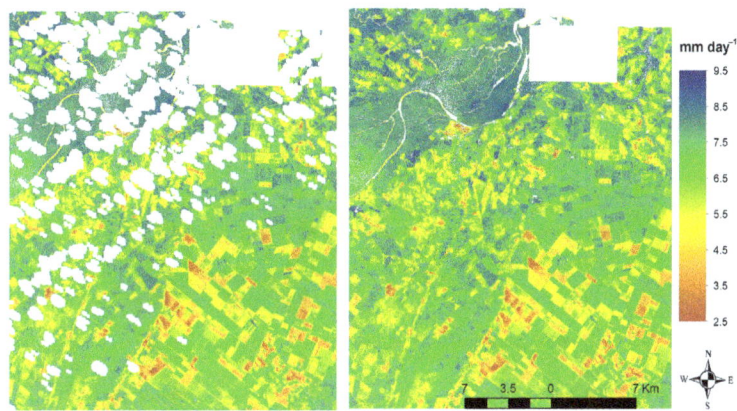

Figure 5. Example of gap-filling cloudy regions in a Landsat 8 ET image for DOY 200. The left image is Landsat-retrieved ET with clouds masked using the Fmask data layer and the right image has been processed through the Landsat gap-filling method.

The gap-filling method described in Sect. 2.2.3 relies on the inputs from both the original Landsat ET and the STARFM predictions, which in turn rely on a filled MODIS image on the target date as well as a MODIS–Landsat image pair on a surrounding date. If the Landsat image in the input pair also has gaps, additional pairs can be used to iteratively fill the target image.

4.2 Evaluation of daily ET retrievals from DisALEXI at the flux tower sites

Modeled and measured instantaneous and daytime integrated surface energy fluxes on Landsat overpass dates are compared in Fig. 7, demonstrating good correspondence. Statistical performance metrics for each flux component for both sites are shown in Table 2, including MAE, root mean square error (RMSE) and mean bias error (MBE). The model performance for each flux is similar between sites, with somewhat lower errors obtained for the clear-cut site (NC3). The latent heat observed at the NC2 site is higher than that at NC3

with or without closure enforcement. As mentioned earlier, closure could not be assessed at NC3 due to failure among the soil heat flux instrumentation. At NC2, closure by residual resulted in an increase in observed ET by approximately 12 % on average.

Time series of ALEXI ET (4 km), Landsat ET retrieved on Landsat overpass dates and Landsat–MODIS fused ET (both at 30 m resolution) are compared in Fig. 8 with ET observed at both the NC2 and NC3 sites from DOY 50 to 330. In addition, daily ET values generated using a simple Landsat-only interpolation scheme are shown for comparison. These were generated using the MODIS and ALEXI gap-filling technique described in Sect. 2.2, conserving the ratio of actual-to-reference ET between Landsat overpass dates. Metrics of statistical performance at daily to seasonal timescales are listed in Table 3.

Figure 8 highlights the value of disaggregation to the tower footprint scale for the purposes of model validation. For NC2, ALEXI 4 km fluxes agree well with tower observations, suggesting that the tower footprint at NC2 is reasonably rep-

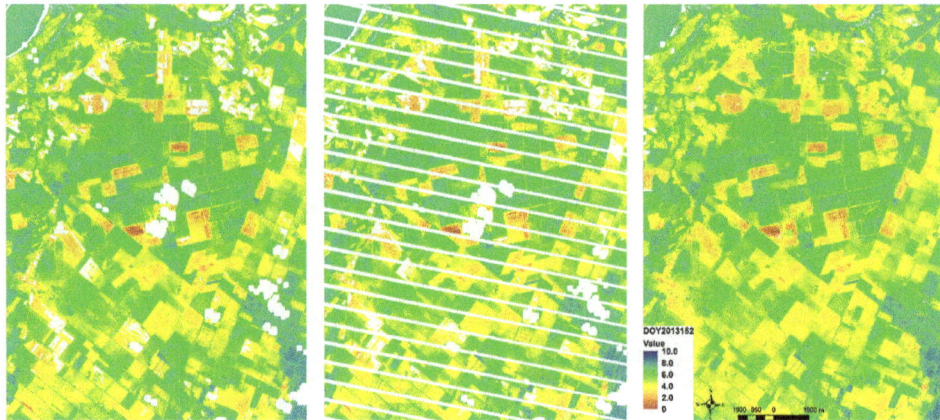

Figure 6. Comparison between the original Landsat ET retrieval for DOY 152 (left panel), an artificially gapped version, imposing SLC gaps from DOY 96 (middle panel), and the gap-filled map (right panel).

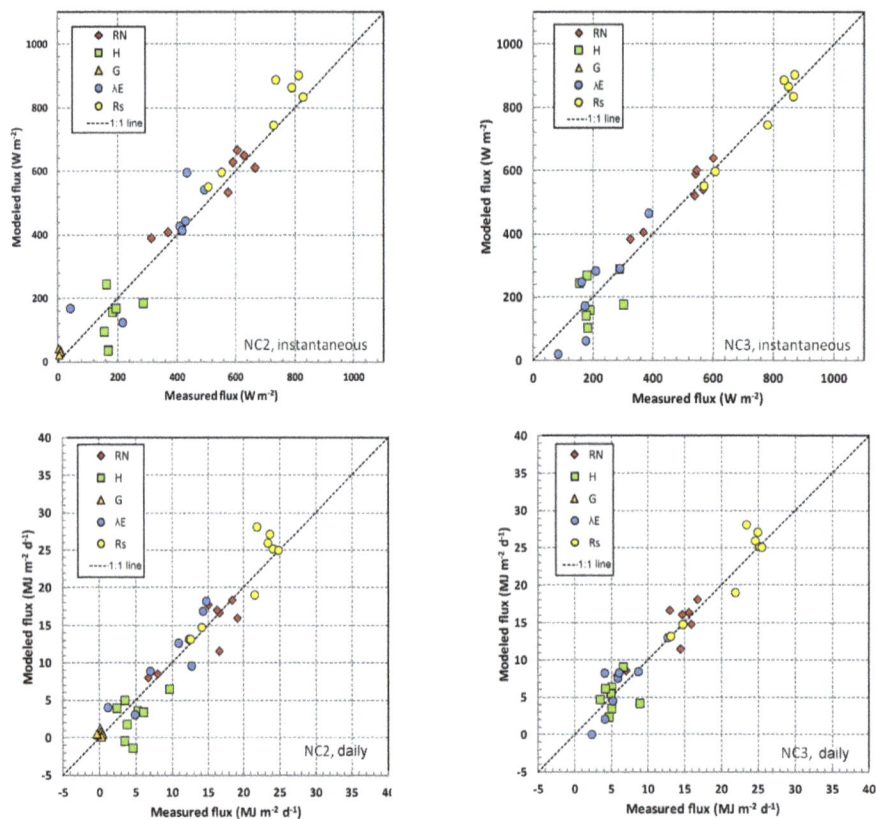

Figure 7. Scatterplot of modeled and measured instantaneous (top row panels) and daily surface fluxes (bottom row panels) on Landsat overpass dates for NC2 (left column panels) and NC3 (right column panels) flux tower sites.

resentative of the surrounding 4 km ALEXI pixel area. The disaggregated 30 m fluxes are also similar to both ALEXI and observations at this site. At NC3, however, the 4 km ALEXI fluxes are notably higher than the observed ET, while the disaggregated fluxes are comparable. The NC3 tower site is not representative at the ALEXI pixel scale, and disaggregation to the tower footprint scale is required to account for local

sub-pixel heterogeneity. Even at 1 km resolution, the MODIS retrieval accuracy was degraded at NC3 in comparison with the Landsat-scale retrievals (Table 3). Recall that NC3 was recently clear-cut with surrounding areas still comprised by more mature forest stands. With the 4 km spatial resolution, ALEXI ET is able to capture the ET status of the major land cover type (i.e., mature forest), but not the particular patch

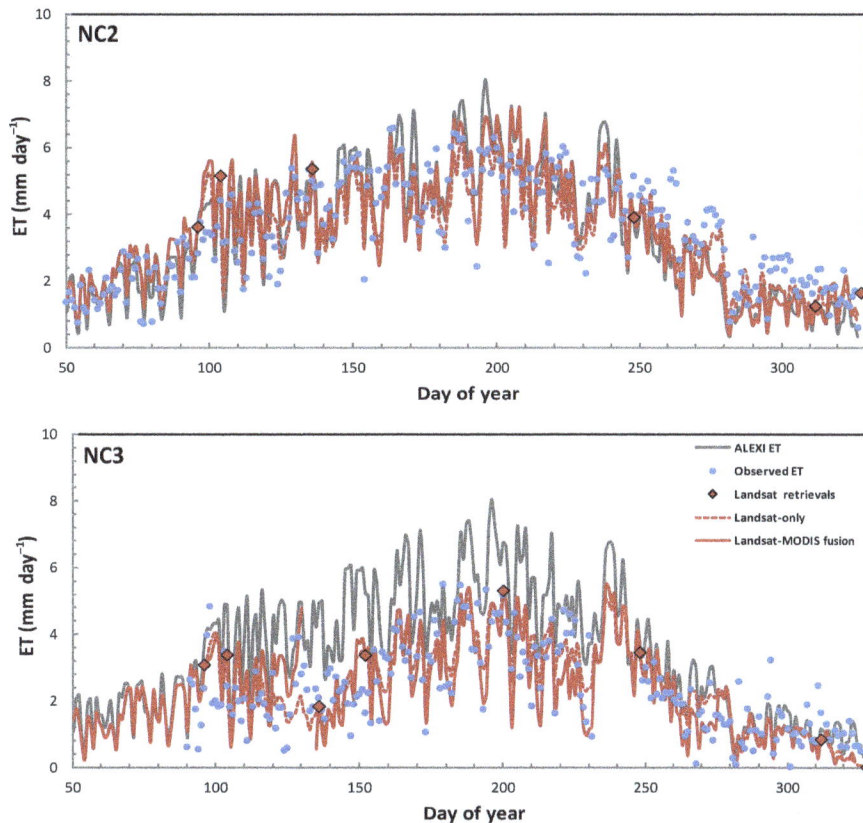

Figure 8. Comparison of time series of ALEXI ET (4 km), observed ET, Landsat ET retrieved on Landsat overpass dates, Landsat-only interpolated ET and Landsat–MODIS fused ET for the NC2 site (and NC3, bottom panel) sites in 2013.

of land where the NC3 tower is located. This underscores the need for appropriate spatial resolution when comparing modeled with observed fluxes, especially for the more heterogeneous land surfaces (e.g., Anderson et al., 2004).

Overall, the performance of the two Landsat retrievals (STARFM and Landsat only) are comparable between sites, with RMSE at daily time steps of ~ 0.8 to $1.0\,\mathrm{mm\,day^{-1}}$ and MAE of 0.6 to $0.8\,\mathrm{mm\,day^{-1}}$ (19–30 % of the mean observed ET). At monthly time steps, performance improves to 11–14 %, due to averaging of random errors – including errors in daily insolation forcings from the NARR meteorological dataset. Fluxes are somewhat underestimated at the end of the growing season at each site due largely to the Landsat retrieval on DOY 312. This highlights the importance of good temporal sampling at the Landsat scale – an additional Landsat scene around DOY 270 during the prolonged gap in coverage may have improved the seasonal water use estimates. A small negative mean model bias is observed for both sites, due primarily to underestimation at the end of growing season of the Landsat retrieval on DOY 312. This is also the reason that Landsat-only interpolated ET performs slightly better than STARFM. More details about the comparison of the two Landsat retrievals can be found in the Discussion section.

The seasonal cumulative ET at NC2 and NC3, calculated for DOY 90–330 from both the observed ET and the Landsat–MODIS fused time series, is shown in Fig. 9. For NC3, the accumulation does not include values of modeled or measured ET during the period DOY 232–248 when flux tower data were not available, so these values do not represent the total seasonal water use at this site. The modeled cumulative water use on DOY 330 agrees with the observed values to within −0.9 % at NC2 and 0.4 % at NC3. Overall, the modeled and measured cumulative ET curves agree well throughout the growing season, indicating the remote sensing method has utility for water use management and assessment at subseasonal timescales.

4.3 Spatiotemporal variability in seasonal water use

4.3.1 ET variations with land cover

Figure 10 shows the land cover types over the study area as described in the NLCD from 2006[2]. The major classes represented in the NLCD over the study area include crop land

---

[2]Note: an updated NLCD map for 2011 was published after the study was implemented, but there was no notable change in land cover types over the study area in comparison with NLCD2006.

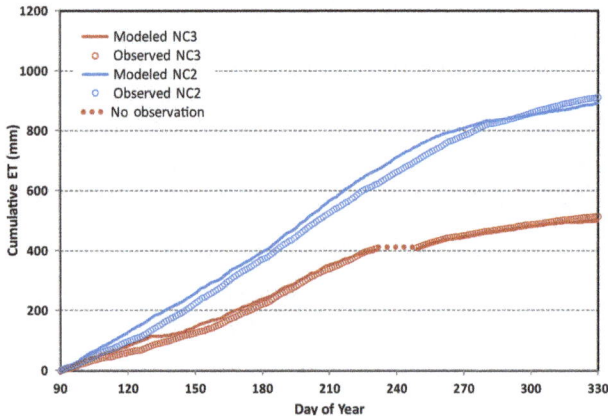

Figure 9. Comparison between the modeled and observed seasonal cumulative ET at NC2 and NC3 during 2013.

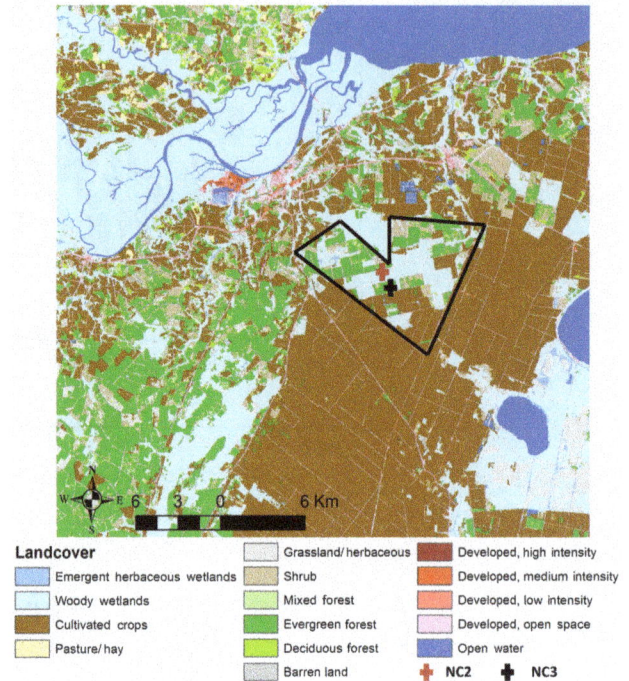

Figure 10. Land cover types over the study area from NLCD 2006. Area in the black outline is the plantation area.

(including corn, cotton and soybeans), forest and woody wetland. The land cover in many plots within the Parker Tract plantation, including the NC2 site, was classified as woody wetland rather than evergreen forest. This misclassification, however, had little impact on the model ET estimates at NC2 due in part to the normalization constraint imposed by the 4 km ALEXI ET output.

Time series maps of monthly and cumulative ET in Fig. 11 over the study area exhibit spatiotemporal water use patterns that are related to land cover type (Fig. 10). The relatively high rates of ET during midseason in the riparian and more densely forested regions are readily apparent. Water use patterns in the cultivated agricultural areas reflect the diversity of crops and water management strategies. Within the Parker Tract plantation, a few fields with persistently low ET may be fresh clear cuts, possibly with a layer of slash to inhibit emergence of new vegetation. In the summer of 2014, after the slash has been collected and piled, these plots may appear more like the recent clear cut near NC3.

Seasonal ET time series were developed for five generalized land cover classes (crop land, natural forest, woody wetland, mature plantation and young plantation) to assess variability in water use with land use/land cover type in the study area. The term "natural forest" is used to describe unmanaged mixed forested areas within the study domain. "Mature plantation" refers to managed stands of loblolly pine within the Parker Tract with ages ranging from 10 to 20 years, while "young plantation" indicates stand ages less than 3 years. Figure 12 shows the time series of modeled field-scale ET averaged from 10 randomly sampled pixels associated with each generalized land cover class for 2013. For the woody wetland class, care was taken to select pixels that were correctly classified by visual inspection of Google Earth imagery. A 7-day moving average was applied to the modeled daily ET to reduce noise and facilitate visual comparison.

Pixels classified as woody wetland, natural forest, agriculture and mature plantation generally showed higher ET than did the young plantation pixels. Water use in the woody wetland areas was the highest among all the different land cover types but was similar to natural forest and mature managed forest during the peak growing season. The seasonal cumulative ET from these five land cover types is shown in Fig. 13. The woody wetlands tended to have higher seasonal ET than the other three classes, slightly exceeding that of natural forest and mature pine plantations. Modeled water use in crop lands exceeded that in young plantation stands, resulting from relatively higher LAI and lower LST observed over the cropped areas.

Figure 14 shows the average cumulative ET between DOY 50 and 330 associated with the five different land cover types, computed from the 10 random samples per class, and the black bars represent the standard deviation among the samples. Natural forests showed the lowest variability in ET (30 mm), while the woody wetland and mature plantation pixels had the highest standard deviations (74 and 73 mm, respectively). The high variability in the latter classes may reflect both management effects and misclassification. Crop lands and young forest plantations showed moderate variability in water use, with standard deviations of 64 and 50 mm, respectively. In terms of coefficient of variation in water use across the modeling domain, the crops and young plantation classes are relatively high at 7.1 %, compared to natural forest at 2.8 %. Crops and young plantations also have higher coefficients of variation in LAI than other land cover types, leading to larger variability in water use demand through transpiration.

Because the two-source land-surface representation in DisALEXI also provides estimates of the evaporation (E) and transpiration (T) components of ET, the model output can also be used to assess variability in E / T partitioning between land cover types and through the season (Fig. 15). In general, soil evaporation losses account for a higher percentage of total ET early in the season, after the spring rains but before the canopies have completely leafed out. On average through the season, the E / T ratio is highest for crop lands, young plantations and woody wetlands. This is reasonable given the lower leaf area characteristic of these classes and the abundant substrate moisture in the case of woody wetlands. Partitioning to T is maximized during the peak growing season (DOY 152, 200 and 248) for all the land cover types. Natural forests and mature plantations tend to have higher rates of transpiration than other land cover types.

4.3.2 ET variations with LAI

Many forest hydrology models (Lu et al., 2003; Scott et al., 2006; Sun et al., 2011a) assume seasonal ET is well correlated with LAI. This assumption was tested over the set of points chosen randomly from different land cover types. Sample points from mature plantations, young plantations and crop lands were all located in drained areas, while sample points from natural forest and woody wetland were in undrained areas. However, LAI values from the sample points were not affected by drainage conditions. High LAI values were obtained from both drained mature plantations and undrained natural forest, while relatively low LAI values were from both drained young plantations and undrained woody wetlands. Figure 16 shows that some (but not all) variability in ET, as predicted by the fusion estimates, is explained by variability in average LAI over the prediction time period ($R^2 = 0.59$). These results indicate that, in 2013, increasing LAI added up to 350 mm of seasonal water use on top of nearly 800 mm from the soil evaporation contribution. The high rates of ET at low LAI are reasonable, given that the study area was fairly wet during 2013 due to plenty of precipitation and shallow water tables.

4.3.3 ET variations with plantation stand age

Within the managed pine plantation at Parker Tract, we also examined variations in seasonal water use with stand age (Fig. 17). This has relevance to forest management practices and their impacts on the water yield of the watershed, which is the difference between precipitation and evapotranspiration over the long term. Higher water yield translates to higher streamflow available for downstream use. Many forest management practices, for example, thinning and reforestation, need to consider the influence of stand age on the hydrological response.

Table 2. Summary of the statistical indices quantifying model performance for instantaneous and daytime integrated surface energy fluxes on Landsat overpass dates.

Variable	Daily fluxes						Instantaneous fluxes					
	$R_{s,d}$	$R_{n,d}$	G_d	H_d	LE_d (closed)	LE_d (unclosed)	R_s	R_n	G_0	H	LE (closed)	LE (unclosed)
Unit	MJ m^{-2} day^{-1}	MJ m^{-2} day^{-1}	MJ m^{-2} day^{-1}	MJ m^{-2} day^{-1}	MJ m^{-2} day^{-1}	MJ m^{-2} day^{-1}	W m^{-2}	W m^{-2}	W m^{-2}	W m^{-2}	W m^{-2}	W m^{-2}
						NC2 site						
n	8	8	8	8	8	8	8	8	8	8	8	8
Ø	20.70	14.60	0.10	4.80	9.70	9.40	714	540	1	214	326	288
MAE	1.10	1.70	0.50	2.80	2.30	2.90	63	48	34	78	68	113
RMSE	4.20	2.80	0.20	4.80	3.00	5.70	78	51	36	89	89	125
MBE	1.60	−0.30	0.50	−1.90	1.10	1.40	63	21	34	−55	41	91
						NC3 site						
n	8	8	8	8	8	8	8	8	0	8	0	8
Ø	22.30	13.80	NA	5.20	NA	6.50	776	507	NA	205	NA	209
MAE	1.50	1.90	NA	2.00	NA	1.70	28	41	NA	64	NA	60
RMSE	2.30	2.20	NA	2.70	NA	2.20	31	44	NA	75	NA	72
MBE	0.70	0.90	NA	−0.10	NA	0.50	2	29	NA	−11	NA	10

$R_{s,d}$ is daytime integrated solar radiation; $R_{n,d}$ is daytime integrated net radiation; G_d is daytime integrated soil flux; H_d is daytime integrated sensible heat; LE_d is daytime integrated latent heat; R_s, R_n, G_0, H and LE are instantaneous fluxes; "closed" indicates energy balance closure by residual, while "unclosed" indicates that energy balance closure was not imposed on the EC measurements. In addition: n is the number of observations; Ø is mean measured flux; MAE is mean absolute error between the modeled and measured quantities; RMSE is root mean square error; MBE is mean bias error. NA means not available.

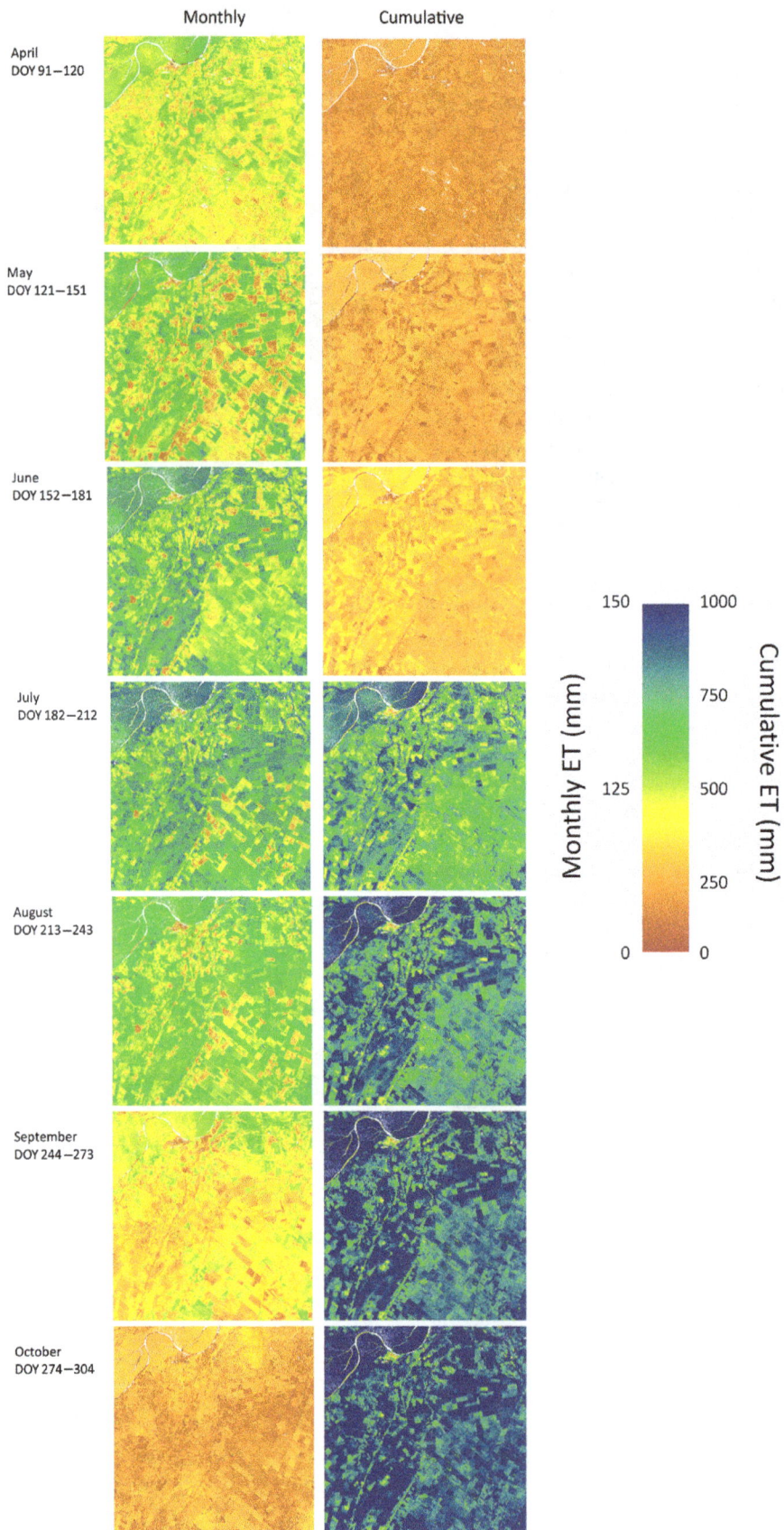

Figure 11. Spatial patterns of monthly cumulative ET (left column panels) from April to October, and cumulative ET on the end day of each month over the study area.

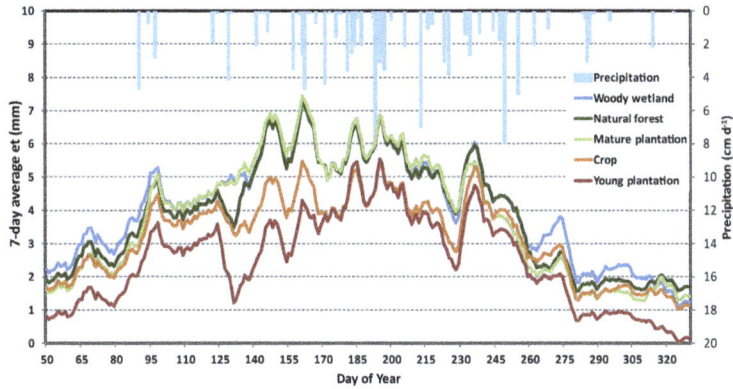

Figure 12. Time series of modeled plot-scale ET (daily values smoothed with a 7-day moving average) associated with different land cover types.

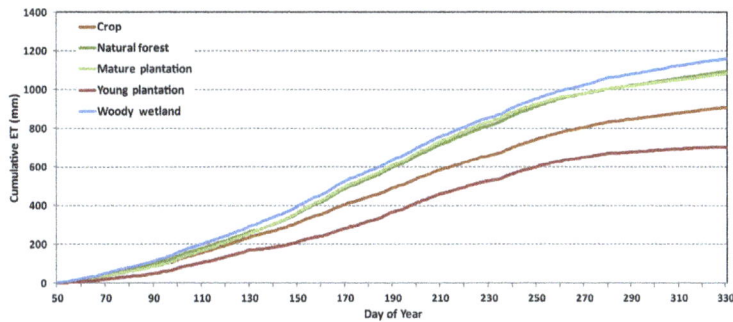

Figure 13. Seasonal cumulative ET for different land cover types.

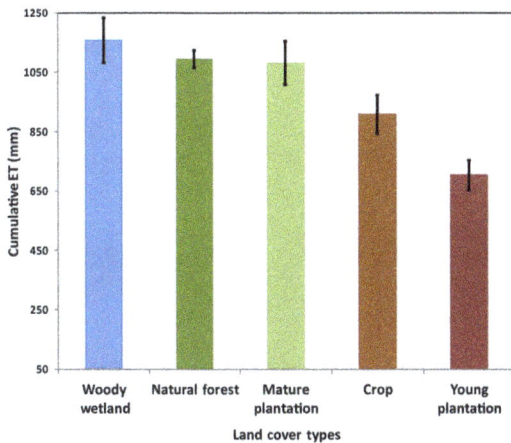

Figure 14. Average cumulative ET at DOY 330 in 2013 over different land cover types, and standard deviations within the sample populations (the black bar).

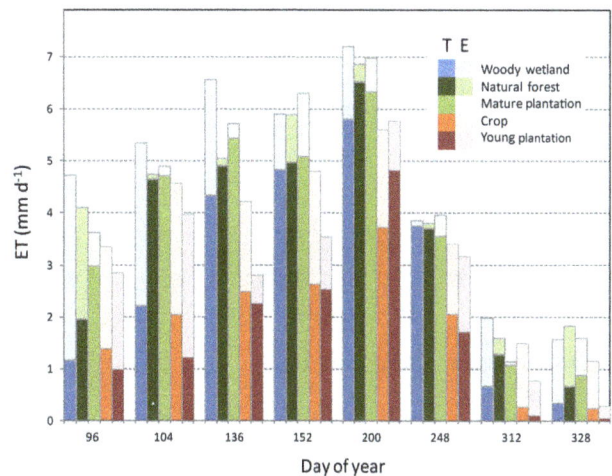

Figure 15. Average evaporation (E) and transpiration (T) components of ET for five land cover types on Landsat overpass days.

Plotting cumulative ET at DOY 330 from various sites against stand age (Fig. 17), there is a clearly positive linear relationship between water use and stand age for the younger stands, between a few years old and around 20 years old (R^2 is 0.82). As the stand age increases, more water is used as ex-pected to sustain larger amounts of biomass. Differences in seasonal cumulative ET curves for different stand ages begin to significantly diverge after DOY 130 – around the middle of May (not shown). When the stand matures beyond 20 years,

Table 3. Statistical metrics describing comparison of retrieved ET time series with tower observations at NC2 and NC3 at daily and monthly time steps, as well as cumulative values over the study period (DOY 50–330).

Site	NC2 site				NC3 site			
Index	STARFM	Landsat only	MODIS smoothed	ALEXI	STARFM	Landsat only	MODIS smoothed	ALEXI
	Daily				Daily			
n	281	216	281	281	223	199	223	223
MAE (mm day^{-1})	0.80	0.64	0.66	0.85	0.83	0.70	2.00	1.45
RMSE (mm day^{-1})	0.99	0.83	0.85	1.04	1.02	0.88	2.27	1.84
MBE (mm day^{-1})	−0.03	−0.29	−0.10	−0.11	−0.05	0.10	1.96	1.31
RE (%)	27.9	19.2	22.0	28.0	30.6	25.1	85.0	56.0
	Monthly				Monthly			
n	12	8	12	12	10	8	10	10
MAE (mm day^{-1})	0.41	0.36	0.48	0.44	0.23	0.33	1.61	1.06
RMSE (mm day^{-1})	0.53	0.39	0.58	0.54	0.29	0.36	1.82	1.33
MBE (mm day^{-1})	0.00	−0.25	−0.11	−0.15	−0.01	0.16	1.61	1.06
RE (%)	13.6	12.1	16.1	14.8	10.9	15.5	76.3	50.1
	Study period				Study period			
n	1	1	1	1	1	1	1	1
MAE (mm day^{-1})	0.03	0.04	0.10	0.11	0.01	0.12	1.88	1.26
MBE (mm day^{-1})	−0.03	0.04	−0.10	−0.11	0.01	0.12	1.88	1.26
RE (%)	0.9	1.3	2.7	3.2	0.4	5.2	81.0	54.3

RE is relative error, which is calculated by dividing MAE by observed average ET.

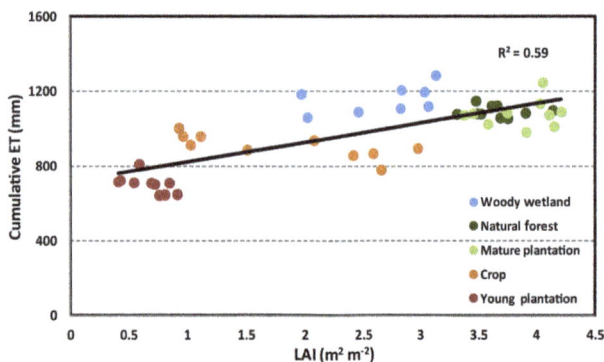

Figure 16. Modeled cumulative ET from DOY 50 to DOY 330 as a function of LAI.

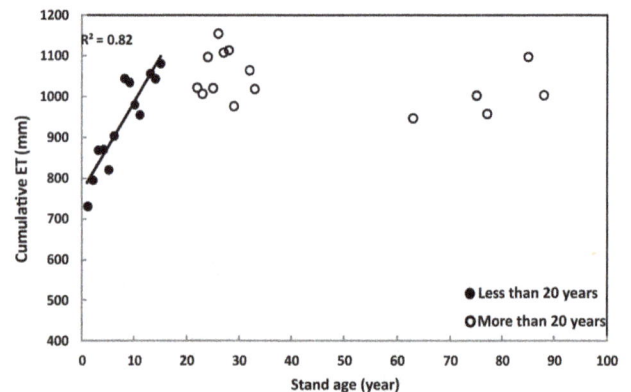

Figure 17. Modeled cumulative ET at DOY 330 as a function of stand age.

the water usage tends to plateau and may actually decrease slightly for stands with trees older than 75 years.

5   Discussion

5.1   Utility of TIR-based data fusion as a daily ET estimation method

In this study, the STARFM modeling approach applied over a full growing season resulted in a relative errors in ET at the daily time step of 28 % for the mid-rotation pine plantation site and 31 % for the clear-cut site, and errors of 13.6 % at the monthly time step. The accuracy of these results is comparable with earlier studies applying the STARFM ET data fusion approach. Over rain-fed and irrigated corn and soybean sites in central Iowa (IA), STARFM yielded a relative error of about 11 % over eight flux towers (Cammalleri et al., 2013). When STARFM was applied to the Bushland, TX, the daily ET estimation had a relative error of 26.6 % for irrigated area and 27 % for rain-fed area (Cammalleri et al., 2014). For a study area near Mead, NE, the relative error was 20.8 % for irrigated crops and 25.4 % for rain-fed crops (Cammalleri et al., 2014). Semmens et al. (2015) estimated daily Landsat-scale ET over California vineyards with relative error of 18 % for an 8-year-old vineyard and 23 % for a 5-year-old vineyard.

In the current study, we found that STARFM yielded similar or marginally lower accuracy than did a simple Landsat-only ET interpolation scheme in estimating fluxes between Landsat overpasses. This is in contrast to results from Cammalleri et al. (2013), who found that STARFM significantly outperformed the Landsat-only approach in comparison with flux measurements acquired over rain-fed crops in central IA during the soil moisture experiments in 2002 (SMEX02). There may be several factors that influence the potential value added by STARFM that can be deduced from this comparison. One factor may be climate and moisture status within the target modeling domain. During the first part of SMEX02 experiment, conditions were becoming quite dry and the crops – particularly the corn fields – were becoming notably stressed, exhibiting leaf curl near the field edges. A rainfall event near the beginning of July (between Landsat overpasses) significantly relieved the crop stress and greatly impacted soil moisture conditions over the SMEX02 study area (Anderson et al., 2013). In contrast, the climate of the Parker Tract study site studied here was very wet during 2013, and vegetation condition was not water limited. Consequently, as long as other factors do not affect the health of the vegetation (e.g., pest/disease infestation) or large variations do not occur in atmospheric demand (large oscillations in radiation, wind and temperature), the evapotranspiration process will remain near a constant fraction of potential and may be reasonably captured by a simple daily interpolation scheme. At the NC sites studied here, there were no major changes in soil moisture status that were additionally captured by the MODIS ET retrievals. This might not be the case in severe drought years. Another factor to be considered is predominant land cover type. The NC sites are dominated by trees, with deeper rooting systems than perennial crops, and are able to extract water that is available much deeper in the soil profile. This further reduces variability in vegetation response during the study period in comparison with the SMEX02 study.

These two factors are consistent with the assumptions in the Landsat-only interpolation, which assumes the ratio between actual ET and reference ET is smoothly varying over time. However, for rain-fed crop land areas (like the central IA sites) that occasionally go through relatively dry and wet periods, STARFM may be better able to capture time variability in f_{RET} in response to changing moisture conditions. Despite the limited value added by STARFM over standard methods in this case, Fig. 8 shows that both methods performed very well.

For real-time applications in water management, STARFM can be used to project water use information beyond the date of the last Landsat overpass within some limited time range, assuming MODIS data are available with low time latency. Practical limits to a viable projection time range may vary with site and season and will depend on the rate of change in weighting factors governing the STARFM fusion process.

5.2 Comparison with prior ET studies over the Parker Tract

Direct measurements of ET within the Parker Tract study area are only available for pine plantations (NC2 and NC3 AmeriFlux sites) (Sun et al., 2010; Domec et al., 2012). Annual ET rates for the NC2 site reported by Sun et al. (2010) vary from 892 mm (a dry year, 2007) to 1226 mm in a normal year (2006). A process-based forest hydrological model (DRAINMOD-FOREST) has been calibrated for the NC2 eddy flux site using drainage and groundwater table data for the period of 2005–2012. ET estimates by the calibrated model vary from 903 mm yr^{-1} in 2008 to 1170 mm yr^{-1} in 2006. These reported annual ET rates are consistent with results from the present study, which indicate ET values of 1000–1200 mm yr^{-1} for pine plantations established in the 1990s.

Empirical models of ET, operative at the monthly timescale, have been developed for the Parker Tract study area (Sun et al., 2011a) and the accuracy is generally uncertain due to the large variability of climate and ecosystem structure and ET processes in general. For example, Sun et al. (2011a) proposed a general predictive model to estimate monthly ET using reference ET, precipitation and LAI over 13 ecosystems, which included the mid-rotation NC2 site and another clear-cut site (NC1 in AmeriFlux database) also in our study area. The monthly ET estimates from that approach had a relative error of 23 % and a RMSE of 15.1 mm month^{-1}, while STARFM relative errors from the current study were 13.6 % at the monthly time step – a substantial improvement in accuracy.

5.3 Water use variations with stand age and land cover type

Previous studies investigating the relationship between land cover and age of forest stands on water usage mainly focused on the resulting impact on the water balance via impacts on streamflow at the watershed scale (Matheussen et al., 2000; Vertessy et al., 2001; Williams et al., 2012). Matheussen et al. (2000) analyzed the hydrological effects of land cover change in the Columbia River basin and found a significant correlation between hydrological change and the tree maturity in the forested areas. In the current study, we found that water use increased linearly with stand age between 1 and 20 years, then plateaued or decreased with age after about

20 years. An investigation of mountain ash forests in Victoria, Australia, found that annual ET from a nearly 35-year stand was 245 mm more than that from a 215-year stand (Vertessy et al., 2001). Similarly, another study of three forest stands, aged 14, 45 and 160 years, found plot transpiration declined from 2.2 mm per day in the 14-year stand and 1.4 mm per day in the 45-year stand to 0.8 mm per day in the 160-year-old forest (Roberts et al., 2001). In another study, Kuczera (1987) found a rapid decrease of mean annual water yield from a mountain ash forest watershed when stand age increased from 1 year to around 25 years, suggesting greater water use by the older forest stands. Murakami et al. (2000) used a Penman–Monteith-based model to simulate an ET–stand age relationship, which also showed a clear upward trend for young forests and a peak in ET at 20 years. The sharp increase of seasonal ET with the increase of stand age from 1 year to around 20 years illustrated in Fig. 17 is consistent with these earlier studies. Plant LAI is closely related to ET (Sun et al., 2011a) and is also an important input in plant physiological and hydrologic/land surface models as well as crop models (Duchemin et al., 2006; Nemani et al., 1993; Tague et al., 2013). As shown in Fig. 16, we also find correlation between cumulative ET and season-average LAI over different land cover types, although LAI explained only 59 % of the modeled variability in ET.

6 Conclusions

This study demonstrates the capability of a multiscale data fusion ET model to estimate daily field-scale ET over a forested landscape. Daily ET retrievals over the growing season of 2013, generated at 30 m spatial resolution, compared well with observed fluxes at AmeriFlux tower sites in a mature pine stand and a recent clear-cut site, demonstrating capability to reasonably capture a range in land-surface conditions within a managed pine plantation. Errors were 29 % at daily time step, 12 % at monthly time step and 0.7 % over the study period.

A new scene gap-filling method was described to maximize the number of Landsat images used for ET retrieval at Landsat scale and will be of benefit in areas with persistent partial cloud cover and for recovering scenes from the Landsat 7 archive that are impacted by the SLC failure. The STARFM data fusion method can help to mitigate the dearth of high spatial–temporal resolution land surface temperature data from currently available satellite systems.

This study suggests that satellite retrievals of ET at the Landsat scale can be used to analyze water use variability over a heterogeneous forested landscape in response to stand age and vegetation composition. The estimates of ET at a high resolution provide insight of seasonal water balances and thus offer useful information for local water resource management. Comparing with traditional forest ET estimation methods, this study describes an accurate, efficient and potentially real-time remote sensing method for estimating landscape-level ET which is suitable for operational applications.

Competing interests. The authors declare that they have no conflict of interest.

Acknowledgements. This work was funded in part by a grant from NASA (NNH14AX36I). We thank the Weyerhaeuser Company for providing stand age data. The US Department of Agriculture (USDA) prohibits discrimination in all its programs and activities on the basis of race, color, national origin, age, disability, and where applicable, sex, marital status, familial status, parental status, religion, sexual orientation, genetic information, political beliefs, reprisal or because all or part of an individual's income is derived from any public assistance program. (Not all prohibited bases apply to all programs.) Persons with disabilities who require alternative means for communication of program information (Braille, large print, audiotape, etc.) should contact USDA's TARGET Center at (202) 720-2600 (voice and TDD). To file a complaint of discrimination, write to USDA, Director, Office of Civil Rights, 1400 Independence Avenue, S.W., Washington, D.C. 20250-9410, or call (800) 795-3272 (voice) or (202) 720-6382 (TDD). USDA is an equal opportunity provider and employer.

Edited by: Y. Chen

References

Abramopoulos, F., Rosenzweig, C., and Choudhury, B.: Improved ground hydrology calculations for global climate models (GCMs): Soil water movement and evapotranspiration, J. Climate, 1, 921–941, 1988.

Allen, R. G., Pereira, L. S., Raes, D., and Smith, M.: Crop evapotranspiration – Guidelines for computing crop water requirements-FAO Irrigation and drainage paper 56, FAO, Rome, 300, D05109, 1998.

Anderson, M. C., Norman, J. M., Diak, G. R., Kustas, W. P., and Mecikalski, J. R.: A two-source time-integrated model for estimating surface fluxes using thermal infrared remote sensing, Remote Sens. Environ., 60, 195–216, 1997.

Anderson, M. C., Norman, J. M., Mecikalski, J. R., Torn, R. D., Kustas, W. P., and Basara, J. B.: A multiscale remote sensing model for disaggregating regional fluxes to micrometeorological scales, J. Hydrometeorol., 5, 343–363, 2004.

Anderson, M. C., Norman, J. M., Mecikalski, J. R., Otkin, J. A., and Kustas, W. P.: A climatological study of evapotranspiration and moisture stress across the continental United States based on thermal remote sensing: 1. Model formulation, J. Geophys. Res., 112, D10117, doi:10.1029/2006JD007506, 2007a.

Anderson, M. C., Kustas, W. P., and Norman, J. M.: Upscaling Flux Observations from Local to Continental Scales Using Thermal Remote Sensing, Agron. J., 99, 240–254, doi:10.2134/agronj2005.0096S, 2007b.

Anderson, M. C., Kustas, W. P., Norman, J. M., Hain, C. R., Mecikalski, J. R., Schultz, L., González-Dugo, M. P., Cammalleri, C., d'Urso, G., Pimstein, A., and Gao, F.: Mapping daily evapotranspiration at field to continental scales using geostationary and polar orbiting satellite imagery, Hydrol. Earth Syst. Sci., 15, 223–239, doi:10.5194/hess-15-223-2011, 2011.

Anderson, M. C., Kustas, W. P., Alfieri, J. G., Gao, F., Hain, C., Prueger, J. H., Evett, S., Colaizzi, P., Howell, T., and Chávez, J. L.: Mapping daily evapotranspiration at Landsat spatial scales during the BEAREX'08 field campaign, Adv. Water Resour., 50, 162–177, doi:10.1016/j.advwatres.2012.06.005, 2012a.

Anderson, M. C., Allen, R. G., Morse, A., and Kustas, W. P.: Use of Landsat thermal imagery in monitoring evapotranspiration and managing water resources, Remote Sens. Environ., 122, 50–65, doi:10.1016/j.rse.2011.08.025, 2012b.

Anderson, M. C., Cammalleri, C., Hain, C. R., Otkin, J., Zhan, X., and Kustas, W.: Using a diagnostic soil-plant-atmosphere model for monitoring drought at field to continental scales, Procedia Environ. Sci., 19, 47–56, doi:10.1016/j.proenv.2013.06.006, 2013.

Bailey, R.: Description of the Ecoregions of the United States, USDA Forest Service, Washington, 1995.

Baldocchi, D., Falge, E., Gu, L., Olson, R., Hollinger, D., Running, S., Anthoni, P., Bernhofer, C., Davis, K., and Evans, R.: FLUXNET: A new tool to study the temporal and spatial variability of ecosystem-scale carbon dioxide, water vapor, and energy flux densities, B. Am. Meteorol. Soc., 82, 2415–2434, 2001.

Beven, K. and Freer, J.: Equifinality, data assimilation, and uncertainty estimation in mechanistic modelling of complex environmental systems using the GLUE methodology, J. Hydrol., 249, 11–29, 2001.

Brutsaert, W. H.: Evaporation in the Atmosphere, D. Reidel pub. Comp, Dordrecht, Boston, London, 1982.

Cammalleri, C., Anderson, M. C., Gao, F., Hain, C. R., and Kustas, W. P.: A data fusion approach for mapping daily evapotranspiration at field scale, Water Resour. Res., 49, 4672–4686, doi:10.1002/wrcr.20349, 2013.

Cammalleri, C., Anderson, M. C., Gao, F., Hain, C. R., and Kustas, W. P.: Mapping daily evapotranspiration at field scales over rainfed and irrigated agricultural areas using remote sensing data fusion, Agr. Forest Meteorol., 186, 1–11, doi:10.1016/j.agrformet.2013.11.001, 2014.

Chen, F. and Dudhia, J.: Coupling an advanced land surface-hydrology model with the Penn State-NCAR MM5 modeling system. Part I: Model implementation and sensitivity, Mon. Weather Rev., 129, 569–585, 2001.

Chen, J., Zhu, X., Vogelmann, J. E., Gao, F., and Jin, S.: A simple and effective method for filling gaps in Landsat ETM+ SLC-off images, Remote Sens. Environ., 115, 1053–1064, 2011.

Cienciala, E. and Lindroth, A.: Gas-exchange and sap flow measurements of Salix viminalis trees in short-rotation forest, Trees, 9, 295–301, 1995.

Cuenca, R. H., Stangel, D. E., and Kelly, S. F.: Soil water balance in a boreal forest, J. Geophys. Res.-Atmos., 102, 29355–29365, 1997.

Domec, J.-C., Sun, G., Noormets, A., Gavazzi, M. J., Treasure, E. A., Cohen, E., Swenson, J. J., McNulty, S. G., and King, J. S.: A comparison of three methods to estimate evapotranspiration in two contrasting Loblolly Pine plantations: age-related changes in water use and drought sensitivity of evapotranspiration components, Forest Sci., 58, 497–512, doi:10.5849/forsci.11-051, 2012.

Duchemin, B., Hadria, R., Erraki, S., Boulet, G., Maisongrande, P., Chehbouni, A., Escadafal, R., Ezzahar, J., Hoedjes, J. C. B., and Kharrou, M. H.: Monitoring wheat phenology and irrigation in Central Morocco: On the use of relationships between evapotranspiration, crops coefficients, leaf area index and remotely-sensed vegetation indices, Agr. Water Manage., 79, 1–27, 2006.

Fang, Y., Sun, G., Caldwell, P., McNulty, S. G., Noormets, A., Domec, J., King, J., Zhang, Z., Zhang, X., and Lin, G.: Monthly land cover specific evapotranspiration models derived from global eddy flux measurements and remote sensing data, Ecohydrology, doi:10.1002/eco.1629, in press, 2015.

Friedl, M. A., McIver, D. K., Hodges, J. C. F., Zhang, X. Y., Muchoney, D., Strahler, A. H., Woodcock, C. E., Gopal, S., Schneider, A., Cooper, A., Baccini, A., Gao, F., and Schaaf, C.: Global land cover mapping from MODIS: algorithms and early results, Remote Sens. Environ., 83, 287–302, doi:10.1016/S0034-4257(02)00078-0, 2002.

Fry, J., Xian, G., Jin, S., Dewitz, J., Homer, C., Yang, L., Barnes, C., Herold, N., and Wickham, J.: Completion of the 2006 national land cover database for the conterminous United States, publ. by: Multi-Resolution L. Charact. Consortium, PE & RS, Vol. 77, 858–864, http//www.mrlc.gov/nlcd2006.php (last access: 24 October 2013), 2011.

Gao, F., Masek, J., Schwaller, M., and Hall, F.: On the blending of the Landsat and MODIS surface reflectance: Predicting daily Landsat surface reflectance, IEEE T. Geosci. Remote, 44, 2207–2218, 2006.

Gao, F., Morisette, J. T., Wolfe, R. E., Ederer, G., Pedelty, J., Masuoka, E., Myneni, R., Tan, B., and Nightingale, J.: An algorithm to produce temporally and spatially continuous MODIS-LAI time series, IEEE T. Geosci. Remote, 5, 60–64, 2008.

Gao, F., Kustas, W. P., and Anderson, M. C.: A data mining approach for sharpening thermal satellite imagery over land, Remote Sens., 4, 3287–3319, 2012a.

Gao, F., Anderson, M. C., Kustas, W. P., and Wang, Y.: Simple method for retrieving leaf area index from Landsat using MODIS leaf area index products as reference, J. Appl. Remote Sens., 6, 63551–63554, 2012b.

Hain, C. R., Crow, W. T., Mecikalski, J. R., Anderson, M. C., and Holmes, T.: An intercomparison of available soil moisture estimates from thermal infrared and passive microwave remote sensing and land surface modeling, J. Geophys. Res., 116, D15107, doi:10.1029/2011JD015633, 2011.

Hain, C. R., Crow, W. T., Anderson, M. C., and Yilmaz, M. T.: Diagnosing neglected soil moisture source/sink processes via a thermal infrared-based two-source energy balance model, J. Hydrometeorol., 16, 1070–1086, 2015.

Hanson, P. J., Amthor, J. S., Wullschleger, S. D., Wilson, K. B., Grant, R. F., Hartley, A., Hui, D., Hunt E Raymond, J., Johnson, D. W., and Kimball, J. S.: Oak forest carbon and water simulations: model intercomparisons and evaluations against independent data, Ecol. Monogr., 74, 443–489, 2004.

Huete, A., Didan, K., Miura, T., Rodriguez, E. P., Gao, X., and Ferreira, L. G.: Overview of the radiometric and biophysical performance of the MODIS vegetation indices, Remote Sens. Environ., 83, 195–213, 2002.

Johnson, L. F. and Trout, T. J.: Satellite NDVI assisted monitoring of vegetable crop evapotranspiration in California's San Joaquin valley, Remote Sens., 4, 439–455, 2012.

Kalma, J. D., McVicar, T. R., and McCabe, M. F.: Estimating land surface evaporation: A review of methods using remotely sensed surface temperature data, Surv. Geophys., 29, 421–469, 2008.

Kannan, N., White, S. M., Worrall, F., and Whelan, M. J.: Sensitivity analysis and identification of the best evapotranspiration and runoff options for hydrological modelling in SWAT-2000, J. Hydrol., 332, 456–466, 2007.

Klein, T., Rotenberg, E., Cohen Hilaleh, E., RazYaseef, N., Tatarinov, F., Preisler, Y., Ogée, J., Cohen, S., and Yakir, D.: Quantifying transpirable soil water and its relations to tree water use dynamics in a water limited pine forest, Ecohydrology, 7, 409–419, 2014.

Kuczera, G.: Prediction of water yield reductions following a bushfire in ash-mixed species eucalypt forest, J. Hydrol., 94, 215–236, 1987.

Kustas, W. P. and Anderson, M.: Advances in thermal infrared remote sensing for land surface modeling, Agr. Forest Meteorol., 149, 2071–2081, doi:10.1016/j.agrformet.2009.05.016, 2009.

Kustas, W. P. and Norman, J. M.: Evaluation of soil and vegetation heat flux predictions using a simple two-source model with radiometric temperatures for partial canopy cover, Agr. Forest Meteorol., 94, 13–29, 1999.

Kustas, W. P. and Norman, J. M.: A two-source energy balance approach using directional radiometric temperature observations for sparse canopy covered surfaces, Agron. J., 92, 847–854, 2000.

Lu, J., Sun, G., McNulty, S. G., and Amatya, D. M.: Modeling actual evapotranspiration from forested watersheds across the southeastern united states, J. Am. Water Resour. Assoc., 39, 886–896, 2003.

Matheussen, B., Kirschbaum, R. L., Goodman, I. A., O'Donnell, G. M., and Lettenmaier, D. P.: Effects of land cover change on streamflow in the interior Columbia River Basin (USA and Canada), Hydrol. Process., 14, 867–885, 2000.

Méndez Barroso, L. A., Vivoni, E. R., Robles Morua, A., Mascaro, G., Yépez, E. A., Rodríguez, J. C., Watts, C. J., Garatuza Payán, J., and Saíz Hernández, J. A.: A modeling approach reveals differences in evapotranspiration and its partitioning in two semiarid ecosystems in northwest Mexico, Water Resour. Res., 50, 3229–3252, 2014.

Murakami, S., Tsuboyama, Y., Shimizu, T., Fujieda, M., and Noguchi, S.: Variation of evapotranspiration with stand age and climate in a small Japanese forested catchment, J. Hydrol., 227, 114–127, 2000.

Mutiibwa, D. and Irmak, S.: AVHRR NDVI based crop coefficients for analyzing long term trends in evapotranspiration in relation to changing climate in the US High Plains, Water Resour. Res., 49, 231–244, 2013.

Myneni, R. B., Hoffman, S., Knyazikhin, Y., Privette, J. L., Glassy, J., Tian, Y., Wang, Y., Song, X., Zhang, Y., Smith, G. R., Lotsch, A., Friedl, M., Morisette, J. T., Votava, P., Nemani, R. R., and Running, S. W.: Global products of vegetation leaf area and fraction absorbed PAR from year one of MODIS data, Remote Sens. Environ., 83, 214–231, doi:10.1016/S0034-4257(02)00074-3, 2002.

Nemani, R. and Running, S.: Estimation of Regional Surface Resistance to Evapotranspiration from NDVI and Thermal-IR AVHRR Data, J. Appl. Meteorol., 28, 276–284, 1988.

Nemani, R., Pierce, L., Running, S., and Band, L.: Forest Ecosystem Processes at the Watershed Scale – Sensitivity to Remotely-Sensed Leaf-Area Index Estimates, Int. J. Remote Sens., 14, 2519–2534, 1993.

Noormets, A., McNulty, S. G., DeForest, J. L., Sun, G., Li, Q., and Chen, J.: Drought during canopy development has lasting effect on annual carbon balance in a deciduous temperate forest, New Phytol., 179, 818–828, 2008.

Noormets, A., Gavazzi, M. J., Mcnulty, S. G., Domec, J.-C., Sun, G., King, J. S., and Chen, J.: Response of carbon fluxes to drought in a coastal plain loblolly pine forest, Global Change Biol., 16, 272–287, doi:10.1111/j.1365-2486.2009.01928.x, 2010.

Norman, J. M., Kustas, W. P., and Humes, K. S.: Source approach for estimating soil and vegetation energy fluxes in observations of directional radiometric surface temperature, Agr. Forest Meteorol., 77, 263–293, doi:10.1016/0168-1923(95)02265-Y, 1995.

Norman, J. M., Anderson, M. C., Kustas, W. P., French, A. N., Mecikalski, J., Torn, R., Diak, G. R., Schmugge, T. J., and Tanner, B. C. W.: Remote sensing of surface energy fluxes at 10 1-m pixel resolutions, Water Resour. Res., 39, 1221, doi:10.1029/2002WR001775, 2003.

Olioso, A., Chauki, H., Courault, D., and Wigneron, J.-P.: Estimation of evapotranspiration and photosynthesis by assimilation of remote sensing data into SVAT models, Remote Sens. Environ., 68, 341–356, 1999.

Pan, M., Sahoo, A. K., Troy, T. J., Vinukollu, R. K., Sheffield, J., and Wood, E. F.: Multisource estimation of long-term terrestrial water budget for major global river basins, J. Climate, 25, 3191–3206, 2012.

Pereira, L. S., Oweis, T., and Zairi, A.: Irrigation management under water scarcity, Agr. Water Manage., 57, 175–206, 2002.

Roberts, S., Vertessy, R., and Grayson, R.: Transpiration from Eucalyptus sieberi (L. Johnson) forests of different age, Forest Ecol. Manage., 143, 153–161, 2001.

Roy, D. P., Ju, J., Lewis, P., Schaaf, C., Gao, F., Hansen, M., and Lindquist, E.: Multi-temporal MODIS-Landsat data fusion for relative radiometric normalization, gap filling, and prediction of Landsat data, Remote Sens. Environ., 112, 3112–3130, 2008.

Schaaf, C. B., Liu, J., Gao, F., and Strahler, A. H.: Aqua and Terra MODIS albedo and reflectance anisotropy products, in: Land Remote Sensing and Global Environmental Change, Springer, 549–561, 2010.

Scott, R. L., Huxman, T. E., Cable, W. L., and Emmerich, W. E.: Partitioning of evapotranspiration and its relation to carbon dioxide exchange in a Chihuahuan Desert shrubland, Hydrol. Process., 20, 3227–3243, 2006.

Semmens, K. A., Anderson, M. C., Kustas, W. P., Gao, F., Alfieri, J. G., McKee, L., Prueger, J. H., Hain, C. R., Cammalleri, C., and Yang, Y.: Monitoring daily evapotranspiration over two California vineyards using Landsat 8 in a multi-sensor data fusion approach, Remote Sens. Environ., 85, 155–170, doi:10.1016/j.rse.2015.10.025, 2015.

Shaw, R. H. and Pereira, A. R.: Aerodynamic roughness of a plant canopy: a numerical experiment, Agr. Meteorol., 26, 51–65, 1982.

Smith, D. M. and Allen, S. J.: Measurement of sap flow in plant stems, J. Exp. Bot., 47, 1833–1844, 1996.

Sun, G., Noormets, A., Gavazzi, M. J., McNulty, S. G., Chen, J., Domec, J.-C., King, J. S., Amatya, D. M., and Skaggs, R. W.: Energy and water balance of two contrasting loblolly pine plantations on the lower coastal plain of North Carolina, USA, Forest Ecol. Manage., 259, 1299–1310, 2010.

Sun, G., Alstad, K., Chen, J., Chen, S., Ford, C. R., Lin, G., Liu, C., Lu, N., Mcnulty, S. G., Miao, H., Noormets, A., Vose, J. M., Wilske, B., Zeppel, M., and Zhang, Y.: A general predictive model for estimating monthly ecosystem evapotranspiration, Ecohydrology, 255, 245–255, doi:10.1002/eco.194, 2011a.

Sun, G., Caldwell, P., Noormets, A., McNulty, S. G., Cohen, E., Moore Myers, J., Domec, J.-C., Treasure, E., Mu, Q., Xiao, J., John, R., and Chen, J.: Upscaling key ecosystem functions across the conterminous United States by a water-centric ecosystem model, J. Geophys. Res., 116, G00J05, doi:10.1029/2010JG001573, 2011b.

Sun, Q., Wang, Z., Li, Z., Erb, A., and Schaaf, C. B.: Evaluation of the global MODIS 30 arc-second spatially and temporally complete snow-free land surface albedo and reflectance anisotropy dataset, Int. J. Appl. Earth Obs. Geoinf., 58, 36–49, doi:10.1016/j.jag.2017.01.011, 2017.

Tague, C. L. and Band, L. E.: RHESSys: Regional hydro-ecologic simulation system-an object-oriented approach to spatially distributed modeling of carbon, water, and nutrient cycling, Earth Interact., 8, 1–42, 2004.

Tague, C. L., Choate, J. S., and Grant, G.: Parameterizing subsurface drainage with geology to improve modeling streamflow responses to climate in data limited environments, Hydrol. Earth Syst. Sci., 17, 341–354, doi:10.5194/hess-17-341-2013, 2013.

Tian, H., Chen, G., Liu, M., Zhang, C., Sun, G., Lu, C., Xu, X., Ren, W., Pan, S., and Chappelka, A.: Model estimates of net primary productivity, evapotranspiration, and water use efficiency in the terrestrial ecosystems of the southern United States during 1895–2007, Forest Ecol. Manage., 259, 1311–1327, doi:10.1016/j.foreco.2009.10.009, 2010.

Vertessy, R. A., Watson, F. G. R., and Sharon, K. O.: Factors determining relations between stand age and catchment water balance in mountain ash forests, Forest Ecol. Manage., 143, 13–26, 2001.

Von Bertalanffy, L.: General systems theory, Braziller, New York, 1968.

Wan, Z., Zhang, Y., Zhang, Q., and Li, Z.-L.: Quality assessment and validation of the MODIS global land surface temperature, Int. J. Remote Sens., 25, 261–274, 2004.

Wang, K. and Dickinson, R. E.: A review of global terrestrial evapotranspiration: observation, modelling, climatology, and climatic variability, Rev. Geophys., 50, 1–54, doi:10.1029/2011RG000373, 2012.

Wickham, J. D., Stehman, S. V, Gass, L., Dewitz, J., Fry, J. A., and Wade, T. G.: Accuracy assessment of NLCD 2006 land cover and impervious surface, Remote Sens. Environ., 130, 294–304, 2013.

Williams, C. A., Reichstein, M., Buchmann, N., Baldocchi, D., Beer, C., Schwalm, C., Wohlfahrt, G., Hasler, N., Bernhofer, C., and Foken, T.: Climate and vegetation controls on the surface water balance: Synthesis of evapotranspiration measured across a global network of flux towers, Water Resour. Res., 48, W00L11, doi:10.1029/2011WR010809, 2012.

Wilson, K. B., Hanson, P. J., Mulholland, P. J., Baldocchi, D. D., and Wullschleger, S. D.: A comparison of methods for determining forest evapotranspiration and its components: sap-flow, soil water budget, eddy covariance and catchment water balance, Agr. Forest Meteorol., 106, 153–168, 2001.

Wullschleger, S. D., Meinzer, F. C., and Vertessy, R. A.: A review of whole-plant water use studies in tree, Tree Physiol., 18, 499–512, 1998.

Yilmaz, M. T., Anderson, M. C., Zaitchik, B. F., Hain, C. R., Crow, W. T., Ozdogan, M., and Chung, J. A.: Comparison of prognostic and diagnostic surface flux modeling approaches over the Nile River Basin, Water Resour. Res., 50, 386–408, 2014.

Zhang, L., Hickel, K., Dawes, W. R., Chiew, F. H. S., Western, A. W., and Briggs, P. R.: A rational function approach for estimating mean annual evapotranspiration, Water Resour. Res., 40, W02502, doi:10.1029/2003WR002710, 2004.

Zhu, Z. and Woodcock, C. E.: Object-based cloud and cloud shadow detection in Landsat imagery, Remote Sens. Environ., 118, 83–94, doi:10.1016/j.rse.2011.10.028, 2012.

Evaporation from cultivated and semi-wild Sudanian Savanna in west Africa

Natalie C. Ceperley[1,2], Theophile Mande[2], Nick van de Giesen[3], Scott Tyler[4], Hamma Yacouba[5], and Marc B. Parlange[1,2]

[1]Department of Civil Engineering, Faculty of Applied Sciences, University of British Columbia, Vancouver, British Columbia, Canada

[2]Laboratory of Environmental Fluid Mechanics and Hydrology, School of Architecture, Civil and Environmental Engineering, Swiss Federal Institute of Technology, Lausanne, Switzerland

[3]Department of Civil Engineering and Geosciences, Delft University of Technology, Delft, the Netherlands

[4]Department of Geological Sciences & Engineering, University of Nevada, Reno, NV, USA

[5]Laboratory Hydrology and Resources in Water, International Institute for Water and Environmental Engineering (2iE), Ouagadougou, Burkina Faso

Correspondence to: Natalie C. Ceperley (natalie.ceperley@unil.ch)

Abstract. Rain-fed farming is the primary livelihood of semi-arid west Africa. Changes in land cover have the potential to affect precipitation, the critical resource for production. Turbulent flux measurements from two eddy-covariance towers and additional observations from a dense network of small, wireless meteorological stations combine to relate land cover (savanna forest and agriculture) to evaporation in a small ($3.5\,\mathrm{km}^2$) catchment in Burkina Faso, west Africa. We observe larger sensible and latent heat fluxes over the savanna forest in the headwater area relative to the agricultural section of the watershed all year. Higher fluxes above the savanna forest are attributed to the greater number of exposed rocks and trees and the higher productivity of the forest compared to rain-fed, hand-farmed agricultural fields. Vegetation cover and soil moisture are found to be primary controls of the evaporative fraction. Satellite-derived vegetation index (NDVI) and soil moisture are determined to be good predictors of evaporative fraction, as indicators of the physical basis of evaporation. Our measurements provide an estimator that can be used to derive evaporative fraction when only NDVI is available. Such large-scale estimates of evaporative fraction from remotely sensed data are valuable where ground-based measurements are lacking, which is the case across the African continent and many other semi-arid areas. Evaporative fraction estimates can be combined, for example, with sensible heat from measurements of temperature variance, to provide an estimate of evaporation when only minimal meteorological measurements are available in remote regions of the world. These findings reinforce local cultural beliefs of the importance of forest fragments for climate regulation and may provide support to local decision makers and rural farmers in the maintenance of the forest areas.

1 Introduction

The Sudanian savanna in southeastern Burkina Faso is a patchwork of savanna, forest, and scrubland with some patches more representative of the drier Sahel and others more representative of the more humid Guinean forests. Vegetation is mainly deciduous according to seasonal moisture availability, but spatial variations in topography, water availability, and plant communities result in some variation in greenness and some evergreen species, for example, near the springs. Historically, people in this region rely on a mix of hunting and gathering, small-scale agriculture, and pastoralism. As land claims and regulations have changed, communities have been forced to rely more on agricultural production as a primary source of food and income, resulting in land conversion for agriculture. Today, small-scale, rain-fed agriculture is the dominant livelihood in large parts of west Africa, despite its high level of dependence on seasonally controlled hydrology. Conversion of the landscape to agriculture involves removing rocks, trees, and natural grasses, and tilling the soil (Swanson, 1978). Transformation of forest land to agriculture has been shown by model simulations to alter global circulation, hydrology, and biogeochemistry both in the present and in predictions of the future (Abiodun et al., 2008; Feddema et al., 2005; Mande et al., 2011; Steiner et al., 2009; Sylla et al., 2015; Vitousek, 1997).

Modification of the land surface results in changes to the physical environment, and specifically hydrological fluxes, by altering the components of the surface energy budget (Pielke et al., 2002). The partition of net radiation into sensible heat, evaporation, and soil heat flux drives global atmospheric processes and is controlled by interacting surface and atmospheric conditions (Foken, 2008; Szilagyi and Parlange, 1999).

The energy balance is challenging to close even in areas and regions with extensive datasets and accessibility, and is particularly challenging in areas with complex surfaces (Burba, 2013; Domingo et al., 2011; Farhadi, 2012; Federer et al., 2003; Foken et al., 2009; Guo et al., 2006; Katul and Parlange, 1992; Krishnan et al., 2012; Kustas et al., 1994; Parlange and Katul, 1992; Williams et al., 2012). Evaporation from vegetated surfaces remains the component of the global distribution of water that is the least frequently measured and thus the least well understood (Brutsaert, 1982; Brutsaert and Parlange, 1992; Burba, 2005; Compaore, 2006; Crago, 1996; Crago and Qualls, 2013), particularly in west Africa, due to limited field observations (Bagayoko et al., 2007; Dolman et al., 1997; Gash et al., 1997; Mande et al., 2011). A connection exists between changes in albedo and the occurrence of drought in west Africa, although the physical processes and direct implications for desertification are debatable (Charney, 1975; Nicholson et al., 1998).

The evaporative fraction, the ratio of latent heat flux to available energy, is useful to estimate total daily evaporation with measurements of a single component of the energy balance and to upscale surface measurements using remote sensing products (Brutsaert and Sugita, 1992; Compaore, 2006; Porte-Agel et al., 2000; Shuttleworth et al., 1989; Szilagyi et al., 1998; Szilagyi and Parlange, 1999). Using evaporative fraction to calculate the total daily evaporation is based on the concept of self-preservation in the diurnal evolution of the surface energy budget (Brutsaert and Sugita, 1992; Porte-Agel et al., 2000), stating that the diurnal cycle of each of the energetic fluxes will resemble that of available energy, even if there is variation in the quantity, allowing for exploiting satellite data that are typically only obtainable once a day at best. Remotely sensed land surface temperature is currently the primary tool for mapping the surface energy budget over a large area (Bateni and Entekhabi, 2012). Evaporative fraction is constant during daytime in fair weather conditions (Gentine et al., 2007) but can be much less constant when

moisture circulation rates are high and available soil moisture increases (Lhomme and Elguero, 1999). Seasonal progression of evaporative fraction response to rainfall and moisture availability can depend on surface conditions. For example, it can respond faster in grassland than in woodland (Farah et al., 2004). These variations are not explained by meteorological conditions, including cloudiness alone, but rather change in surface resistance and moisture advection and availability (Farah et al., 2004; Lohou et al., 2010, 2014).

We measured the energetic and hydrologic fluxes in two sites of a semi-arid, mixed-use catchment over a year and a half, capturing both the greening and dry-down phases. The land is used as an agroforestry parkland that is farmed every 2–3 years, and contains a forested area made up of evergreen trees arranged in a gallery forest surrounding a spring and an open wooded savanna (savanna forest) on a plateau about 100 m above the surrounding land. These land covers are representative of the surrounding region and capture the range from more to less anthropogenic land uses. The multi-use comparison over multiple seasonal cycles puts this study among the few recent, long-term studies in this region (Bagayoko et al., 2007; Brümmer et al., 2008; Ezzahar et al., 2009; Guichard et al., 2009; Guyot et al., 2009, 2012, Lohou et al., 2010, 2014; Mamadou et al., 2014, 2016; Mauder et al., 2006; Ramier et al., 2009; Timouk et al., 2009; Velluet et al., 2014). We calculate the evaporative fraction over the study period and compare it with land cover and atmospheric controls, in order to provide estimation based on the physical basis of fluxes. Our measurements are significant because they allow calibration and comparison for calculation of the components of the energy budget from lower cost and more easily maintained stations (Nadeau et al., 2009; Simoni et al., 2011) and corresponding data from satellite imagery. This study has implications for development priorities, as it takes place in context where local livelihood can be dramatically affected by slight changes in the water balance and land cover.

2 Measurements and calculations

2.1 Site description

Observations were made in a small catchment (3.5 km^2 area) neighboring the village of Tambarga in the commune of Madjoari, in the Gourma province, in Burkina Faso, west Africa (Fig. 1). The ephemeral stream defining the catchment (outlet located at 11°26′29.7″ N 1°12′57.7″ E) flows into the Singou River, which joins the Pendjari River and eventually flows into the White Volta of the Volta River basin, the third largest river basin by area in west Africa, after the Niger and the Senegal. A rocky escarpment defines the catchment with a plateau on average some 100 m above the lower agricultural fields, and the soil is predominantly sandy loam (Ceperley, 2014). These fields are the "house" farms and are smaller

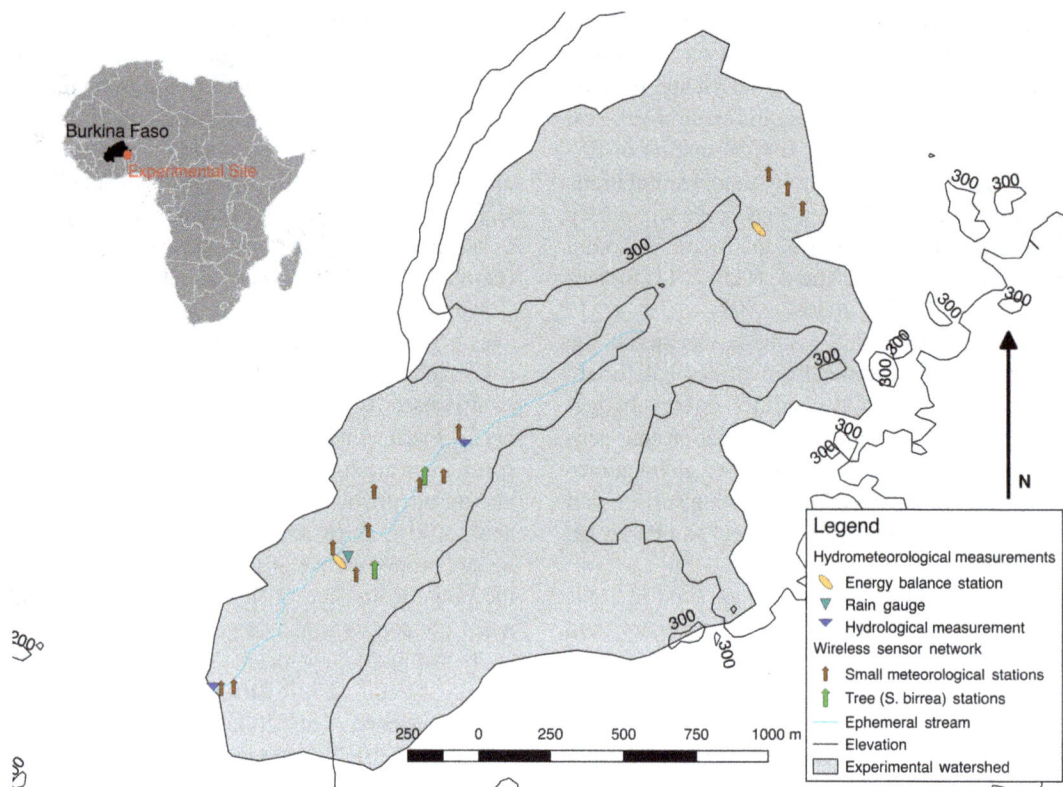

Figure 1. Map of experimental watershed. The site is located next to the village of Tambarga, in the southeastern corner of Burkina Faso, west Africa. Energy balance stations, small meteorological stations, including those near agroforestry trees, and hydrologic monitoring stations are shown. Springs are located at the source of the ephemeral stream, at the base of the rocky escarpment.

than the main revenue farms. In 2009, they were farmed with short rotation millet and in 2010 they were left fallow. Plowing occasionally uses animal-drawn plows, but is primarily done by hand – this is not intensive agriculture. The open wooded savanna (savanna forest) on top of the rocky escarpment and rain-fed grain (corn or millet) are the two dominant land covers of the catchment according to area. At opposite ends of the catchment, there is a dense gallery forest in the valley that grows near perennial springs and an ephemeral wetland used for rice cultivation near the point considered the outlet of the watershed. The existence of a raised plateau with perennial springs suggests that there may be lateral subsurface water transfer. Farming is the main livelihood in the village, and crops include millet, sorghum, cotton, and rice. Agroforestry trees in the fields are common and consist most often of the tree species *Vitellaria paradoxa*, *Sclerocarya birrea*, and *Ficus* sp. (Bordes, 2010; see Table 2). Burning mostly occurs between November and January, but occasionally there is a fire in February or March. The village is made up of a majority of people from the Gourmantché ethnic group, though there is a significant population of Fulani (Peulh), and some migrants from other areas of Burkina Faso or neighboring countries (Ceperley, 2014).

The watershed falls in the Sudanian zone of the west African monsoon climate system, defined by alternating wet and dry seasons, with the rain falling between May and September. The natural vegetation is Sudanian wooded savanna, composed of a mix of deciduous woody trees, shrubs, and tall grasses. In addition, due to the variation in topography and water availability, there are gallery forests near streams or rivers that contain many species endemic further south in the Guinean zone. The surrounding area is a patchwork of hunting reserves and national parks and thus has a higher level of vegetation cover than most of the country. This watershed is a prime location to study the consequences of land use change from Sudanian savanna to agricultural fields, since it contains both open wooded savanna that has not been memorably farmed and regularly farmed fields. Agriculture is primarily rain-fed and not mechanized corn and millet cultivation. In addition, the surrounding Sudanian savanna, which is characterized by fire-selected grasses ranging from 20 cm to 1.5 m in height, also includes patches of woody scrubland, open forests, gallery forests, and riparian stands (Arbonnier, 2004). Inventory of woody species taller than breast height in the two major land covers was inventoried by family, chorology, and life form (according to Adamou, 2005) and is reported in Table 2.

Table 1. Inventory of instruments used for energy balance analysis. The name of the sensor or instrument is followed by the measurement it performs, the height and depth of each sensor, the total number used, and the interval of measurement. The heights were identical at both measurement points.

Instrument	Measurement	Height/depth	Number	Interval	Time span
CSAT-3 sonic anemometer (Campbell Scientific, Logan, UT, USA)	3-D wind speed and direction, air temperature	2.2 m	3	20 Hz, proc. 30 min	May 2009–October 2010
LI-7500 infrared gas analyzer (LICOR, Lincoln, NE, USA)	H_2O concentration	2.2 m	3	20 Hz, proc. 30 min	May 2009–October 2010
CNR2 radiometer (Kipp & Zonen, Delft, the Netherlands)	SW LW radiation	2.1 m	2	1 min	October 2009–October 2010
HMP450 (Campbell Scientific, Logan, UT, USA) with radiation shield	Air temperature, air humidity	2.25 m	2	1 min	May 2009–October 2010
Pluviometer 3029 (Précis Mécanique, Bezons CEDEX, France)	Precipitation	1 m	1	0.1 mm	May 2009–January 2015
Davis Instruments (Hayward, CA, USA)	Precipitation	0.2 m	12	1 min	May 2009–January 2015
Davis Instruments (Hayward, CA, USA)	Shortwave solar radiation, incoming	1.8m	12	1 min	May 2009–January 2015
Infrared thermometer TN901 (Zytemp, Taiwan, R.O.C.)	Surface temperature	1.1 m	12	1 min	May 2009–January 2015
SHT7 (Sensiron AG, Staefa Zurich, Switzerland)	Air temperature and humidity	1.7 m	12	1 min	May 2009–January 2015
5TM, 5TE, ECTM (Decagon, Pullman, WA, USA)	Soil humidity	5–30 cm	∼ 24 (varied)	1 min	May 2009–January 2015

2.2 Field measurements

Two energy balance stations were installed from May 2009 to September 2010. One was situated in an agricultural field planted with short season millet in 2009 and left fallow in 2010, and the second one measured over the gallery forest when the wind came from the west ($90° \pm 45°$) and over the open wooded savanna when the wind came from the south ($180° \pm 45°$). They were equipped with sonic anemometers, infrared open-path gas analyzers, net radiometers, and air temperature and humidity sensors (Table 1). Eddy-covariance equipment was placed facing two opposite directions (46 and 226°) on the lower station over the field and in the dominant wind direction on the upper station over the gallery forest. The distance between the two measurement points was approximately 1 km and 100 m of height difference. Near the station in the field was a high-precision rain gauge measuring at a resolution of 0.1 mm. In addition, a network of up to 12 small meteorological stations (Ingelrest et al., 2010) was distributed across the watershed with sensors to measure incoming solar radiation, wind direction and speed, air temperature and relative humidity, rainfall, soil moisture and temperature, and surface temperature. In this analysis, we use data from May 2009 to October 2010, the 15 months with both towers operational, but when possible, we present the longest time series possible for climatic context. We attempted to measure ground heat flux using heat flux plates but ultimately rejected the observations because of irregularities in the land surrounding the plates.

2.3 Flux calculation

The surface energy budget is written in Eq. (1):

$$R_n = L_e E + H + G, \tag{1}$$

where $L_e E$ is latent heat flux, H is sensible heat flux, R_n is the net radiation, and G is the soil heat flux, all in watts per square meter ($W\,m^{-2}$).

The sensible heat is expressed in Eq. (2):

$$H = \rho c_p \overline{w'T'}, \tag{2}$$

where ρ is the air density ($kg\,m^{-3}$), c_p is the specific heat ($J\,kg^{-1}\,K^{-1}$), and $w'T'$ is the covariance of fluctuations of

Table 2. Inventory of species found in both agricultural field site and savanna forest site.

Inventory of woody vegetation in savanna forest (1 ha)			
Acacia macrostachya	Leg.–Mim.	mph	S
Burkea africana	Leg.–Caes.	mph	SZ
Combretum nigricans	Combretaceae	mph	S
Daniellia oliveri	Leg.–Caes.	MPh	SZ
Detarium microcarpum	Leg.–Caes.	mph	S
Gardenia erubescens	Rubiaceae	nph	S
Grewia flavescens	Tiliaceae	Lmph	GC
Guiera senegalensis	Combretaceae	nph	SZ
Hymenocardia acida	Euphorbiaceae	mph	SZ
Lannea acida	Anacardiaceae	mPh	S
Parkia biglobosa	Leg.–Mim.	mPh	S
Prosopis africana	Leg.–Mim.	mPh	S
Pteleopsis suberosa	Combretaceae	mph	SZ
Pterocarpus erinaceus	Leg.–Pap.	mPh	S
Sclerocarya birrea	Anacardiaceae	mph	S
Sterculia setigera	Sterculiaceae	mph	S
Strychnos spinosa	Loganiaceae	LmPh	PAL
Terminalia avicennioides	Combretaceae	mph	S
Terminalia laxiflora	Combretaceae	mPh	S
Terminalia schimperiana (glaucescens)	Combretaceae	mph	S
Terminalia mollis	Combretaceae	mph	PRA
Vitellaria paradoxa	Sapotaceae	mPh	S
Ximenia americana	Olacaceae	nph	Pt

Inventory of woody vegetation in agricultural fields			
Acacia sieberiana	Leg.–Mim.	mph	SZ
Bombax costatum	Bombacaceae	mph	S
Detarium microcarpum	Leg.–Caes.	mph	S
Ficus sp.	Moraceae	MPh	
Lannea sp.	Anacardiaceae	mPh	S/SZ
Piliostigma reticulatum	Leg.–Caes.	mph	SG
Sclerocarya birrea	Anacardiaceae	mph	S
Terminalia schimperiana (glaucescens)	Combretaceae	mph	S
Terminalia laxiflora	Combretaceae	mPh	S
Terminalia mollis	Combretaceae	mph	PRA
Ziziphus mauritiana	Rhamnaceae	mph	PAL

Chorology		No.
S	Sudanian	19
SZ	Sudano-Zambezian	6
GC	Guineo-Congolian	1
PAL	Paleotropical	1
Pt	Pantropical	1
SG	Sudano/Guinean transition	1
PRA	Pluriregional African	1

Abbreviations	
Leg.–Caes.	Leguminosae–Caesalpinioideae
Leg.–Mim.	Leguminosae–Mimosoideae
Leg.–Pap.	Leguminosae–Papilionoideae

Life forms		
mph	microphanerophyte	2–8 m
MPh	megaphanerophyte	> 30 m
nph	nanophanerophyte	0.5–2 m
L	Liana	
mPh	mesophanerophyte	8–30 m

vertical wind speed (m s^{-1}) and temperature (K). Latent heat flux is expressed in Eq. (3):

$$L_e = L_e \rho \overline{w'q'}, \tag{3}$$

where L_e is the latent energy of vaporization (J g^{-1}) and $w'q'$ is covariance of fluctuations of vertical wind speed (m s^{-1}) and humidity (g m^{-3}).

All measurements were taken at 10 Hz, and fluxes of sensible (H) and latent ($L_e E$) heat were calculated at a half-hour time step using the covariance calculations as written above. Only daylight measurements, consistently between sunrise and sunset (08:00–16:00 UTC), corresponding to when energetic fluxes were of significant magnitude, were used for the comparison. The total of turbulent fluxes, $H + L_e E$, was subtracted from the net radiation, R_n, to give an indicator of ground heat flux (G; see Eq. 1), any unaccounted for flux transfers, and the error (Brutsaert, 1982; Higgins, 2012).

We used a planar fit correction that effectively tilted measurements of the three components of the wind field perpendicular to the direction of flow, so that the vertical wind was equal to zero over 1-month averaging periods (Aubinet et al., 2012; Burba, 2005; Oldroyd et al., 2015; Rebmann et al., 2012; Wilczak et al., 2001). We then performed a linear regression using the mean wind vectors to obtain a matrix that we used to adjust wind vectors and stress tensors in a new coordinate system with a z axis perpendicular to the mean streamline. Finally, we rotated the intermediate winds and stress tensors. The Webb–Pearman–Leuning equation (Foken et al., 2012; Leuning, 2007; Webb et al., 1980):

$$E = (1 + \mu\sigma)\left(\overline{\omega'\rho'_v}\right) + \frac{\overline{T}}{\rho_v}\overline{\omega'T'} \tag{4}$$

was used to correct for any influence of trace gas concentrations on temperature and humidity fluctuations.

2.4 Evaporative fraction

Evaporative fraction (EF) was calculated for each half hour of data, separately over the savanna forest and the agricultural area by dividing the latent heat flux by the available energy, which is equivalent to the sum of sensible and latent heat fluxes. Although the true measure of available energy would be the difference between the net radiation and the ground heat flux, the sum of the turbulent fluxes, H and $L_e E$, was deemed more accurate given the rejection of our ground heat flux measurement. The midday average (10:00–14:00 UTC) was used as the EF for a given day as that is when it was the most stable over the year (Fig. 3):

$$EF = \frac{L_e E}{H + L_e E}. \tag{5}$$

2.5 Volumetric water content

Volumetric water content (VWC, m^3 m^{-3}) in the soil was monitored at 15 and 30 cm depths in 2009 and at 5 and 20 cm

depths thereafter at some of the small meteorological stations representative of the various land covers (Table 1) using a measure of soil dielectric permittivity and converted to VWC (Topp et al., 1980). Measurements were averaged on the half-hour time step for comparison with EC measurements and by day for comparison with EF. Gaps in measurement were due to sensor malfunction. A vertical spatial average of measurements at various depths was used to obtain a continuous record for the watershed.

2.6 Cloud cover

Cloud cover was calculated by dividing the incoming shortwave radiation (W m^{-2}) measured with a radiometer (Table 1) at each small meteorological station by the theoretical incoming radiation (W m^{-2}) calculated with a simple model (Whiteman and Allwine, 1986) for each of the small meteorological stations operating on any given day:

$$CC = \frac{SW_{measured}}{SW_{Whitman \& Allwine}}. \tag{6}$$

The cloud cover was calculated independently for all stations and then averaged to give a single value per day.

2.7 Vegetation index

Normalized difference vegetation index (NDVI) is based on the amount of infrared radiation absorbed, which is related to the amount of photosynthesis taking place. It is considered a measure of the density of chlorophyll. NDVI is a ratio of the near-infrared (NIR) to red wavelengths:

$$NDVI = \frac{NIR - red}{NIR + red}. \tag{7}$$

Seasonal change in NDVI was observed by extraction of the area of interest from the 250 m resolution west Africa eMODIS 10-day temporally smoothed data (USGS FEWS NET). It has been corrected for molecular scattering, ozone absorption, and aerosols, and then smoothed using a least squares linear regression (Swets et al., 1999). The NDVI values are validated with in situ observation and photographs. The pixels that contained our stations were extracted to give a catchment-wide seasonal impression of the vegetation change. All of the pixels that cover our catchment are composed of multiple vegetation types, given the relatively coarse resolution, but they have been sorted by dominant land cover. An inventory of woody species (diameter at breast height of 1.3 m, greater than 10 cm), was performed for 1 ha of the savanna forest and the entirety of the agricultural area within the catchment (Table 2). Vegetation was classified according chorology and life form (Adomou, 2005).

2.8 Wind sector partition

Dominant wind direction for each half-hour covariance measurement was used to sort the sensible and latent energy

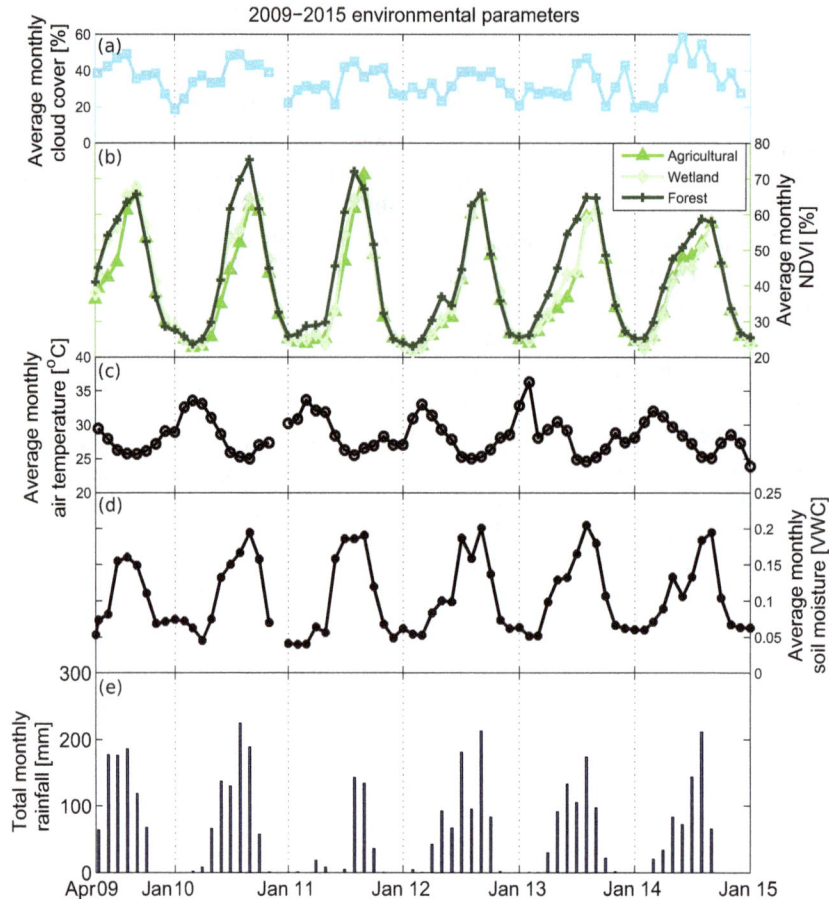

Figure 2. Environmental parameters at the study site for the monitoring period (2009–2015): (**a**) cloud cover measured with shortwave radiometers and averaged over all stations; (**b**) average NDVI from MODIS satellite data (250 m resolution, 10-day composite) averaged over each land cover of pixels containing stations; (**c**) monthly average air temperature (red line), also averaged for all stations; (**d**) volumetric soil moisture averaged over all stations and all land covers between 5 and 30 cm depth (blue line); and (**e**) monthly rainfall (missing bars indicate lack of rain).

fluxes (Fig. 6). Computation of the mean flux according to wind direction allowed for examination of the effect of wind direction, and corresponding land surface, on the flux magnitudes.

3   Results

3.1   Seasonality

A total of 1600 mm of precipitation was measured over the period of intensive monitoring (2009–2010) – 789 mm in 2009 and 811 mm in 2010 – and it fell almost entirely during the period from May to October. As seen in Fig. 2, average monthly air temperature, cloud cover, soil moisture, and NDVI followed the seasonal cycle of the rain: temperature was higher in the dry season (November–March) than in the wet season (May–October); cloud cover was lower in the

dry season and increased starting in March and April peaking both years in July; soil moisture was highest in the wet season; and vegetation, as shown by NDVI, increased over the course of the wet season, starting in May and peaking in September and declining afterwards as grasses senesced. The high level of seasonality is characteristic of semi-arid environments. Separation between the lower and upper parts of the catchment is apparent in the NDVI time series, where the savanna forest consistently stays more green, with the field only surpassing it due to its delayed senescence in September 2010. Plowing, early season harvests, and late season harvests are visible earlier in some years over only the agricultural land. However, since these are averages over a few potentially mixed pixels, which were composed of multiple crops, the individual behaviors are not visible. The land use differentiation is visible even at 250 m pixel resolution.

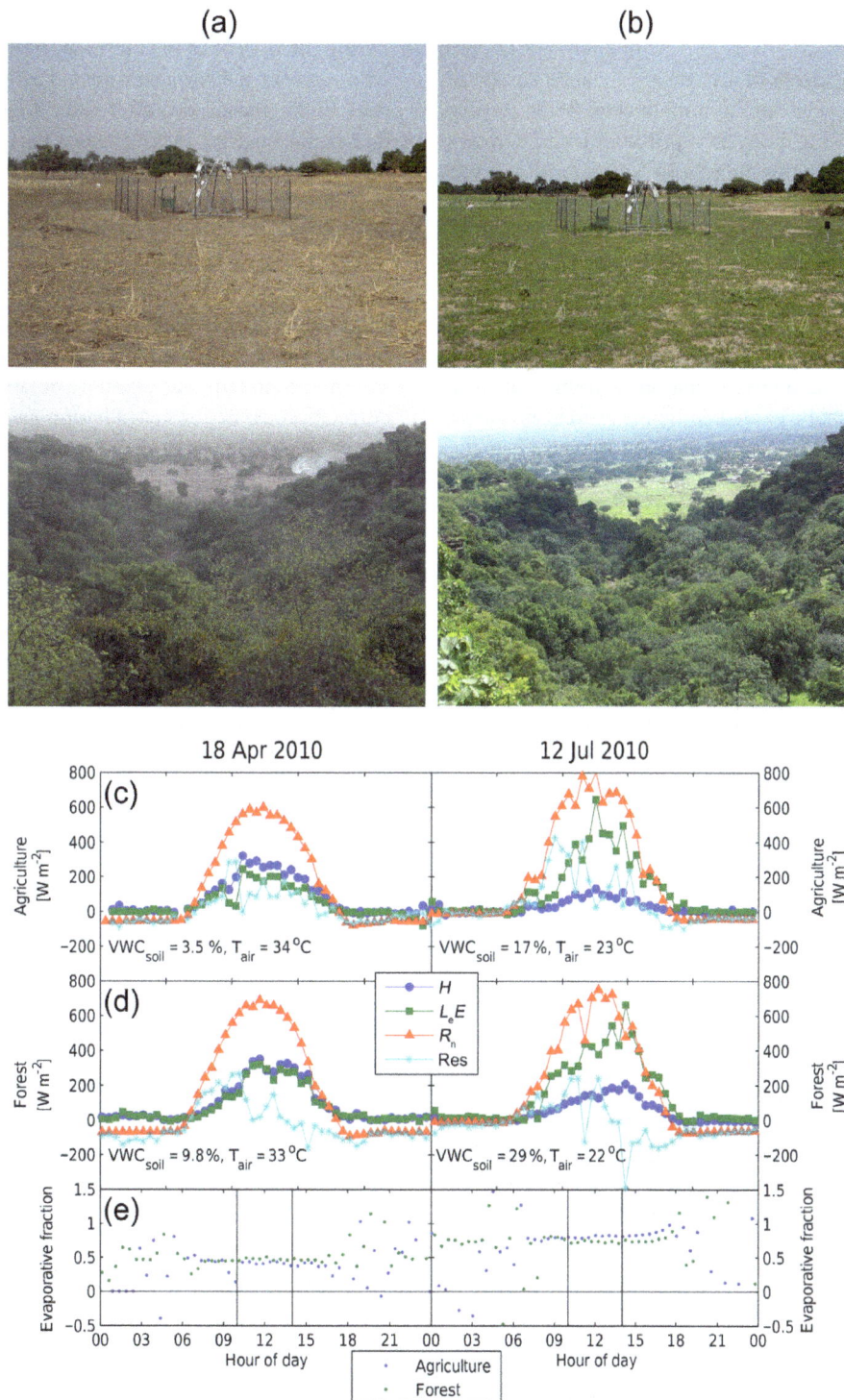

Figure 3. The diurnal cycle of the energy balance. The upper four photographs correspond to the four subplots – the date of the photograph is the same month as the representative plot of diurnal energy budget. Note that April **(a)** was the dry season and the atmosphere was very hazy, in part due to fires. July was the start of the rainy or wet season **(b)**. The energy budget is made up of the sensible heat (H, blue), latent heat ($L_e E$, green), net radiation (R_n, red) on the y axis ($\mathrm{W\,m^{-2}}$) over the savanna forest **(d)**, and agricultural land **(c)** according to the time of day (x axis, 24 h). The residual of the energy budget is also shown in turquoise. The final row shows the half-hour calculation of evaporative fraction. Daily averages were taken between 10:00 and 14:00 UTC, shown with the vertical lines, with the savanna forest in green and agriculture in blue.

3.2 Components of energy balance

3.2.1 By day

The average diurnal cycle of the energy balance varied according to diurnal cycles and by month (Fig. 3). In this figure, we compare a single day in April with a single day in July. The wet season begins after April, and prior to any vegetation growth in the agricultural field, however, some of the evergreen forest canopy is already visible in the photos. By July, crops or fallow are growing in the field and the canopy is greener. This change in vegetation cover and moisture availability is apparent in the diurnal patterns of the energy balance for both sites. Net radiation is slightly higher during the wet season for both land covers, and none of the fluxes are as smooth, which can be explained by the presence of atmospheric humidity and cloud cover. The sensible heat is higher for both land covers in the dry season; however, over the savanna forest, there is still latent flux that nearly matches the sensible heat flux even in the dry season. By July, the latent heat flux surpasses the sensible heat flux for both land covers. This can be explained by moisture availability, as the soil moisture content is much higher in both cases. Over the savanna forest, we can see that the latent heat flux does not decline until the late afternoon, suggesting that it is radiation limited and not moisture limited, whereas over the field it peaks closer to midday. The residual, which is a combination of the ground heat flux and any error, is lower over the savanna forest and is negative in the afternoon. For the month of April, the noon median residual is 20 % of the noon net radiation for the agricultural field and 13 % for the savanna forest, and for the month of July, it is 31 % for the agricultural field and 25 % over the savanna forest. The evaporative fraction is correspondingly higher in the rainy season than in the dry season and, as we would expect, has a higher value over the savanna forest in the dry season but over the agricultural field in the wet season.

Net radiation was the most similar component between the two sites, although during the "dust" season of March and April it was lower over the agricultural field. Sensible heat was greater over the savanna forest for all months, with the greatest difference in the "hot" period of March through May, which is an important period for the triggering of early convective storms.

There is a scale discrepancy between the eddy-covariance measurements and the net radiometer measurements since the latter only senses exchanges directly above and below it, whereas the former's range of detection can span a larger area depending on the wind speed. To account for this, we modeled the net radiation at each small station and then compared it to that measured with net radiometers with acceptable results (see the Supplement).

Latent heat flux was also observed to be greater over the savanna forest compared to the agricultural field. The point in the day when the latent heat flux peaks signals the moment when the system becomes moisture limited; during the early part of the year, the dry season through May, the diurnal cycle of latent heat flux peaked over the agricultural field during the middle of the morning, from 09:00 to 10:00 UTC, whereas over the savanna forest, during the same period, the peak in the diurnal was after noon. In general, our data show that latent heat flux was greatest early in the diurnal cycle during the dry season, which suggests depletion of all available moisture early in the day.

3.2.2 Over the entire study period

Time series of turbulent fluxes, soil moisture, normalized difference vegetation index, and rainfall demonstrate the highly seasonal moisture and energy availability (Fig. 4). A high correlation exists for net radiation and the sum of turbulent fluxes (Fig. 5) between the savanna forest and the field, with more scatter occurring when soil was wetter (blue). However, the sum of the turbulent fluxes is higher over the savanna forest than over the field. Since there are equal amounts of net radiation, we can deduce that there is a greater ground heat flux in the field. The lack of shading in the field, and the greater abundance of trees, with a high level of productivity, and rocks support this observation. Examination of the two components of net radiation – net longwave and net shortwave – shows that soil moisture exerts much greater control on net longwave radiation, with the change in net longwave radiation according to changes in soil moisture in the field much more apparent. Although there is more scatter in net shortwave when soil is wetter, it is less uniformly a response to the two land covers. The savanna forest's net longwave radiation is greater when the soil is dry, whereas the agricultural area has greater net longwave radiation when it is wet. Sensible heat over all land covers is greater under dry conditions than wet (blue), but both sensible and latent heat fluxes are greater over the savanna forest than over the agricultural field regardless of soil moisture.

Furthermore, in Fig. 6, we see that each wind sector has a distinct signature of when latent heat flux is greater than sensible heat flux. This variation can be explained because certain features, such as the ephemeral wetland and the gallery forest (shown in the top left of the forest plot, 16–46°), contributed to higher fluxes that have access to moisture that persists longer into the dry season. Over the agricultural field, there is more scatter, whereas over the savanna forest, there is a minimal level of about $-170\,\mathrm{W\,m^{-2}}$ for net longwave radiation. In this case, the tree canopy buffers the bare ground from the radiation loss.

3.2.3 Month by month

Figure 7 shows the two contrasting trends in surface heat fluxes by comparison of the month-by-month ratios between the measurements over the savanna forest and agricultural land. The savanna forest contributed more sensible and latent

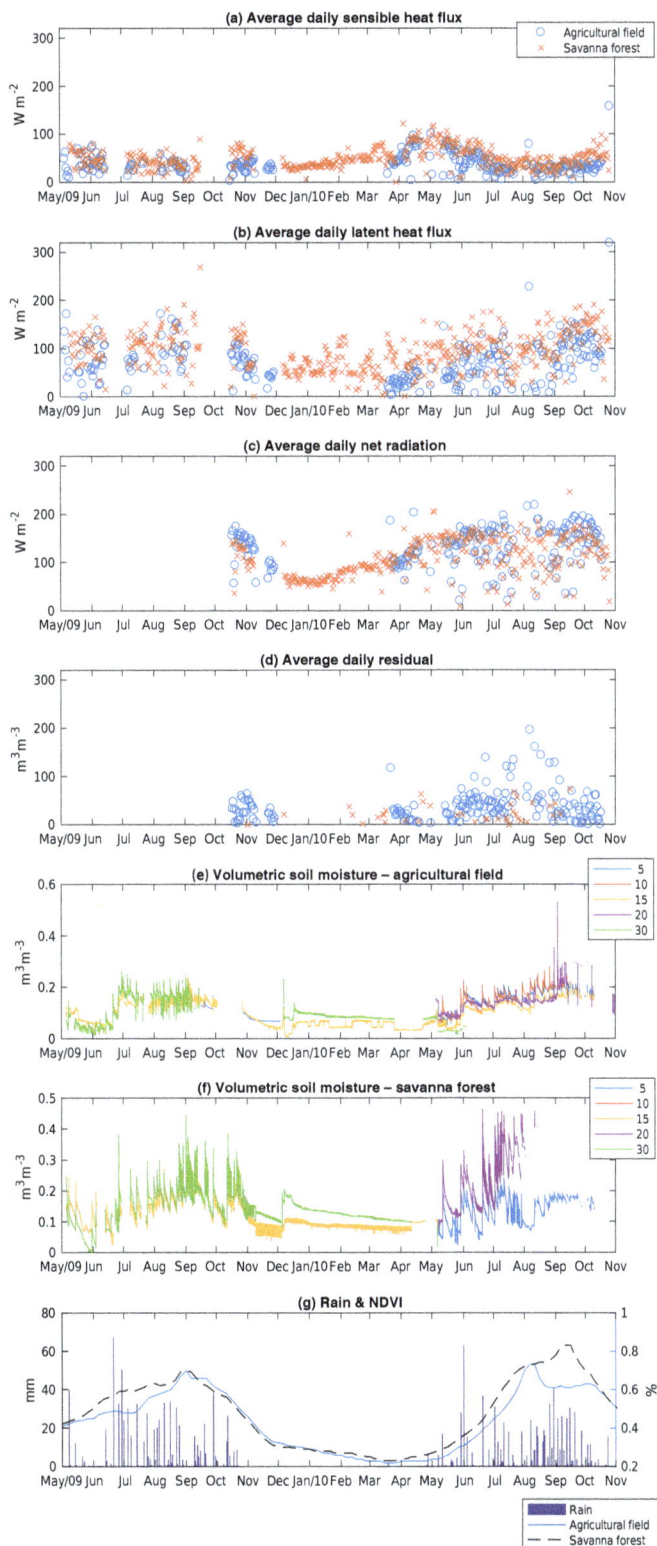

Figure 4. Time series of average daily sensible heat flux (**a**), average daily latent heat flux (**b**), average daily net radiation (**c**), average daily residual (**d**), volumetric soil moisture in the agricultural field (**e**), volumetric soil moisture in the savanna forest (**f**), and daily rain and NDVI (**g**). Panels (**a–d**) show the data from the energy balance stations over the agricultural field (blue circles) and savanna forest (red crosses) in watts per square meter. Panels (**e–f**) show the volumetric soil moisture at five different depths over the same time period. Finally, panel (**g**) shows the NDVI for the pixels containing the energy balance station in the field (blue dashed) and in the savanna forest (red, solid). The daily rain in millimeters is shown as a bar graph.

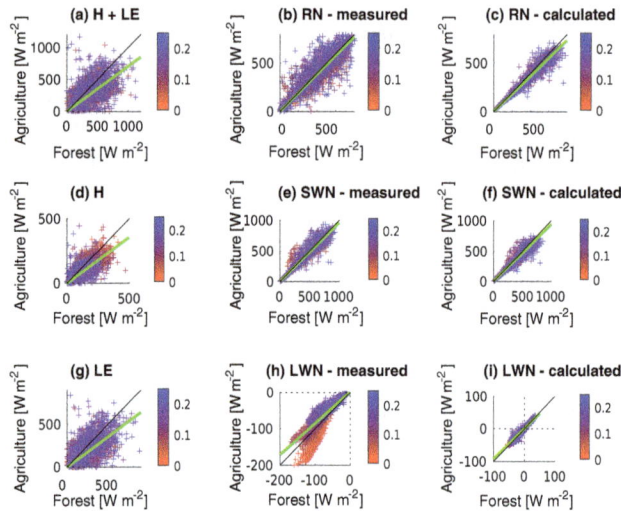

Figure 5. Comparison between fluxes measured over the savanna forest and the field. In each plot, the color indicates soil volumetric water content: red is dry and blue is wet. The least squares regression lines are shown in green and the 1 : 1 lines are in black. Measurements over the savanna forest are on the x axis and those over agriculture are on the y axis. All fits were significant ($p < 0.005$). Panels (**b, e, h**) show the components of net radiation calculated with the net radiometers. Panels (**c, f, i**) show the components of net radiation calculated using parameters measured at the small meteorological stations: air temperature (Ta), incoming shortwave radiation (SW), and soil temperature (Ts); see the Supplement.

heat fluxes throughout the year (ratio less than 1), but the difference in sensible heat flux between the two sites was greatest at the end of the year, the beginning of the dry season, and the difference latent heat flux between the two sites was greatest at the beginning of the year, after land was cleared by burning. These trends were consistent over the 2 years of measurement; however, there were some months (July and August 2010) when the sensible heat was close to equal in the two sites. The higher level of similarity in sensible heat between the two sites in 2009 can be explained by the crop choice that year; that field was planted with early (60–95 days) maturing pearl millet crop compared to its usual late variety (130–150 days), requiring a unusually late tilling and an unusually early harvesting, resulting in bare ground during the growing season. These differences are also visible in the NDVI (Figs. 2 and 4). Net radiation was more similar than the other fluxes; the greatest difference occurred when there was bare ground in the field, at end of the dry season, suggesting a higher albedo during this time.

3.3 Evaporative fraction

3.3.1 Correlations with surface and atmospheric conditions

Examination of the relationship between evaporation and the environmental variables that dominate in various models shows that, for our site, soil moisture and vegetation cover have the strongest positive correlation with evaporative fraction (Fig. 8). Over both the savanna forest and the field, we see that landscape moisture availability, expressed as both NDVI and soil moisture (VWC), exert a strong influence on the evaporative fraction, with higher rates of evaporation occurring at higher levels of soil moisture and vegetation cover or, in other words, moisture availability from either plant or soil. Total net radiation does not show a strong influence, suggesting that this is not a radiation-limited system.

Wind speed shows a strong negative correlation with more evaporation occurring at lower wind speeds, contrary to standard evaporation models. Evaporative fraction and the cloud cover exhibit a positive correlation both over the field and the savanna forest and could be explained by a two-part discontinuous function, with a break at 0.4 (Brutsaert, 1982). In a radiation-limited system, cloud cover would reduce evaporative fraction, but in this case, since it is positive, we can deduce that cloud cover is an indicator of high rates of evaporation and moisture availability, thus further supporting our hypothesis that this is a moisture-limited system.

3.3.2 Explanatory model

The relationship between soil moisture, vegetation index, and evaporative fraction can be fitted with a linear regression (Fig. 9):

$$\begin{cases} \text{EF}_{\text{agriculture}} = 0.41 \times \text{NDVI} + 1.4 \times \text{VWC} + 0.27 \\ \text{EF}_{\text{savanna forest}} = 0.48 \times \text{NDVI} + 0.35 \times \text{VWC} + 0.34. \end{cases} \quad (8)$$

Evaporative fraction depends on both soil moisture and vegetation index over agriculture, whereas over the savanna forest it responds more directly to vegetation index, as shown by the direction of the evaporative fraction color gradient. The evaporative fraction is more variable over the agricultural field, explaining the less good fit of the regression ($R^2 = 59\%$) compared to that over the savanna forest ($R^2 = 66\%$). Inclusion of net radiation, cloud cover, downwelling radiation, and wind speed in this model did not significantly change the quality of the regression. This further supports our understanding that this is a moisture-limited system. This linear regression model provides an estimation that confirms our understanding of the physical basis of the fluxes.

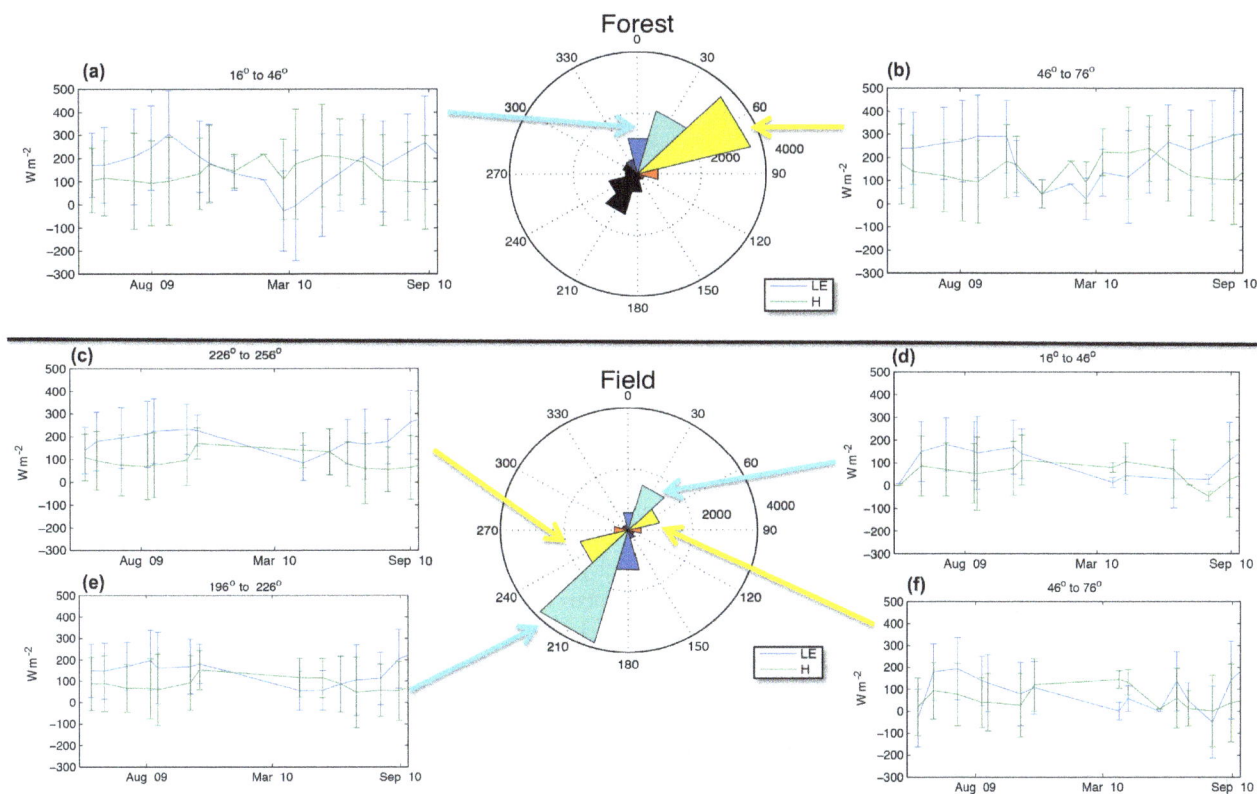

Figure 6. Two dominant wind sectors for each eddy-covariance setup are plotted: over the savanna forest (**a, b**) and over the agricultural field (**c–f**). Mean latent heat flux for each month with standard deviations is shown in blue and sensible heat in green. Note that for some months there were no data.

4 Discussion

4.1 Energy balance

The most striking observation is that the savanna forest had consistently higher levels of both sensible and latent heat fluxes across all months (Figs. 4 and 5). Sensible heat fluxes over the two surfaces showed the greatest difference in November and October (Fig. 6). Latent heat flux is the most different between the two land uses in August and May, which are transition times when the access to water in the catchment is not uniform (Fig. 6). The greatest similarity between the two land covers was during the wet season (May through September). The difference between the energy balance of the two sites was the most accentuated in the transition from the wet to dry season that occurred in the month of October for sensible heat flux and in the early wet season for latent heat flux (Figs. 4, 5, and 7). Because of this observation, we can expect land use changes to have the most impact during these transition periods, due to differences in growth patterns and rooting depths. In particular, since agricultural crops are planted, their germination and development is determined by agricultural decisions above water and energy availability. The behavior during the growing season from June through August varied so dramatically between 2009

and 2010 because in 2009 short season millet was planted, whereas in 2010, it was left fallow. We can imagine that this difference is due to the high growth rates of the agricultural crop through August and the subsequent harvest, whereas the fallow crops grow quickly, transpiring the most in July and then stabilizing into August (Fig. 4). The sensible and latent energy fluxes from the fallow were lower than those of the forest in 2010, whereas when short season millet was growing, they were more similar. By comparing the changes in soil moisture along the diurnal cycle, we see that even in the dry season there is some variation, particularly at shallow depths; thus, in contrast to other authors, we cannot attribute the early peak purely to stomata behavior during the dry season (Mamadou et al., 2016). In contrast, during the rainy season, it declined, following the cycle of available radiation.

The net radiation was very similar over both land surfaces, with the greatest differences occurring in the dry season. However, early in the dry season, net radiation was higher over the agricultural field, whereas early in the dry season, it was higher over savanna forest. This second counter-intuitive finding is supported by other work in the subregion that found that net radiation was higher over woody vegetation during the dry season and explained the difference because

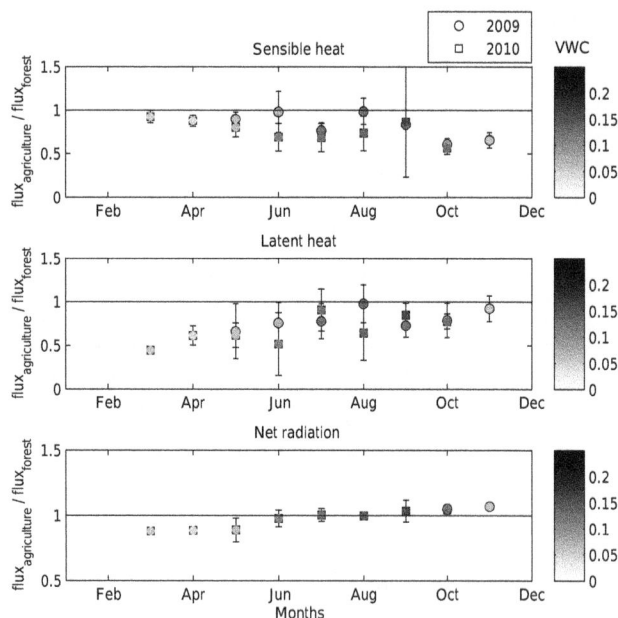

Figure 7. Examination of ratio of the daily fluxes by month and year. The ratios between flux measurements over the field and savanna forest are shown according to the month (x axis) and year (in color: blue indicates 2009 and red indicates 2010). The soil moisture value is shown with shading. Error bars show the standard deviation around the average of daily ratios. Points below the 1 line indicate when the savanna-forest flux is higher than the agricultural field flux. It is important to note that the field was farmed in 2009 until the end of July but left fallow in 2010.

surface temperature was lower and net longwave radiation was higher (Guyot et al., 2009). The residual of the energy budget showed fewer clear patterns, but across seasons, it was greater over the field, likely due to larger ground heat flux into the bare soil.

Two contrasting trends explain why the sensible heat fluxes becomes less similar as the year progresses, whereas the latent heat fluxes becomes more similar. First, at the start of the year, the agricultural field is covered with bare ground and the rocks are exposed above the forested area, and throughout the growing season, the bare ground is progressively covered with grass, whereas on the hill, the rocks remain exposed. At the end of the wet season, the grass senesces and remains until it is burned in late December or January. The contrast of the annual cycles of bare ground and bare rock drives the difference in sensible heat flux. The bare ground has a high level of albedo compared to the rocks, creating a difference in available energy for turbulent fluxes.

Second, at the start of the year, the upper trees have access to water coming from the springs at the base of the rocks. Although the level of water availability and vegetation increases during the wet season and declines during the following dry season, the spring is permanent perhaps due to subsurface lateral water transfer that also produces shallow groundwater

that is available to the trees (Mande, 2014). In contrast, the water availability in the field and the corresponding greenness closely follow the annual cycle of precipitation, driving the variation in latent heat. The ephemeral stream stops flowing at the end of December, when the grasses dry up and the latent heat flux returns to being drastically different between the two sites (Mande, 2014).

Our observations of H and $L_e E$ are higher than fluxes previously measured in the region (Bagayoko et al., 2007; Dolman et al., 1997; Gash et al., 1997; Guichard et al., 2009; Guyot et al., 2009; Mauder et al., 2006; Schüttemeyer et al., 2006; Timouk et al., 2009), which can be explained by our site's location inside a semi-protected area with regionally relative high amount of vegetation, increasing $L_e E$, and an abundance of rocks, raising H. Furthermore, annual cycles of ratios between H and $L_e E$ vary by wind sector (Fig. 6), demonstrating that small variations in land cover, topography, and moisture availability can lead to dramatic differences in evaporation and evaporative fraction. This is consistent with local land management philosophy, which emphasizes the importance of maintaining the gallery forest, and springs therein, as a common moisture reservoir in the dry season and in the case of drought. More continuous, long-term measurements during extreme years would reinforce the validity of this local belief. Our results also emphasize that the forest, even though it is primarily an open wooded savanna, has a higher level of productivity than the rain-fed, hand-farmed fields.

Our values of sensible and latent heat fluxes are most similar to those measured in Ejura, Ghana, in 2002 (Schüttemeyer et al., 2006). Ejura is about 500 km from our site, and though quite far and typically placed in a different category of climate, we measure similar values compared to other areas of west Africa (Guyot et al., 2009), perhaps because measurements took place in November, when vegetation there would be most similar to that at our site. Kompienga is the most similar site and is closest to ours; though measurements display the same seasonality as those at our site, they are still lower than ours (Bagayoko et al., 2007). The scale incongruity between the turbulent flux (sensible and latent) measurements and the net radiation may explain our lack of closure instead of, for example, surface heat water storage during floods (Guyot et al., 2009). There is a strong topographic difference close to the forested area; however, we are confident that our planar fit correction effectively corrected for the corresponding effect on turbulent fluxes.

The high magnitude of turbulent fluxes can be explained by Tambarga–Madjoari's location in the midst of a natural park and hunting concession as well as our measurement of a gallery forest, which would demonstrate the importance of the nearby vegetation cover. On the whole, our measurements are comparable to those elsewhere in west Africa, given that most of these sites (Mali, Niger) are more Sahelian, and thus have less moisture availability, and others (Nigeria, Benin,

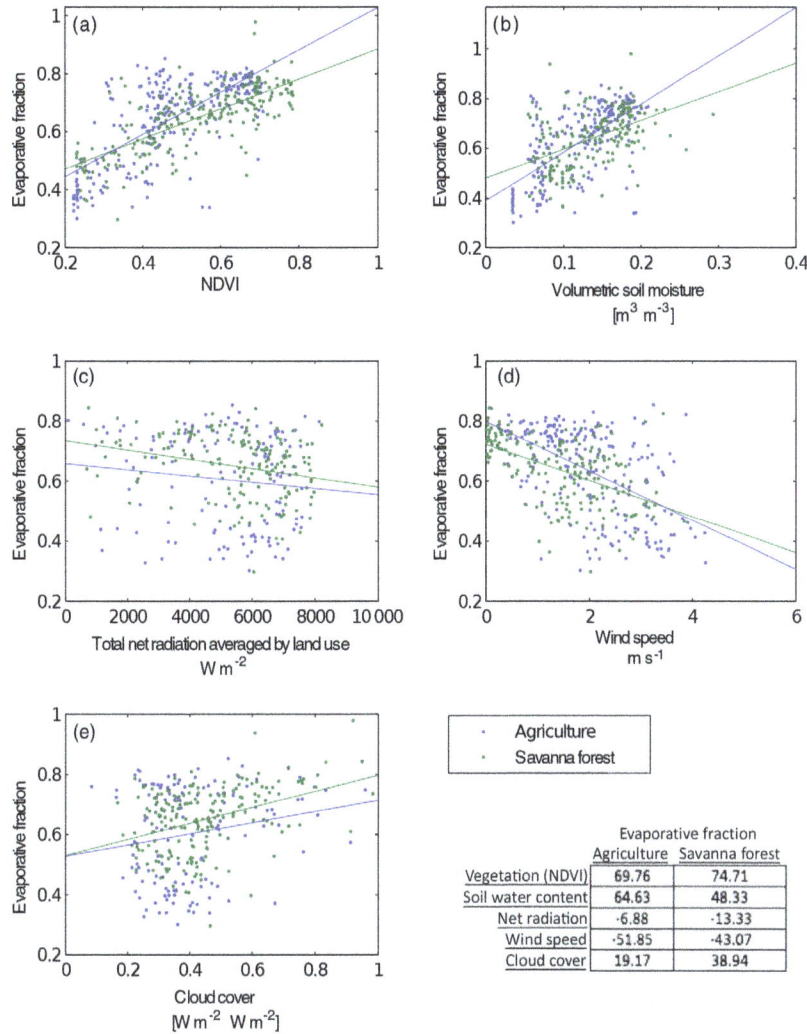

Figure 8. Daily evaporative fraction over study period for the agricultural field (blue) and the savanna forest (green) compared with the observed **(a)** NDVI, **(b)** VWC, **(c)** wind speed, **(d)** net radiation, and **(e)** cloud cover. In all cases, environmental variables are from the average of stations with the same land cover with the energy balance station. These plots show a better correlation with NDVI and VWC **(a, b)**, suggesting a moisture-limited and not a radiation-limited system. The least squares regression lines are shown for each plot. Correlations between variables are in the table.

Ghana) are considerably further south and thus have denser vegetation (White, 1986).

4.2 Evaporative fraction

The median monthly evaporative fraction in the savanna forest was lowest in April at 0.47 and highest in September at 0.77, and for the agricultural fields, it was also lowest in April at 0.31 and highest in September at 0.79. These values are in general higher than previously found in similar environments during the rainy season (Bagayoko et al., 2007; Brümmer et al., 2008; Guyot et al., 2012; Mamadou et al., 2016). In contrast, our site has a lower dry season evaporative fraction than nearby sites, which suggests that moisture cycling during the wet season is more complete and that there is less storage

into the dry season. This is likely due to a combination of vegetation, soil, and topographic characteristics.

We can compare our values of evaporative fraction with other environmental conditions. Over west Africa, the self-preservation concept of the evaporative fraction could be used together with variables such as albedo, temperature at the surface, and a vegetation index to obtain a reasonable estimate of evaporation (Compaore, 2006). A high correlation between midday evaporation and the evaporative fraction exists in Kenyan grasslands (Farah et al., 2004). Evaporative fraction might be affected by cloud cover which alters incoming radiation (Brutsaert and Sugita, 1992); however, cloudiness was not related to the stability of evaporative fraction in 2005–2006 in Brazil (Santos et al., 2010). Evaporative fraction may also be related to other environmental param-

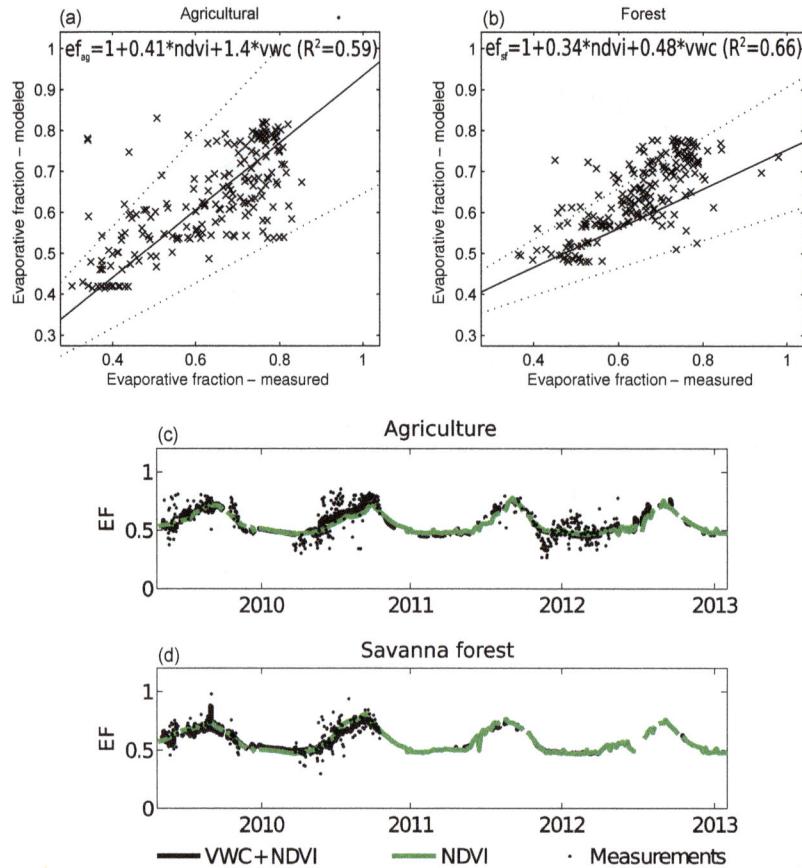

Figure 9. Further examination of relationship between soil moisture, vegetation index, and evaporative fraction. Panels (**a, b**) show the quality of fit of the linear regression model relating soil volumetric water content, vegetation index, and evaporative fraction over agriculture (**a**) and over the savanna forest (**b**). The 1 : 1 line is in red and the 95 % confidence interval is shown with dotted lines for the range of available soil and vegetation. Evaporative fraction over the field (**b**) and the savanna forest (**d**) are shown. Measured data points are in black (points), calculated evaporative fraction based on soil moisture and NDVI are shown in blue, and NDVI alone is shown in green.

eters that are increasingly available and reliable, which are obtained remotely through satellites, such as soil moisture (Crago, 1996; Hall et al., 1992). One limitation of this approach is that of the temporal scale since evaporative fraction is computed by day but surface conditions and moisture availability can change instantaneously. However, it is extremely valuable to be able to estimate evaporative fraction when only NDVI or other remotely sensed parameters are available. For this reason, our estimation of evaporative fraction with a simple model based on the physical basis of fluxes linked to vegetation and soil moisture is useful (Fig. 9). This model can be refined with longer and more varied datasets to increase its accuracy and suitability regionally.

4.3 Social context

The gallery forest over which we measure fluxes includes two springs that were the main water source for the village at its founding until a generation ago (Ceperley, 2014). Its important role in local history means that special institutions ex-

isted in local tradition for its conservation. For example, wetlands and other areas with abundant water, fall into a Gourmantché category for land called "Tinjali" or land that was not farmed because of cultural taboos until recent development projects and the introduction of rice farming (Swanson, 1978). In addition, since it is forested, it is protected because Gourmantché believe that trees have spirits, which may prohibit them from being cut or used, and that some are considered good and some bad. So it is reasonable to conclude that the presence of the forest is not only because the water availability provides the habitat but also because the village has protected it on some level. Additionally, it is reasonable to think that the institutions that protect this forest are not only protecting it but also the ecosystem services that it preserves. Our work suggests that these ecosystem services include the cycling of moisture into the atmosphere and the eventual generation of rainfall. In this light, the Gourmantché myth explains that there is a sack of water above the atmosphere that spirits can pierce, bringing rain delivered by clouds (Alves, 2012). If the gods pierced those clouds with "trees", then our research seems to be right on target in terms of validating what traditions have long known. This is an important tool: as land use becomes more and more contested, the validation of local institutions or land uses can ensure the continuation of these practices.

5 Conclusions

Sensible and latent heat fluxes were higher over a savanna forest than a semi-cultivated (millet-fallow) field according to our measurements in a Sudanian ecosystem of west Africa in 2009 and 2010. The sensible heat and latent heat flux are generally higher over the savanna forest because of its more permanent water availability and corresponding greenness, higher productivity, and the amount of rocky terrain. For example, the diurnal cycle of latent heat flux peaked earlier in the day during the dry season in the agricultural field, suggesting depletion of all available moisture by late morning. These observations of sensible and latent heat fluxes are higher than fluxes previously measured in the region, potentially due to this site's location inside of a semi-protected area. Analysis of wind sectors separately revealed that particular sectors, corresponding to the location of particular features, for example, over the ephemeral wetland and the gallery forest, contributed higher amounts to the flux. Local land management emphasizes the importance of maintaining the gallery forest, and springs therein, as a common humidity reservoir in the dry season and in the case of drought. Changes in land cover may even have consequences for local rainfall triggering, causing cascading effects that transform both the energy and water budgets.

Continuous, long-term measurements during drought and moist years are essential to prove the long-term validity of our observations. Additionally, variations in exchanges according to small landscape features could result in enormous underestimation for upscaling. We recommend using eddy-covariance measurements such as these to improve estimates with more easily maintained and obtained meteorological station and satellite data. The evaporative fraction is dependent on NDVI, which is an important finding for modeling and upscaling. Efforts focused on preserving hydrologic services need to take anomalies into account and reinforce cultural institutions that protect wetlands and gallery forests.

The development of a simple NDVI-based indicator of evaporative fraction is transferable to other semi-arid systems, agroforestry parklands, and open wooded savannas around the globe. Globally, the African continent and often semi-arid environments represent a gap in observations of land–atmosphere interactions. This study is an important step in filling this gap and proposing a tool with still greater potential.

Our results point to the necessity for ground measurements for eventual upscaling from point to regional evaporation measurements in remote and less-studied regions of the globe. We began this work with discussion with community partners and, to bring it full circle, we conclude this paper by relating it back to the cultural context.

Competing interests. The authors declare that they have no conflict of interest.

Special issue statement. This article is part of the special issue "Observations and modeling of land surface water and energy exchanges across scales: special issue in Honor of Eric F. Wood". It is a result of the Symposium in Honor of Eric F. Wood: Observations and Modeling across Scales, Princeton, New Jersey, USA, 2–3 June 2016.

Acknowledgements. The Velux and 3rd Millennium foundations funded a large part of this research, in addition to support from the young researcher's (KFPE) grant from the Swiss Agency for Development and Cooperation. Alexandre Repetti, Jean-Claude Bolay, and the center for cooperation at EPFL initiated and continued to this project. We would like to thank all of our colleagues, assistants, and students based both in Switzerland and in Burkina Faso who helped us with fieldwork. In particular, collaboration with the International Institute of Environment and Water Engineering facilitated this research. We give additional gratitude to the commune of Madjoari, its residents, and its government, who hosted us and our equipment for the duration of this project. The manuscript was greatly improved by feedback from reviewers and the scientific community; for that, we are grateful. The final analysis was completed with the support of the NSERC discovery grant.

Edited by: Reed Maxwell

References

Abiodun, B. J., Pal, J. S., Afiesimama, E. A., Gutowski, W. J., and Adedoyin, A.: Simulation of West African monsoon using RegCM3 Part II: impacts of deforestation and desertification, Theor. Appl. Climatol., 93, 245–261, 2008.

Adomou, A.: Vegetation patterns and environmental gradients in Benin, PhD thesis, Wageningen University, Wageningen, the Netherlands, 2005.

Alves, J. P. G.: Anthropologie et écosystèmes au Niger: humains, lions et esprits de la forêt dans la culture gourmantché, Editions Harmattan, Paris, France, 448 pp., 2012.

Arbonnier, M.: Trees, Shrubs and lianas of West African dry Zones, CIRAD, MNHN, Montpellier, France, 2004.

Aubinet, M., Feigenwinter, C., Heinesch, B., Laffineur, Q., Papale, D., Reichstein, M., Rinne, J., and Gorsel, E. V.: Nighttime Flux Correction, in: Eddy Covariance, edited by: Aubinet, M., Vesala, T., and Papale, D., Springer Netherlands, 133–157, https://doi.org/10.1007/978-94-007-2351-1, 2012.

Bagayoko, F., Yonkeu, S., Elbers, J., and van de Giesen, N.: Energy partitioning over the West African savanna: Multi-year evaporation and surface conductance measurements in Eastern Burkina Faso, J. Hydrol., 334, 545–559, https://doi.org/10.1016/j.jhydrol.2006.10.035, 2007.

Bateni, S. M. and Entekhabi, D.: Relative efficiency of land surface enrgy balance components, Water Resour. Res., 48, W04510, https://doi.org/10.1029/2011WR011357, 2012.

Bordes, C.: La Gestion Des Arbres Par Les Paysans: Etude d'une enclave au milieu de reserves forestieres au sud-est du Burkina Faso, Ingenieur, ISTOM, Cergy-Pontoise, France, 2010.

Brümmer, C., Falk, U., Papen, H., Szarzynski, J., Wassmann, R., and Brüggemann, N.: Diurnal, seasonal, and interannual variation in carbon dioxide and energy exchange in shrub savanna in Burkina Faso (West Africa), J. Geophys. Res., 113, 1–11, https://doi.org/10.1029/2007JG000583, 2008.

Brutsaert, W.: Evaporation into the Atmosphere: Theory, History, and Applications, Kluwer Academic Publishers, Dordrecht, the Netherlands, 1982.

Brutsaert, W. and Parlange, M. B.: The Unstable Surface Layer Above the Forest: Regional Evaporation and Heat Flux, Water Resour. Res., 28, 3129–3134, 1992.

Brutsaert, W. and Sugita, M.: Application of Self-Preservation in the Diurnal Evolution of the Surface Energy Budget to Determine Daily Evaporation, J. Geophys. Res., 97, 18377–18382, 1992.

Burba, G.: Eddy Covariance Method for Scientific, Industrial, Agricultural and Regulatory Applications, Li-COR Biosciences, Lincoln, Nebraska, USA, 2005.

Burba, G.: Eddy Covariance Method for Scientific, Industrial, Agricultural and Regulatory Applications: A Field Book on Measuring Ecosystem Gas Exchange and Areal Emission Rates, LI-Cor Biosciences, Lincoln, Nebraska, USA, 2013.

Ceperley, N. C.: Ecohydrology of a Mixed Savanna-Agricultural Catchment in South-East Burkina Faso, West Africa, Swiss Federal Institute of Technology, Lausanne, available at: http://infoscience.epfl.ch/record/195232/files/EPFL_TH6040.pdf (last access: 5 February 2015), 2014.

Ceperley, N. C., Mande, T., Parlange, M. B., Tyler, S., van de Giesen, N., and Yacouba, H.: Energy Balance, Tambarga, Burkina Faso, 2009–2010, https://doi.org/10.4121/uuid:0dbbaf01-bea4-4520-aee9-c3ebd354b27c, 2017.

Charney, J. G.: Dynamics of deserts and drought in the Sahel, Q. J. Roy. Meteor. Soc., 101, 193–202, 1975.

Compaore, H.: The impact of savannah vegetation on the spatial and temporal variation of the actual evapotranspiration in the Volta Basin, Navrongo, Upper East Ghana, Ecol. Dev. Ser., edited by: Denich, M., Martius, C., Rogers, C., van de Giesen, N., and Compaoré, H., Cuvillier Verlag, Göttingen, Germany, 1–144, 2006.

Crago, R. D.: Conservation and variability of the evaporative fraction during the daytime, J. Hydrol., 180, 173–194, https://doi.org/10.1016/0022-1694(95)02903-6, 1996.

Crago, R. D. and Qualls, R.: The value of intuitive concepts in evaporation research, Water Resour. Res., 49, 6100–6104, https://doi.org/10.1002/wrcr.20420, 2013.

Dolman, A. J., Gash, J. H. C., Goutorbe, J.-P., Kerr, Y., Lebel, T., Prince, S. D., and Stricker, J. N. M.: The role of the land surface in Sahelian climate: HAPEX-Sahel results and future research needs, J. Hydrol., 188–189, 1067–1079, https://doi.org/10.1016/S0022-1694(96)03183-6, 1997.

Domingo, F., Serrano-Ortiz, P., Were, A., Villagarcía, L., García, M., Ramírez, D. A., Kowalski, A. S., Moro, M. J., Rey, A., and Oyonarte, C.: Carbon and water exchange in semi-arid ecosystems in SE Spain, J. Arid Environ., 75, 1271–1281, https://doi.org/10.1016/j.jaridenv.2011.06.018, 2011.

Ezzahar, J., Chehbouni, A., Hoedjes, J., Ramier, D., Boulain, N., Boubkraoui, S., Cappelaere, B., Descroix, L., Mougenot, B., and Timouk, F.: Combining scintillometer measurements and an aggregation scheme to estimate area-averaged latent heat flux during the AMMA experiment, J. Hydrol., 375, 217–226, https://doi.org/10.1016/j.jhydrol.2009.01.010, 2009.

Farah, H. O., Bastiaanssen, W. G. M., and Feddes, R. A.: Evaluation of the temporal variability of the evaporative fraction in a tropical watershed, Int. J. Appl. Earth Obs., 5, 129–140, https://doi.org/10.1016/j.jag.2004.01.003, 2004.

arhadi, L.: Estimation of Land Surface Water and Energy Balance Flux Components and Closure Relation Using Conditional Sampling, MIT, Cambridge, Mass., USA, 2012.

Feddema, J. J., Oleson, K. W., Bonan, G. B., Mearns, L. O., Buja, L. E., Meehl, G. A., and Washington, W. M.: The importance of land-cover change in simulating future climates, Science, 310, 1674–1678, https://doi.org/10.1126/science.1118160, 2005.

Federer, C. A., Vörösmarty, C., Fekete, B., and Olume, V.: Sensitivity of Annual Evaporation to Soil and Root F Properties in Two Models of Contrasting Complexity, J. Hydrometeorol., 4, 1276–1290, https://doi.org/10.1175/1525-7541(2003)004<1276:SOAETS>2.0.CO;2, 2003.

Foken, T.: The energy balance closure problem: An overview, Ecol. Appl., 18, 1351–1367, https://doi.org/10.1890/06-0922.1, 2008.

Foken, T., Mauder, M., Liebethal, C., Wimmer, F., Beyrich, F., Leps, J.-P., Raasch, S., DeBruin, H. A. R., Meijninger, W. M. L., and Bange, J.: Energy balance closure for the LITFASS-2003 experiment, Theor. Appl. Climatol., 101, 149–160, https://doi.org/10.1007/s00704-009-0216-8, 2009.

Foken, T., Leuning, R., Oncley, S. R., Mauder, M., and Aubinet, M.: Corrections and Data Quality Control, in: Eddy Covariance: A Practical Guide to Measurement and Data Analysis, edited by: Aubinet, M., Vesala, T., Papale, D., Springer Atmospheric Sciences, https://doi.org/10.1007/978-94-007-2351-1, 2012.

Gash, J. H., Kabat, P., Monteny, B. A., Amadou, M., Bessemoulin, P., Billing, H., Blyth, E. M., Elbers, J. A., Friborg, T., Harrison, G., and Holwill, C. J. T.: The variability of evaporation during the HAPEX-Sahel intensive observation period, J. Hydrol., 188, 385–399, 1997.

Gentine, P., Entekhabi, D., Chehbouni, A., Boulet, G., and Duchemin, B.: Analysis of evaporative fraction diurnal behavior, Agr. Forest Meteorol., 143, 13–29, 2007.

Guichard, F., Kergoat, L., Mougin, E., Timouk, F., Baup, F., Hiernaux, P., and Lavenu, F.: Surface thermodynamics and radiative budget in the Sahelian Gourma: Seasonal and diurnal cycles, J. Hydrol., 375, 161–177, https://doi.org/10.1016/j.jhydrol.2008.09.007, 2009.

Guo, Z., Dirmeyer, P. A., Koster, R. D., Sud, Y. C., Bonan, G., Oleson, K. W., Chan, E., Verseghy, D., Cox, P., Gordon, C. T., McGregor, J. L., Kanae, S., Kowalczyk, E., Lawrence, D., Liu, P., Mocko, D., Lu, C.-H., Mitchell, K., Malyshev, S., McAvaney, B., Oki, T., Yamada, T., Pitman, A., Taylor, C. M., Vasic, R., and Xue, Y.: GLACE: The Global Land Atmosphere Coupling

Experiment. Part II: Analysis, J. Hydrometeorol., 7, 611–625, https://doi.org/10.1175/JHM511.1, 2006.

Guyot, A., Cohard, J.-M., Anquetin, S., Galle, S., and Lloyd, C. R.: Combined analysis of energy and water balances to estimate latent heat flux of a sudanian small catchment, J. Hydrol., 375, 227–240, https://doi.org/10.1016/j.jhydrol.2008.12.027, 2009.

Guyot, A., Cohard, J.-M., Anquetin, S., and Galle, S.: Long-term observations of turbulent fluxes over heterogeneous vegetation using scintillometry and additional observations: A contribution to AMMA under Sudano-Sahelian climate, Agr. Forest Meteorol., 154–155, 84–98, https://doi.org/10.1016/j.agrformet.2011.10.008, 2012.

Hall, F. G., Huemmrich, K. F., Goetz, S. J., Sellers, P. J., and Nickeson, J. E.: Satellite Remote Sensing of Surface Energy Balance Success, Failures, and Unresolved Issues in FIFE, J. Geophys. Res., 97, 19061–19089, 1992.

Higgins, C. W.: A-posteriori analysis of surface energy budget closure to determine missed energy pathways, Geophys. Res. Lett., 39, 1–5, https://doi.org/10.1029/2012GL052918, 2012.

Ingelrest, F., Barrenetxea, G., Schaefer, G., Vetterli, M., Couach, O., and Parlange, M.: SensorScope, ACM Trans. Sens. Netw., 6, 1–32, https://doi.org/10.1145/1689239.1689247, 2010.

Katul, G. G. and Parlange, M. B.: A Penman-Brutsaert Model for Wet Surface Evaporation, Water Resour. Res., 28, 121–126, 1992.

Krishnan, P., Meyers, T. P., Scott, R. L., Kennedy, L., and Heuer, M.: Energy exchange and evapotranspiration over two temperate semi-arid grasslands in North America, Agr. Forest Meteorol., 153, 31–44, https://doi.org/10.1016/j.agrformet.2011.09.017, 2012.

Kustas, W. P., Rango, A., and Uijlenhoet, R.: A simple energy budget algorithm for the snowmelt runoff model, Water Resour. Res., 30, 1515–1527, https://doi.org/10.1029/94WR00152, 1994.

Leuning, R.: The correct form of the Webb, Pearman and Leuning equation for eddy fluxes of trace gases in steady and non-steady state, horizontally homogeneous flows, Bound.-Lay. Meteorol., 123, 263–267, 2007.

Lhomme, J.-P. and Elguero, E.: Examination of evaporative fraction diurnal behaviour using a soil-vegetation model coupled with a mixed-layer model, Hydrol. Earth Syst. Sci., 3, 259–270, https://doi.org/10.5194/hess-3-259-1999, 1999.

Lohou, F., Saïd, F., Lothon, M., Durand, P., and Serça, D.: Impact of Boundary-Layer Processes on Near-Surface Turbulence Within the West African Monsoon, Bound.-Lay. Meteorol., 136, 1–23, https://doi.org/10.1007/s10546-010-9493-0, 2010.

Lohou, F., Kergoat, L., Guichard, F., Boone, A., Cappelaere, B., Cohard, J.-M., Demarty, J., Galle, S., Grippa, M., Peugeot, C., Ramier, D., Taylor, C. M., and Timouk, F.: Surface response to rain events throughout the West African monsoon, Atmos. Chem. Phys., 14, 3883–3898, https://doi.org/10.5194/acp-14-3883-2014, 2014.

Mamadou, O., Cohard, J. M., Galle, S., Awanou, C. N., Diedhiou, A., Kounouhewa, B., and Peugeot, C.: Energy fluxes and surface characteristics over a cultivated area in Benin: daily and seasonal dynamics, Hydrol. Earth Syst. Sci., 18, 893–914, https://doi.org/10.5194/hess-18-893-2014, 2014.

Mamadou, O., Galle, S., Cohard, J.-M., Peugeot, C., Kounouhewa, B., Biron, R., Hector, B., and Zannou, A. B.: Dynamics of water vapor and energy exchanges above two contrasting Sudanian climate ecosystems in Northern Benin (West Africa): WATER VAPOR AND ENERGY EXCHANGES, J. Geophys. Res.-Atmos., 121, 11269–11286, https://doi.org/10.1002/2016JD024749, 2016.

Mande, T.: Hydrology of the Sudanian Savannah in West Africa, Burkina Faso, PhD thesis, Swiss Federal Institute of Technology, Lausanne, Lausanne, Switzerland, 17 January, available at: https://infoscience.epfl.ch/record/195231/files/EPFL_TH6011.pdf (last access: 4 October 2015), 2014.

Mande, T., Ceperley, N., Barrenetxea, G., Repetti, A., and Niang, D.: Rainfall-Runoff Processes in a Mixed Sudanian Savanna Agriculture Catchment?: Use of a distributed sensor network, in: Geophysical Research Abstracts, vol. 13, p. 1, Vienna, Austria, 2011.

Mauder, M., Jegede, O. O., Okogbue, E. C., Wimmer, F., and Foken, T.: Surface energy balance measurements at a tropical site in West Africa during the transition from dry to wet season, Theor. Appl. Climatol., 89, 171–183, https://doi.org/10.1007/s00704-006-0252-6, 2006.

Nadeau, D. F., Brutsaert, W., Parlange, M. B., Bou-Zeid, E., Barrenetxea, G., Couach, O., Boldi, M.-O., Selker, J. S., and Vetterli, M.: Estimation of urban sensible heat flux using a dense wireless network of observations, Environ. Fluid Mech., 9, 635–653, https://doi.org/10.1007/s10652-009-9150-7, 2009.

Nicholson, S. E., Tucker, C. J., and Ba, M. B.: Desertification, drought, and surface vegetation: An example from the West African Sahel, B. Am. Meteorol. Soc., 79, 815–830, 1998.

Oldroyd, H. J., Pardyjak, E. R., Huwald, H., and Parlange, M. B.: Adapting Tilt Corrections and the Governing Flow Equations for Steep, Fully Three-Dimensional, Mountainous Terrain, Bound.-Lay. Meteorol., 159, 539–565, 2015.

Parlange, M. B. and Katul, G. G.: Estimation of the diurnal variation of potential evaporation from a wet bare soil surface, J. Hydrol., 132, 71–89, https://doi.org/10.1016/0022-1694(92)90173-S, 1992.

Pielke, R. A., Marland, G., Betts, R. A., Chase, T. N., Eastman, J. L., Niles, J. O., Niyogi, D. D. S., and Running, S. W.: The influence of land-use change and landscape dynamics on the climate system: relevance to climate-change policy beyond the radiative effect of greenhouse gases, Philos. T. R. Soc. A, 360, 1705–1719, 2002.

Porte-Agel, F., Parlange, M. B., Cahill, A. T., Gruber, A., and Porte, F.: Mixture of Time Scales in Evaporation?: Desorption and Self-Similarity of Energy Fluxes, Agron. J., 92, 832–836, 2000.

Ramier, D., Boulain, N., Cappelaere, B., Timouk, F., Rabanit, M., Lloyd, C. R., Boubkraoui, S., Métayer, F., Descroix, L., and Wawrzyniak, V.: Towards an understanding of coupled physical and biological processes in the cultivated Sahel – 1. Energy and water, J. Hydrol., 375, 204–216, https://doi.org/10.1016/j.jhydrol.2008.12.002, 2009.

Rebmann, C., Kolle, O., Heinesch, B., Queck, R., Ibrom, A., and Aubinet, M.: Data Acquisition and Flux Calculations, in: Eddy Covariance, edited by: Aubinet, M., Vesala, T., and Papale, D., Springer Netherlands, 59–83, https://doi.org/10.1007/978-94-007-2351-1, 2012.

Santos, C. A. C. D., Silva, B. B. D., and Rao, T. V. R.: Analysis of the evaporative fraction using eddy covariance and remote sensing techniques, Rev. Bras. Meteorol., 25, 427–436, https://doi.org/10.1590/S0102-77862010000400002, 2010.

Schüttemeyer, D., Moene, A. F., Holtslag, A. A. M., Bruin, H. A. R., and Giesen, N. V.: Surface Fluxes and Characteristics of Drying Semi-Arid Terrain in West Africa, Bound.-Lay. Meteorol., 118, 583–612, https://doi.org/10.1007/s10546-005-9028-2, 2006.

Shuttleworth, W. J., Gurney, R. J., Hsu, A. Y., and Ormsby, J. P.: FIFE: the variation in energy partition at surface flux sites, Remote Sensing and Large-Scale Global Processes, Proceedings of the IAHS Third Int. Assembly, Baltimore, MD, USA, May 1989, IAHS Publ. no. 186, 1989.

Simoni, S., Padoan, S., Nadeau, D. F., Diebold, M., Porporato, A., Barrenetxea, G., Ingelrest, F., Vetterli, M., and Parlange, M. B.: Hydrologic response of an alpine watershed: Application of a meteorological wireless sensor network to understand streamflow generation, Water Resour. Res., 47, W10524, https://doi.org/10.1029/2011WR010730, 2011.

Steiner, A. L., Pal, J. S., Rauscher, S. A., Bell, J. L., Diffenbaugh, N. S., Boone, A., Sloan, L. C., and Giorgi, F.: Land surface coupling in regional climate simulations of the West African monsoon, Clim. Dynam., 33, 869–892, 2009.

Swanson, R. A.: Gourmantche agriculture, Ouagadougou USAID, Development Anthropology Technical Assistance Component, Integrated Rural Development Project, Eastern ORD, BAEP, Upper Volta Contract AID-686-049-78, USAID, Washington, D.C., USA, 1978.

Swets, D. L., Reed, B. C., Rowland, J. D., and Marko, S. E.: A weighted least-squares approach to temporal NDVI smoothing, in: Proceedings of the 1999 ASPRS Annual Conference, 17–21 May 1999, Portland, Oregon, USA, available at: https://phenology.cr.usgs.gov/pubs/ASPRSSwetsetalSmoothing.pdf (last access: 16 August 2017), 1999.

Sylla, M. B., Pal, J. S., and Wang, G.: Impact of land cover characterization on regional climate modeling over West Africa, Clim. Dynam., 46, 637–650, 2015.

Szilagyi, J. and Parlange, M. B.: Defining Watershed-Scale Evaporation Using a Normalized Difference Vegetation Index, J. Am. Water Resour. As., 35, 1245–1255, https://doi.org/10.1111/j.1752-1688.1999.tb04211.x, 1999.

Szilagyi, J., Rundquist, D. C., Gosselin, D. C., and Parlange, M. B.: NDVI relationship to monthly evaporation, Geophys. Res. Lett., 25, 1753–1756, 1998.

Timouk, F., Kergoat, L., Mougin, E., Lloyd, C. R., Ceschia, E., Cohard, J.-M., Rosnay, P., Hiernaux, P., Demarez, V., and Taylor, C. M.: Response of surface energy balance to water regime and vegetation development in a Sahelian landscape, J. Hydrol., 375, 178–189, 2009.

Topp, G. C., Davis, J. L., and Annan, A. P.: Electromagnetic determination of soil water content: Measurements in coaxial transmission lines, Water Resour. Res., 16, 574–582, https://doi.org/10.1029/WR016i003p00574, 1980.

Velluet, C., Demarty, J., Cappelaere, B., Braud, I., Issoufou, H. B.-A., Boulain, N., Ramier, D., Mainassara, I., Charvet, G., Boucher, M., Chazarin, J.-P., Oï, M., Yahou, H., Maidaji, B., Arpin-Pont, F., Benarrosh, N., Mahamane, A., Nazoumou, Y., Favreau, G., and Seghieri, J.: Building a field- and model-based climatology of local water and energy cycles in the cultivated Sahel – annual budgets and seasonality, Hydrol. Earth Syst. Sci., 18, 5001–5024, https://doi.org/10.5194/hess-18-5001-2014, 2014.

Vitousek, P. M.: Human Domination of Earth's Ecosystems, Science, 277, 494–499, https://doi.org/10.1126/science.277.5325.494, 1997.

Webb, E., Pearman, G., and Leuning, R.: Correction of flux measurements for density effects due to heat and water vapour transfer, Q. J. Roy. Meteorol. Soc., 106, 85–100, 1980.

White, F.: La vegetation de l'Afrique: Memoire accompagnant la carte de vegetation de l'Afrique, vol. 20, IRD Editions, Paris, France, 1986.

Whiteman, C. D. and Allwine, K. J.: Extraterrestrial solar radiation on inclined surfaces, Environ. Softw., 1, 164–169, 1986.

Wilczak, J. M., Oncley, S. P., and Stage, S. A.: Sonic anemometer tilt correction algorithms, Bound.-Lay. Meteorol., 99, 127–150, 2001.

Williams, C. A., Reichstein, M., Buchmann, N., Baldocchi, D., Beer, C., Schwalm, C., Wohlfahrt, G., Hasler, N., Bernhofer, C., Foken, T., Papale, D., Schymanski, S., and Schaefer, K.: Climate and vegetation controls on the surface water balance: Synthesis of evapotranspiration measured across a global network of flux towers, Water Resour. Res., 48, W06523, https://doi.org/10.1029/2011WR011586, 2012.

5

Stochastic generation of multi-site daily precipitation focusing on extreme events

Guillaume Evin, Anne-Catherine Favre, and Benoit Hingray

Univ. Grenoble Alpes, CNRS, IRD, Grenoble INP, Grenoble, France

Correspondence: Guillaume Evin (guillaume.evin@irstea.fr)

Abstract. Many multi-site stochastic models have been proposed for the generation of daily precipitation, but they generally focus on the reproduction of low to high precipitation amounts at the stations concerned. This paper proposes significant extensions to the multi-site daily precipitation model introduced by Wilks, with the aim of reproducing the statistical features of extremely rare events (in terms of frequency and magnitude) at different temporal and spatial scales. In particular, the first extended version integrates heavy-tailed distributions, spatial tail dependence, and temporal dependence in order to obtain a robust and appropriate representation of the most extreme precipitation fields. A second version enhances the first version using a disaggregation method. The performance of these models is compared at different temporal and spatial scales on a large region covering approximately half of Switzerland. While daily extremes are adequately reproduced at the stations by all models, including the benchmark Wilks version, extreme precipitation amounts at larger temporal scales (e.g., 3-day amounts) are clearly underestimated when temporal dependence is ignored.

1 Introduction

Stochastic precipitation generators are often employed in risk assessment studies to estimate the return periods of very rare flooding events (e.g., 10 000-year events). The observed series of streamflows are too short to produce reliable estimations of very rare and large floods. Typically, extreme hydrological events can be reproduced using long series of simulated precipitation data as input to hydrological models (Lamb et al., 2016).

In the last two decades, a number of precipitation models have been proposed to deal with the temporal and spatial properties of daily precipitation, for both intermittency and amount, and all have different strengths and weaknesses. Many of these models use exogenous variables to predict the statistical properties of precipitation using generalized linear models (Chandler and Wheater, 2002; Mezghani and Hingray, 2009; Serinaldi and Kilsby, 2014b), atmospheric analogs (Lafaysse et al., 2014), or modified Markov models (Mehrotra and Sharma, 2010). Introducing a link between exogenous atmospheric variables can be used to reconstruct past events, make predictions, or downscale global-climate-model-based simulations of future climate. Such models are classically referred to as statistical downscaling models (see Maraun et al., 2010, for a review.) Closely related to this approach, weather "types" or "regimes" (Ailliot et al., 2015) can be used to specifically account for different atmospheric circulation patterns. Using a hidden Markov model (HMM) with transitions between these weather states, stochastic weather generators can then simulate various aspects of the precipitation process (Rayner et al., 2016).

Alternatively, purely stochastic precipitation models can be used. These can be broadly classified into three main types.

– *Resampling methods.* The stochastic generation of precipitation fields can be performed using resampling techniques such as the K-nearest neighbors algorithm (Buishand, 1991; Yates et al., 2003). Unobserved precipitation amounts can be obtained using perturbation techniques (Sharif and Burn, 2007).

– *Random fields.* Spatiotemporal precipitation models can simulate precipitation fields over a regular grid. This approach is particularly useful for hydrological applications, since areal precipitation values over a basin are obtained directly. Poisson cluster-based models (Burton et al., 2008, 2010; Leonard et al., 2008; McRobie et al., 2013) randomly simulate rain disk cells, with random centers, radius and intensity, over the study area. Meta-Gaussian models (Vischel et al., 2009; Kleiber et al., 2012; Allard and Bourotte, 2015; Baxevani and Lennartsson, 2015; Bennett et al., 2017) are based on truncated and transformed random Gaussian fields. Closely related, the turning band method can be used to simulate intermittent precipitation fields with different types of advection (Leblois and Creutin, 2013). These model structures are appealing since they are able to simulate realistic precipitation fields at fine spatial scales. However, their complexity leads to numerous technical issues during parameter estimation and simulation, notably in terms of computational cost. Moreover, they are usually unable to represent large regions comprising very distinct precipitation regimes.

– *Statistical multi-site models.* In this last type of weather generator, the properties of precipitation are directly fitted at a limited number of stations using different statistical structures. This type of generator preserves the interdependency between all pairs of stations, even when the area under study exhibits different precipitation regimes. Bárdossy and Pegram (2009) and Rasmussen (2013) combine a multivariate autoregressive process and transformations (V-transform, power transformation) to simultaneously model precipitation occurrence and amount. More precisely, with these models, transformed precipitation amounts follow truncated distributions. Alternatively, Wilks (1998) proposes a multi-site model in which precipitation occurrence and amount are handled separately. Several extensions to this popular structure have been proposed in the literature. Thompson et al. (2007) reformulate the Wilks model as a hidden Markov model, inferring three precipitation states ("dry", "light", and "heavy"). Mehrotra and Sharma (2007b) apply semi-parametric techniques to add more flexibility to the spatial structure of precipitation occurrence and amount. Srikanthan and Pegram (2009) propose a modified version in which daily, monthly, and annual amounts are nested such that precipitation statistics are preserved for all these levels of aggregation.

Mehrotra et al. (2006) compare three different precipitation models, the Wilks model, a HMM and a resampling approach, and they provide strong arguments in favor of the Wilks model in terms of performance, computation time, model, and level of complexity of the model structure. Furthermore, as indicated above, this model offers a flexible structure, which can be applied to a large number of stations

with very different precipitation regimes (like in mountainous areas). This paper presents several significant extensions of the Wilks precipitation model, referred to as GWEX versions, which will be used to generate long scenarios. These extensions aim to fit the most extreme precipitation amounts at different temporal (1- and 3-day amounts) and spatial scales. Novel components are thus introduced in GWEX, including robust estimation methods (regionalization methods) for critical parameters impacting directly on the behavior of extreme precipitation at each station. Recent advances in the choice of the marginal distributions for daily precipitation amounts are also included. Using 15 029 long daily precipitation records (> 50 years) from around the world, Papalexiou et al. (2013) conclude that heavy-tailed distributions are generally in better agreement with observed precipitation extremes. Follow-up studies (Papalexiou and Koutsoyiannis, 2013; Serinaldi and Kilsby, 2014a) apply extreme value theory to annual maxima and "peaks over threshold" (POTs) of a large subset of these records and confirm that extreme daily precipitation is not adequately represented by light-tailed distributions. Based on statistical tests on 90 000 station records of daily precipitation, Cavanaugh et al. (2015) also come to the same conclusion. These findings have important implications for precipitation models.

– Light-tailed distributions such as exponential, Gamma, and Weibull distributions, which are applied in the vast majority of the existing precipitation models, often lead to an underestimation of extreme daily precipitation amounts.

– While nonparametric densities with Gaussian kernels (Mehrotra and Sharma, 2007a, 2010) offer the flexibility of fitting the observed range of precipitation amounts, their tail also belongs to the domain of attraction of the Gumbel distribution and suffers from the same drawbacks.

Alternatively, current statistical procedures consisting of fitting a flexible distribution to the bulk of the observations and using it for extrapolation are highly questionable, as major assumptions are usually violated (Klemeš, 2000a, b). Since the tail of the distribution on precipitation amounts at each station will dictate the generation of the most extreme precipitation events, important features of GWEX are as follows:

– the application of a heavy-tailed distribution to precipitation amounts at each station (Naveau et al., 2016),

– the determination of robust estimates of the shape parameter of this distribution, which indicates the heaviness of the tail, using a regionalization approach, as in Evin et al. (2016).

Furthermore, following Bárdossy and Pegram (2009), GWEX also employs the copula theory to introduce a tail dependence between the precipitation amounts simulated at the different stations. The second version of the GWEX model includes a disaggregation method, the observed precipitation amounts being fitted at a 3-day scale in a first step. This paper compares the performance of the different model versions and assesses the impact of the different statistical components (e.g., heavy-tailed distribution, tail dependence).

We first describe the study area in Sect. 2. The features of different multi-site precipitation models are then described in Sect. 3. The evaluation framework, presented in Sect. 4, aims to assess the performance of these models at different spatial and temporal scales. Section 5 presents an application of these daily precipitation models to 105 stations located in Switzerland, with a summary of the results focusing on the reproduction of extreme events. Finally, Sect. 6 presents our conclusions.

2 Data and study area

The Aare River basin covers the northern part of the Swiss Alps and has an area of $17\,700\,\mathrm{km}^2$. Basin elevations approximately range from 310 m.a.s.l. in Koblenz (next to the German border in the north) to 4270 m.a.s.l. at the Finsteraarhorn summit (in the south of the basin). The mean annual precipitation for the basin as a whole is 1300 mm. The basin can be divided into five main sub-basins, with different hydrometeorological regimes highly governed by regional terrain features (Jura Mountains in the northwest, northern Alps in the south of the basin, and lowlands in the middle).

Figure 1 shows the location of the 105 precipitation stations used for the development and evaluation of weather generators. Located within or close to the Aare River basin, they correspond to the stations for which long daily time series of observations with less than 3 years of missing data are available over the period 1930–2014. The 105 precipitation stations cover the Aare River basin relatively well.

The proposed precipitation models are designed to simulate flood scenarios, via a conceptual hydrological model, for the whole Aare River basin and for its different sub-basins. For Switzerland, Froidevaux et al. (2015) show that the generation of floods is mainly influenced by areal precipitation amounts accumulated over short periods (e.g., 1 to 3 days). These results are obtained by analyzing a wide variety of basins, their areas ranging from 10 to $12\,000\,\mathrm{km}^2$. Therefore, the properties of the weather scenarios must be evaluated at different spatial and temporal scales, from the high resolutions required to simulate the hydrological behavior of the system (e.g., sub-daily, $100\,\mathrm{km}^2$) to lower resolutions relevant at the scale of the entire basin (e.g., n-days, $17\,700\,\mathrm{km}^2$). In this study, the performance of the different precipitation models is evaluated at the station scale, at the scale of 15 and 5 sub-basins partitioning the Aare River basin, and at the scale of the entire study area (see Sect. 5). Note that for those evaluations, areal estimates of precipitation are ob-

tained from the precipitation amounts at the stations using the Thiessen polygon method.

3 Multi-site precipitation model

As indicated above, GWEX refers to multi-site precipitation models that rely strongly on the structure proposed by Wilks (1998). At each location k, let $P_t(k)$ be a random variable representing the accumulated precipitation over day t. The structure proposed by Wilks considers a hidden occurrence process $X_t(k)$ that can be represented by a two-state Markov chain as follows:

$$X_t(k) = \begin{cases} 0, & \text{if day } t \text{ is dry at location } k. \\ 1, & \text{if day } t \text{ is wet at location } k. \end{cases} \quad (1)$$

Precipitation amount $P_t(k)$ is then defined as

$$P_t(k) = Y_t(k)X_t(k), \quad (2)$$

where $Y_t(k)$ is a random variable describing the non-zero precipitation amounts. Non-zero precipitation amounts $Y_t(k)$ are thus modeled independently of precipitation occurrences $X_t(k)$, which act as a mask.

3.1 Precipitation occurrence process

3.1.1 At-site occurrence process

At each location, the temporal persistence of dry and wet events is introduced with a p-order Markov chain model for $X_t(k)$ so that

$$\begin{aligned} \Pr\{X_t(k) &= 1 | X_{t-1}(k), \ldots, X_1(k)\} \\ &= \Pr\{X_t(k) = 1 | X_{t-1}(k), \ldots, X_{t-p}(k)\}; \end{aligned} \quad (3)$$

i.e., the probability of having a wet day at time t depends only on the p previous states, for days $t-1, \ldots, t-p$. While many authors suppose that a first-order Markov is sufficient (e.g., Wilks, 1998; Keller et al., 2015), Srikanthan and Pegram (2009) apply a fourth-order Markov chain and show that it improves the reproduction of dry/wet period lengths. In this study, different orders for this Markov chain are considered.

At each site, the probability of having a wet day on day t is given by the transition probability $\Pr\{X_t(k) = 1 | X_{t-1}(k) = i_1, \ldots, X_{t-p}(k) = i_p\}$, where $i_1, \ldots, i_p$ are equal to 0 or 1. This Markov chain is thus fully characterized by a transition matrix $\mathbf{\Pi}$ with dimension 2^p.

3.1.2 Spatial occurrence process

The spatial dependence of the precipitation states $X_t(k)$ is modeled using an unobserved Gaussian stochastic process $\mathbf{U}_t = \{U_t(1), \ldots, U_t(K)\}$, where K is the number of stations. Here, Gaussian random variables $U_t(k), k = 1, \ldots, K$, are

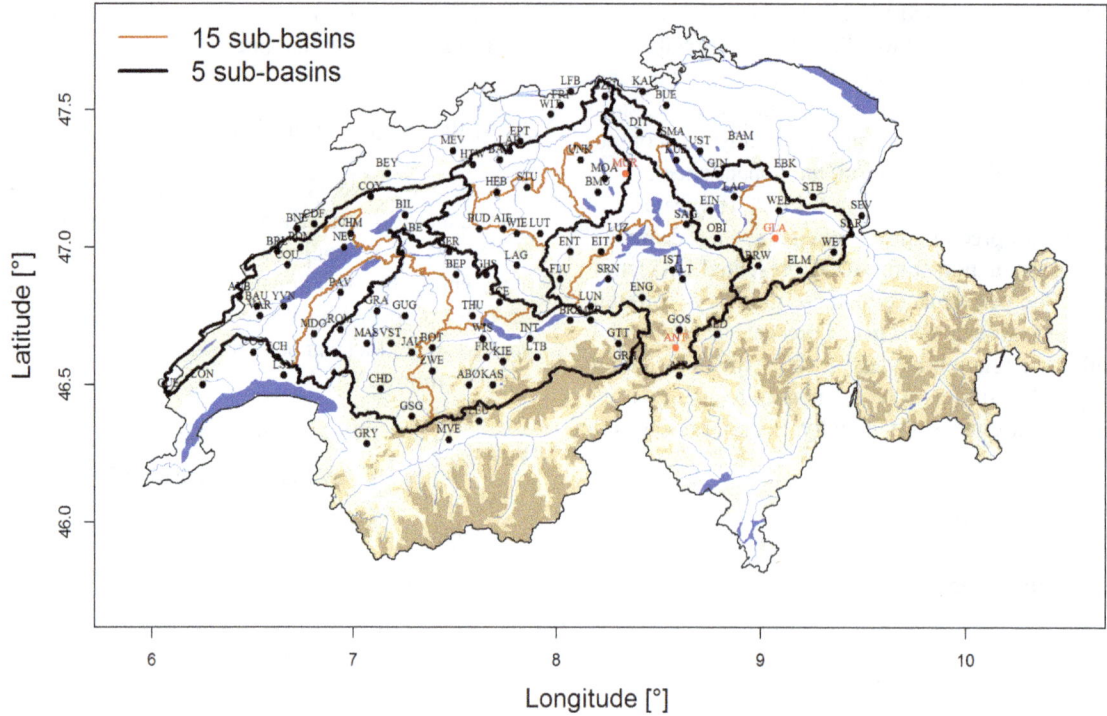

Figure 1. Location of the 105 precipitation stations in Switzerland. Different partitions of the Aare River basin into 5 and 15 sub-basins are shown.

temporally independent and $\mathbf{U}_t$ follows a multivariate normal distribution:

$$\mathbf{U}_t \sim N(0, \mathbf{\Omega}_X), \tag{4}$$

where $\mathbf{\Omega}_X = \{\omega_{kl}\}$ is a positive-definite correlation matrix. At any location k, the precipitation state $X_t(k)$ is assumed to be completely determined by $U_t(k)$ and the previous p states at the same location. Specifically, if $X_{t-1}(k) = i_1, \ldots, X_{t-p}(k) = i_p$, and $p_1 = \Pr\{X_t(k) = 1 | X_{t-1}(k) = i_1, \ldots, X_{t-p}(k) = i_p\}$, then

$$X_t(k) = \begin{cases} 1, & \text{if } U_t(k) \le \Phi^{-1}(p_1). \\ 0, & \text{otherwise}, \end{cases} \tag{5}$$

where $\Phi[.]$ indicates the standard Gaussian cumulative distribution function.

Let $\rho_{kl} = \mathrm{Corr}(X_t(k), X_t(l))$ denote the inter-site correlation between the states $X_t(k)$ and $X_t(l)$. Following Srikanthan and Pegram (2009), ρ_{kl} can be expressed as

$$\rho_{kl} = \frac{\pi_{00}(k,l) - \pi_0(k)\pi_0(l)}{\sqrt{\pi_0(k)\pi_1(k)}\sqrt{\pi_0(l)\pi_1(l)}}, \tag{6}$$

where $\pi_0(s) = \Pr\{X_t(s) = 0\}$ and $\pi_1(s) = \Pr\{X_t(s) = 1\}$ denote the probabilities of having dry and wet states at location s, respectively, and $\pi_{00}(k,l) = \Pr\{X_t(k) = 0, X_t(l) = 0\}$ denotes the joint probability of having dry states at both locations k and l.

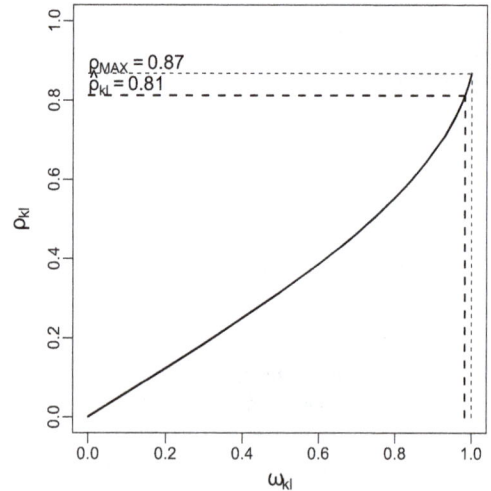

Figure 2. Illustration of the relationship between ω_{kl} and ρ_{kl} for the month of July and for stations GOS and ANT. A Markov chain of order 4 is considered in this example. The correlation between the observed states is $\hat{\rho}_{kl} = 0.81$ and can be reproduced using a bivariate Gaussian distribution with a correlation parameter of $\omega_{kl} = 0.98$. The maximum correlation ρ which can be obtained if $\omega_{kl} = 1$ is $\rho_{\mathrm{MAX}} = 0.87$.

The relationship between ω_{kl} and ρ_{kl} is not direct since the temporal persistence of dry and wet events introduced at each station with a Markov chain also influences ρ_{kl} (Wilks,

1998). Figure 2 illustrates this relationship, obtained for the month of July via Monte Carlo simulations, for two close stations, GOS and ANT. In a first step, transition probabilities with a Markov chain of order 4 are estimated for these two stations. Given these transition probabilities, stochastic simulations of occurrence are then generated for different values of ω_{kl}, leading to different values of ρ_{kl}. Since this relationship is monotonic (see Fig. 2), it can be used to identify the value ω_{kl} leading to a specific $\hat{\rho}_{kl}$, namely the empirical value obtained from the observed time series of occurrence. The estimate of ω_{kl} is found by iterating until the evaluation of the correlation between the simulated precipitation states, ρ_{kl}, matches $\hat{\rho}_{kl}$. Note that a very high value for $\hat{\rho}_{kl}$ cannot always be reached, even if $\omega_{kl} = 1$. This is, however, a situation which rarely occurs in practice.

3.2 Precipitation intensity process

Given the occurrence of precipitation $X_t(k)$ at different locations k, GWEX models generate the amounts of precipitation $Y_t(k)$ using

- marginal heavy-tailed distributions,

- a tail-dependent spatial distribution,

- an autocorrelated temporal process.

3.2.1 Marginal distributions

At a given location k, daily precipitation has often been modeled by light-tailed distributions: exponential and Weibull distributions (Bárdossy and Pegram, 2009); gamma distributions (Srikanthan and Pegram, 2009; Mezghani and Hingray, 2009); a mixture of exponential distributions (Wilks, 1998; Keller et al., 2015); and a mixture of gamma distributions (Chen et al., 2014). However, as shown by many recent studies on a very large number of daily precipitation series (Papalexiou et al., 2013; Serinaldi and Kilsby, 2014a; Cavanaugh et al., 2015), exponentially decaying tails often result in a severe underestimation of extreme event probabilities. The introduction of a heavy-tailed distribution is thus crucial for the reproduction of the most extreme precipitation events (Hundecha et al., 2009).

In this work, the distribution representing the precipitation intensity at each location, $Y_t(k)$, is the E-GPD distribution. This distribution was first proposed by Papastathopoulos and Tawn (2013), who referred to it as an extended GP-Type III distribution, and it has since been shown to adequately model the whole range of precipitation intensities (Naveau et al., 2016). Compared to other heavy-tailed distributions applied to daily precipitation amounts (e.g., mixtures of GPD and gamma distribution; see Vrac and Naveau, 2007), the E-GPD is parsimonious and provides a very good compromise between flexibility and stability, which is an essential feature for extrapolation.

This distribution can be described by a smooth transition between a gamma-like distribution and a heavy-tailed Generalized Pareto distribution (GPD). This transition is obtained via a transformation function, $G(\nu)$, such that the whole range of precipitation intensities is modeled without a threshold selection (Naveau et al., 2016):

$$F_Y\{Y_t(k)\} = G\Big[H_\xi\big\{Y_t(k)/\sigma\big\}\Big], \tag{7}$$

where

$$H_\xi(z) = \begin{cases} 1 - (1 + \xi z)_+^{-1/\xi} & \text{if } \xi \neq 0, \\ 1 - e^{-z} & \text{if } \xi = 0, \end{cases} \tag{8}$$

where $a_+ = \max(a, 0)$ is the standard cumulative distribution function of the GPD, $\sigma > 0$ is a scale parameter, and $G(\nu) = \nu^\kappa, \kappa > 0$. Thus, a three-parameter set $\{\sigma, \kappa, \xi\}$ needs to be estimated at each station.

3.2.2 Spatial and temporal dependence of precipitation amounts

Spatial and temporal dependence of precipitation amounts is represented using a multivariate autoregressive model of order 1 (MAR(1)). A MAR(1) process has been used by different authors (Bárdossy and Pegram, 2009; Rasmussen, 2013) to simultaneously represent spatial and temporal dependences. Let $\mathbf{Z}_t$ denote a vector of K Gaussian random variables with mean 0 defined as

$$Z_t(k) = \Phi^{-1}\big[F_Y\{Y_t(k)\}\big]. \tag{9}$$

The stochastic Gaussian process $\mathbf{Z}_t$ is assumed to follow a MAR(1) process defined as follows:

$$\mathbf{Z}_t = \mathbf{A}\mathbf{Z}_{t-1} + \boldsymbol{\epsilon}_t, \tag{10}$$

where $\mathbf{A}$ is a $K \times K$ matrix and $\boldsymbol{\epsilon}_t$ is an innovation term described by a random $K \times 1$ noise vector. The elements of $\boldsymbol{\epsilon}_t$ have zero means and are independent of the elements of $\mathbf{Z}_{t-1}$. The covariance matrix of $\boldsymbol{\epsilon}_t$ is denoted by $\boldsymbol{\Omega}_Z$. Following Bárdossy and Pegram (2009), $\mathbf{A}$ is taken to be a diagonal matrix with diagonal elements that are the lag-1 serial correlation coefficients of the intensity process $Y_t(k)$. The matrix $\boldsymbol{\Omega}_Z$ can be expressed as

$$\boldsymbol{\Omega}_Z = \mathbf{M}_0 - \mathbf{A}\mathbf{M}_0'\mathbf{A}, \tag{11}$$

where $\mathbf{M}_0$ is the covariance matrix of $\mathbf{Z}_t$, which indicates the degree of spatial dependence between each pair of stations, and $\mathbf{M}_0'$ is its transpose.

Innovations $\boldsymbol{\epsilon}_t$ are often assumed to follow a standard multivariate normal distribution. However, the upper tail dependence of the multivariate normal distribution is 0, which means that extreme precipitation amounts simulated at the different sites are not spatially dependent. To introduce a tail dependence between at-site extremes, a possibility is to use a Student copula to represent the dependence structure of ϵ_t, providing an additional parameter, ν, related to the tail dependence. Both dependence structures will be considered in the following.

3.3 Parameter estimation

3.3.1 Occurrence process

Following Wilks (1998), parameters related to the occurrence process $X_t(k)$ are estimated using the method of moments, i.e., using the empirical counterparts of the parameters. Observed states are first obtained using a low precipitation threshold (e.g., 0.2 mm). The matrix $\mathbf{\Pi}$ of transition probabilities is then estimated directly by the proportion of wet days $X_t(k) = 1$ following observed sequences $\{X_{t-1}(k), ..., X_{t-p}(k)\}$. Concerning the spatial occurrence process, $\hat{\rho}_{kl}$ estimates are obtained using the empirical counterparts of π_{00}, π_0, and π_1 (see Eq. 6), which correspond respectively to the proportion of days for which dry states are observed simultaneously at two locations ($\hat{\pi}_{00}$) and to the proportions of dry days $\hat{\pi}_0$ and wet days $\hat{\pi}_1$. The correlation matrix $\hat{\mathbf{\Omega}}_X$ is then composed of the cross-correlations $\hat{\omega}_{kl}$ obtained for all possible pairs of stations. If $\hat{\mathbf{\Omega}}_X$ is not positive-definite, the closest positive-definite matrix is considered (Rousseeuw and Molenberghs, 1993; Rebonato and Jaeckel, 2011). Furthermore, the seasonality of the occurrence process is taken into account by estimating these parameters on a monthly basis.

3.3.2 Intensity process

E-GPD distributions are first fitted to precipitation amounts available at each location k. As local estimations of the GPD tail exhibit a lack of robustness, we propose estimating the ξ parameter of the E-GPD (see Eq. 8) using a regionalization method similar to that of Evin et al. (2016), which can be summarized as follows.

1. Following Burn (1990), for each station, a region-of-influence (RoI) is delimited by a circle around the site, the radius being determined using homogeneity tests. All the stations inside this RoI are then considered homogeneous up to a scale factor.

2. The ξ parameters are then estimated with the maximum likelihood method using the precipitation observations from all the stations inside the RoI.

This regionalization method is applied to the precipitation data available from 666 stations in Switzerland, for four different seasons:

- *winter:* December, January, and February;

- *spring:* March, April, and May;

- *summer:* June, July, and August;

- *autumn:* September, October, and November.

In this work, the estimation of the ξ parameter is bounded below by 0. When $\xi < 0$, the E-GPD distribution has an upper bound. As shown by many recent studies (e.g., Serinaldi and Kilsby, 2014a), negative estimates of ξ are usually due to parameter uncertainty and are not realistic. The two remaining parameters of the E-GPD, the scale parameter σ and the parameter of the transformation κ, are estimated from the observations available at that station. Here, we use a method of moments based on probability weighted moments (see Naveau et al., 2016, for further details).

Concerning the spatial and temporal dependence of precipitation amounts, direct estimates of $\mathbf{M}_0$ and $\mathbf{A}$ cannot be obtained since non-zero precipitation amounts $Y_t(k)$ are not observed. Here, we follow the methodology proposed by Wilks (1998) and Keller et al. (2015). For each pair of stations, we generate long sequences of precipitation amounts $P_t(k)$ using the estimated parameters of the occurrence process ($\hat{\mathbf{\Pi}}$ and $\hat{\omega}_{kl}$), the parameters of the marginal distributions, and a correlation coefficient $m_0(k, l)$, indicating the degree of spatial dependence. Similarly to the occurrence process, $\hat{m}_0(k, l)$ is then found iteratively by matching the correlation between these long random streams with the observed correlation $\mathrm{Corr}(P_t(k), P_t(l))$ (see Wilks, 1998; Keller et al., 2015, for further details). The correlation matrix $\hat{\mathbf{M}}_0$ is then composed of the cross-correlations $\hat{m}_0(k, l)$ obtained for all possible pairs of stations. For each station, the estimates of the lag-1 serial correlation coefficients of the matrix $\mathbf{A}$ are obtained using the same simulation approach.

The matrix $\hat{\mathbf{\Omega}}_Z$, i.e., the estimate of the covariance matrix of the innovations ϵ_t, is then obtained using Eq. (11). Since $\hat{\mathbf{\Omega}}_Z$ is not necessarily positive-definite (see Eq. 11), the closest positive-definite matrix is taken as the covariance matrix of ϵ_t if necessary. Given $\hat{\mathbf{\Omega}}_Z$, the parameter ν is estimated by maximizing the likelihood, as described in McNeil et al. (2005, Sect. 5.5.3.).

Similarly to the occurrence process, the seasonal aspect of the precipitation intensity is taken into account by performing the parameter estimation for each month, on a 3-month moving window.

3.4 Model versions

Different versions of the proposed multi-site precipitation model are considered in this paper, each corresponding to different extensions of the Wilks model. A flowchart summarizing the increasing complexity of these models is presented in Fig. 3.

3.4.1 Wilks

A first benchmark version of the multi-site model, referred to here as "Wilks", is considered. It closely matches the multi-site model proposed by Wilks (1998), detailed in particular as follows.

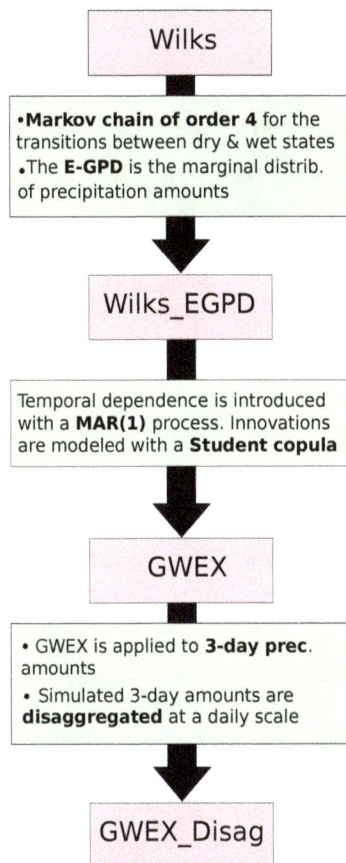

Figure 3. Flowchart of the different model versions. The differences between the models are summarized inside green boxes.

- The at-site occurrence process is a Markov chain of order 1.

- The marginal distribution on precipitation amounts is a mixture of exponential distribution, for which the probability density function is defined as

$$f(x) = \frac{w}{\beta_1} \exp\left(-\frac{x}{\beta_1}\right) + \frac{1-w}{\beta_2} \exp\left(-\frac{x}{\beta_2}\right). \quad (12)$$

The parameters w, β_1 and β_2 are estimated using the expectation-maximization (EM) method (Dempster et al., 1977).

- Precipitation amounts are not considered to be temporally correlated; i.e., the matrix **A** in Eq. (10) is a zero matrix. Furthermore, innovations ϵ_t follow a standard multivariate normal distribution and represent the spatial correlations.

3.4.2 Wilks_EGPD

A modified Wilks version is considered, for which the at-site occurrence process is a Markov chain of order 4 and

the mixture of exponential distributions is replaced by the E-GPD distribution. As indicated above, Srikanthan and Pegram (2009) show that a fourth-order Markov chain improves the reproduction of dry/wet period lengths. This direct extension of the Wilks model is used to illustrate the impact of using a Markov chain of order 4 compared to order 1. Differences in performance between a heavy-tailed distribution (E-GPD) and a low-tailed distribution (mixture of exponentials) will be highlighted.

3.4.3 GWEX

The initial GWEX model has the following characteristics.

- The at-site occurrence process is a Markov chain of order 4.

- The marginal distribution for precipitation amounts is the E-GPD distribution.

- Precipitation amounts follow a MAR(1) process with innovations modeled by a Student copula.

3.4.4 GWEX_Disag

In this paper, an alternative version, referred to as GWEX_Disag, is also proposed. GWEX_Disag is applied to 3-day precipitation amounts and has the same characteristics as GWEX, except the following.

- The at-site occurrence process is a Markov chain of order 1.

- A threshold of 0.5 mm separates dry and wet states.

With GWEX_Disag, daily scenarios are first generated at a 3-day scale and then disaggregated at a daily scale using a method of fragments (e.g., Wójcik and Buishand, 2003). Simulated 3-day amounts are disaggregated using the temporal structures of the closest observed 3-day amounts, in terms of similarity of the spatial fields. The same observed 3-day sequence is thus used to disaggregate the 3-day amounts simulated at the 105 stations, which ensures the spatial coherence of these disaggregated amounts. Details of the disaggregation method are provided in Appendix A. Compared to GWEX, GWEX_Disag offers the following advantages.

- The 3-day precipitation amounts are directly modeled and have a better chance of being adequately reproduced.

- The disaggregation of 3-day precipitation amounts creates an inherent link between the occurrence and the intensity processes. For very extreme precipitation events, we can expect these processes to be dependent (higher chance of being in a wet state over the whole Aare River basin, as well as large and persistent precipitation amounts).

4   Multi-scale evaluation

The proposed stochastic models intend to preserve the most critical properties of precipitation at different spatial and temporal scales, especially extreme precipitation amounts. For hydrological applications, it can be assumed that a precipitation model preserving these properties has a better chance of adequately reproducing flood properties for small sub-basins as well as for large basins. This statement is supported by empirical evidence provided by Froidevaux (2014) and Froidevaux et al. (2015) for our study area (i.e., Switzerland). Using 60 years of gridded precipitation data, Froidevaux et al. (2015) show that, in Switzerland, high discharge events are usually triggered by meteorological events with a duration of several days, in late summer and autumn. Typically, the 2-day precipitation sum before floods is most correlated with flood frequency and flood magnitude.

The performance of the different multi-site precipitation models is thus assessed for multiple spatial and temporal scales. We investigate whether or not the statistical properties of precipitation data are adequately reproduced at the scale of the stations and for different partitions of the Aare River basin (see Fig. 1). In order to achieve this, 100 daily precipitation scenarios are generated, each scenario having a length of 100 years.

For the different evaluated statistics, performance is categorized according to the comprehensive and systematic evaluation (CASE) framework proposed by Bennett et al. (2017). The CASE framework enables a systematic comparison of stochastic models and offers a consistent way of computing the performance metrics, which is important in order to obtain a fair assessment of the strengths/weaknesses of the different model versions. This approach consists in assigning one of three categories: "good", "fair", and "poor" performance, to each metric, according to the agreement between the observed metric and the simulated metrics computed from the 100 scenarios. Table 1 summarizes the tests leading to each performance category. A good performance is obtained when the observed metric is inside the 90 % probability limits of the 100 simulated metrics (case 1). It indicates that simulated metrics are in good agreement with the observed metric. However, an observed metric can obviously lie outside these limits without necessarily indicating a failure of the model. In this case, fair performance may be assigned if either of the following two rules is satisfied.

1. *Case 2.* The observed metric is outside the 90 % probability limits but within 3 standard deviations (SD) of the simulated mean, which corresponds to the 99.7 % probability limits if we assume that the uncertainty in the statistics is normally distributed. This case covers the situation where we could expect that the observed metric is outside the 90 % limits due to sampling uncertainty.

Table 1. Performance categorization criteria from Bennett et al. (2017).

Performance classification	Key	Test
Good	■	Observed metric inside 90 % limits (case 1)
Fair	■	Observed metric outside 90 % limits but within the 99.7 % limits (case 2) OR absolute relative difference between the observed metric and the average simulated metrics is 5 % or less (case 3)
Poor	■	Otherwise (case 4)

Table 2. Hydrological regimes and characteristics of extreme floods in Switzerland (Froidevaux, 2014).

	Mean elevation (m)	Season	Triggering events
Glacial	> 1900	summer	showers and snow melt
Nival	1200–1900	summer, spring	showers, long rain
Pluvial	< 1200	summer	long rain

2. *Case 3.* The absolute relative difference $|(S_{obs} - \overline{S}_{sim})/S_{obs}|$ between the observed metric S_{obs} and the mean of the simulated metrics $\overline{S}_{sim}$ is 5 % or less. If the variability of the simulated metrics is very small, it can happen that the observed metrics lie outside the 99.7 % limits without being too far from the simulated mean in terms of relative difference.

Otherwise, we consider that performance is poor, indicating that the model fails to reproduce this particular statistical properly.

In summary, good performance represents cases for which the observed metric is clearly well reproduced by the model, whereas fair performance indicates a reasonable match between the observed and the simulated metrics. The number of metrics for which poor performance is obtained is thus the first criterion indicating the overall performance of a model.

For illustration purposes, we also present the results of the evaluation for three precipitation stations corresponding to different hydrological regimes (see Table 2). Figure 1 shows the 3 (out of 105) selected precipitation stations. Station ANT (at Andermatt) is located in a glacial basin, station GLA (at Glarus) in a nival basin, and station MUR (at Muri) in a pluvial basin.

5 Results

This section presents the results of the multi-scale evaluation framework (see Sect. 4) for several metrics related to the occurrence process of the precipitation events, daily amounts, and precipitation extremes. Summary assessments are provided, with several statistics provided for all the spatial scales of interest.

The precipitation observations are split into two sets. (1) A total of 45 years randomly chosen among the period 1930–2014 are used to estimate the parameters, and (2) the 40 remaining years are used to evaluate the performance of the models. This separation between an estimation set and a validation set is crucial to test the ability of the model to adequately represent the statistical properties of events which have not been used during the fitting procedure. In this study, the multi-scale evaluation is only applied to the 40-year validation set.

5.1 Parameter estimation and generation of scenarios

The different model parameters are estimated with the 45-year estimation set of observations, following the methodology described in Sect. 3.3, except for the ξ parameter of the E-GPD, which is estimated using all available precipitation data in Switzerland. This approach ensures that robust estimates are obtained for this parameter, which is crucial in our context since extreme simulated precipitation amounts are highly sensitive to the ξ parameter.

For GWEX, the estimation of the ξ parameter is performed at a daily scale. In order to highlight spatial patterns of ξ over Switzerland, we show the maps of the interpolated parameter estimates in Fig. 4. Fat tails are obtained in the southern and eastern parts of the Aare River basin, particularly during spring and summer seasons. In the south of Switzerland, a region with high estimates ($\xi \sim 0.2$), highlighted in red, is obtained for the summer and autumn seasons. These high ξ estimates are consistent with the presence of strong convective storms in this mountainous region during this period of the year (Rudolph and Friedrich, 2012).

For GWEX_Disag, the regionalization method is applied at a 3-day scale (see Fig. 5). The resulting estimates are similar to the ones obtained at a daily scale. However, note that the very high estimates obtained during the summer season at a daily scale are lower at a 3-day scale. This seems to confirm the interpretation of these high ξ estimates; i.e., the relationship between summer convective storms and high ξ estimates is not as strong at a 3-day scale since storms of this type usually have a shorter duration. Note that non-zero ξ estimates in Figs. 4 and 5 (in green, yellow, and red) indicate that low-tailed distributions lead to an underestimation of extreme precipitation in these regions.

Figure 6 compares empirical and fitted distributions (mixture of exponentials and E-GPD) at a daily scale, for three illustrative stations and for the months of January, April, July, and October. Both distributions fit the observed precipitation amounts reasonably well. Concerning the highest precipitation intensities, it is hard to draw conclusions on a significant over- or underestimation. Indeed, local assessments of precipitation extremes are often inconclusive due to insufficient information on the distribution tails (Papalexiou and Koutsoyiannis, 2013).

For each multi-site precipitation model investigated in this paper (Wilks, Wilks_EGPD, GWEX and GWEX_Disag), we generate 100 daily precipitation scenarios with these parameter estimates, each scenario having a length of 100 years. These scenarios are compared to the precipitation observed for the 40-year validation period.

5.2 Occurrence process

The monthly number of wet days obtained from observed and simulated precipitation data are compared in Fig. 7. The average number of wet days is adequately reproduced by all models, with approximately 30 % of cases with poor performance. These poor performance cases seem to occur mainly during the winter and spring seasons. The SD of the monthly number of wet days indicates the inter-annual variability of this metric. While the magnitudes of the SD from the simulated precipitation roughly match the corresponding observed SD, it seems that the highest observed variabilities are underestimated by all the models, most markedly by the Wilks model.

Figures 8 and 9 show the distributions of observed and simulated dry and wet spells, respectively, for the three illustrative stations. Concerning the distributions of dry spell lengths, the Wilks_EGPD, GWEX, and GWEX_Disag models lead to adequate performance, the performance being classified as good in 48, 48, and 49 % of the cases, respectively. The performance of the Wilks model is slightly lower because of an imprecise reproduction of the frequency of the shortest dry spells. This difference in performance is explained by the order of the Markov chain used to simulate the transitions between dry and wet states, which is the only difference between the occurrence processes of Wilks and Wilks_EGPD or GWEX. The fourth-order Markov chain of the Wilks_EGPD and GWEX models seems to provide a more adequate representation of these transitions than the first-order Markov chain of the Wilks model, confirming previous findings (Srikanthan and Pegram, 2009).

The frequencies of wet spell lengths are adequately reproduced by the Wilks, Wilks_EGPD, and GWEX models, with more than 50 % of good performance. The lower overall performance of GWEX_Disag for this metric is due to a slight underestimation of the longest wet spells for some stations (which is however not the case for the stations shown in Fig. 9).

Season 1: Dec, Jan, Feb   Season 2: Mar, Apr, May

Season 3: Jun, Jul, Aug   Season 4: Sep, Oct, Nov

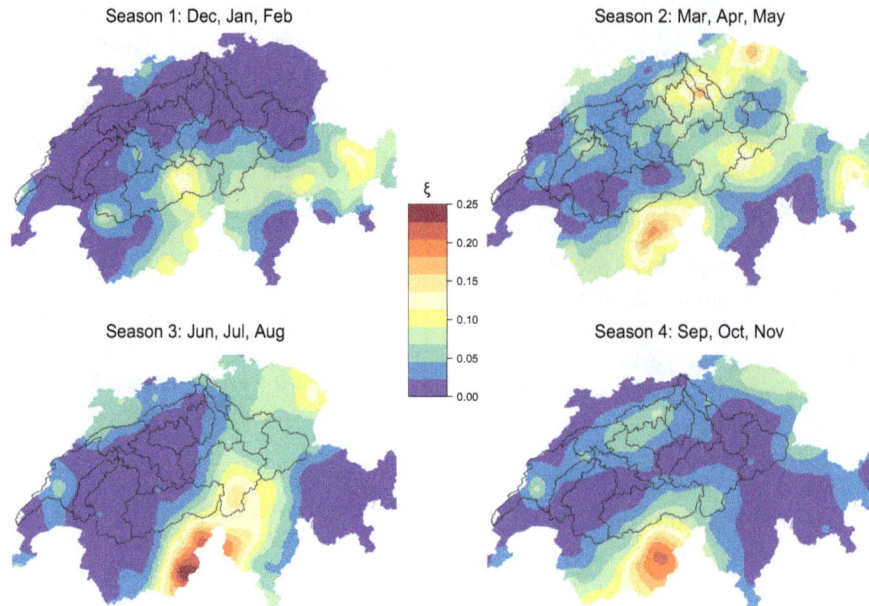

Figure 4. Regionalized ξ parameters at a daily scale, for the different seasons. Here, we present the spatial interpolation of at-site estimates for a better readability of their variability.

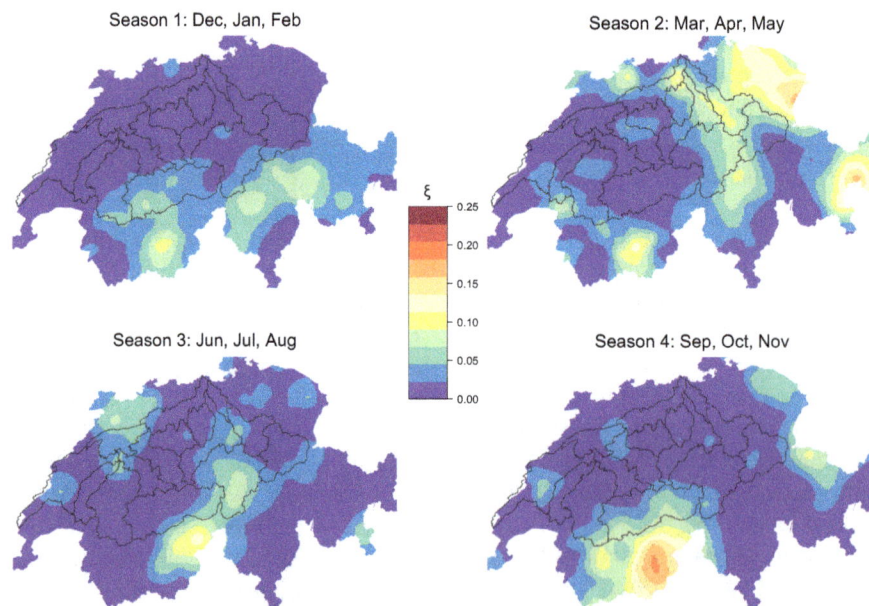

Season 1: Dec, Jan, Feb   Season 2: Mar, Apr, May

Season 3: Jun, Jul, Aug   Season 4: Sep, Oct, Nov

Figure 5. Regionalized ξ parameters at a 3-day scale, for the different seasons. Here, we present the spatial interpolation of at-site estimates for a better readability of their variability.

5.3 Inter-site correlations of precipitation amounts

Figure 10 compares observed and simulated inter-site correlations for the different model versions. Unlagged cross-correlations, which represent the spatial dependence, are close to the 1 : 1 diagonal line, as expected given that these correlations are explicitly taken into account by all model versions. However, a slight underestimation can be observed,

especially concerning correlations above 0.8. This underestimation is a side effect of the transformation applied to obtain a positive-definite matrix (see Sect. 3.3).

An adequate reproduction of lag-1 inter-site correlations is important for the reproduction of persistent precipitation events. Simulated lag-1 cross-correlations are close to 0 for the Wilks and Wilks_EGPD models, as expected given that these versions ignore the temporal dependence. Con-

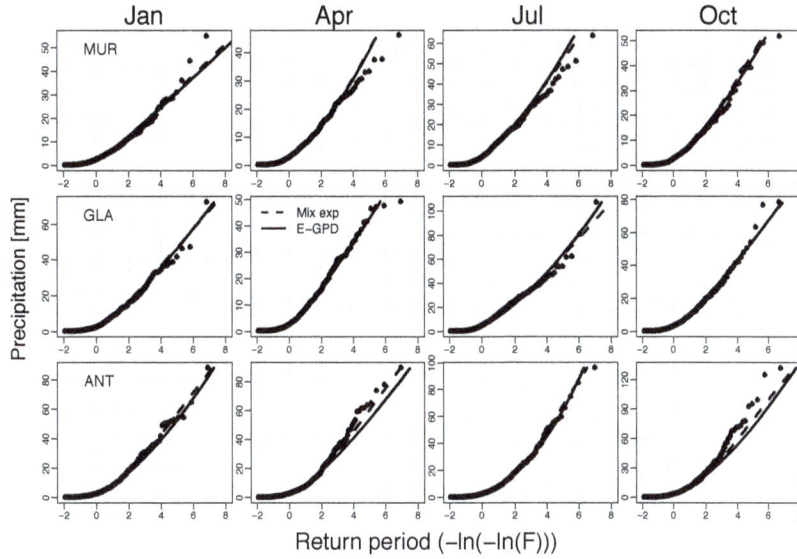

Figure 6. Empirical and fitted distributions (dashed curves for mixture of exponentials and solid curves for E-GPD) at a daily scale, for the three illustrative stations and for the months of January, April, July, and October.

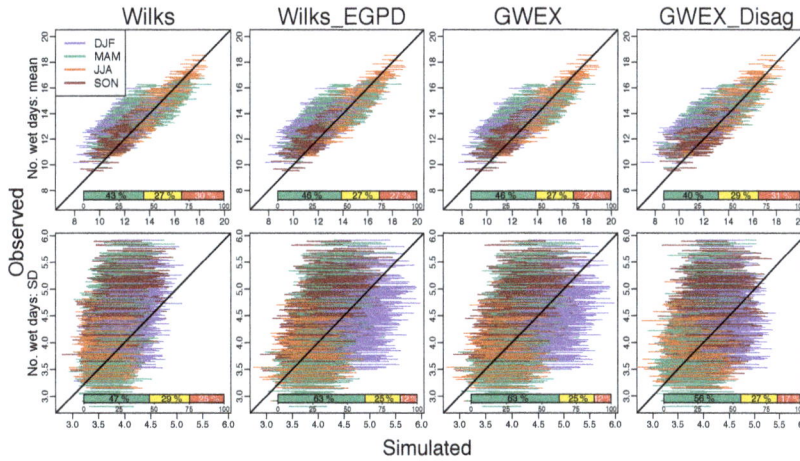

Figure 7. At-site number of wet days for all sites and months: inter-annual mean and standard deviation (SD). The 90 % probability limits are shown for the different seasons. Overall performance is represented by the indicated percentages of good, fair, and poor performance for all sites and months ($105 \times 12 = 1260$ cases).

sequently, these two model versions significantly underestimate observed lag-1 cross-correlations, which range between 0 and 0.4. Concerning GWEX, lag-1 serial autocorrelations at the stations (black points in the bottom plots) are perfectly aligned along the 1 : 1 line, as expected given that they are explicitly fitted by the MAR(1) process. Simulated and observed lag-1 cross-correlations are roughly in agreement, though the largest observed cross-correlations are underestimated. This is also the case, to a lesser extent, for GWEX_Disag. However, the agreement between observed and simulated cross-correlations is much stronger.

5.4 Daily amounts

The reproduction of precipitation amounts at a daily scale is assessed in Fig. 11, for all spatial scales and months. For all models, we obtain a reasonable agreement between observed and simulated average daily amounts (90 % limits close to the 1 : 1 line), with more than 40 % of good cases and less than 30 % of poor cases. The SD of these daily amounts are also adequately reproduced (Fig. 11, bottom plots).

5.5 Extreme precipitation amounts

Figures 12 and 13 show the relative differences, expressed as a percentage, between observed and simulated 10- and

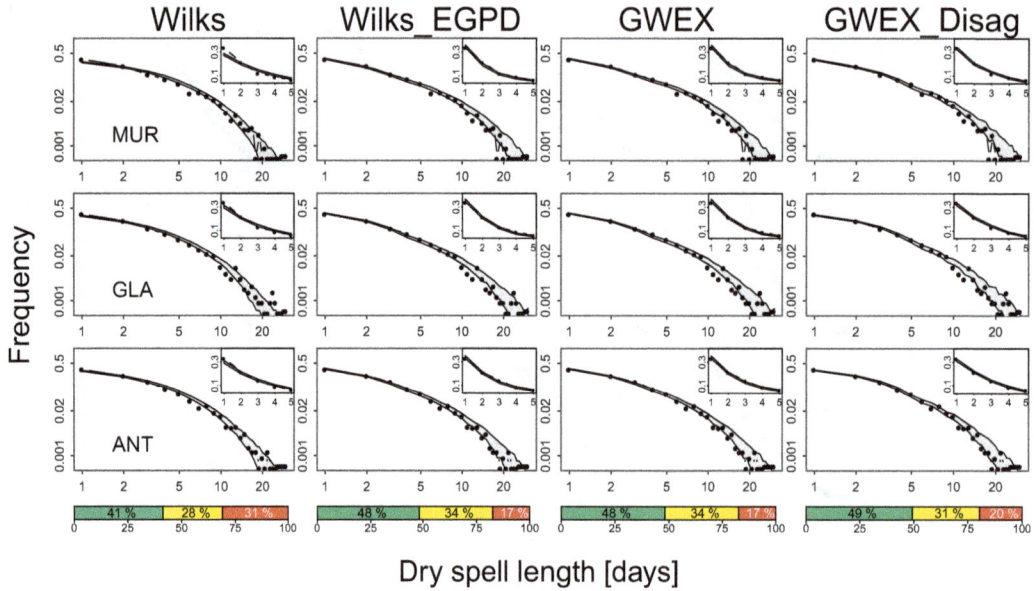

Figure 8. Distribution of dry spell lengths at the stations: the 90 % probability limits are shown. Overall performance is represented by the indicated percentages of good, fair, and poor performance for all sites. Inset plots provide a zoom for durations of 1 to 5 days.

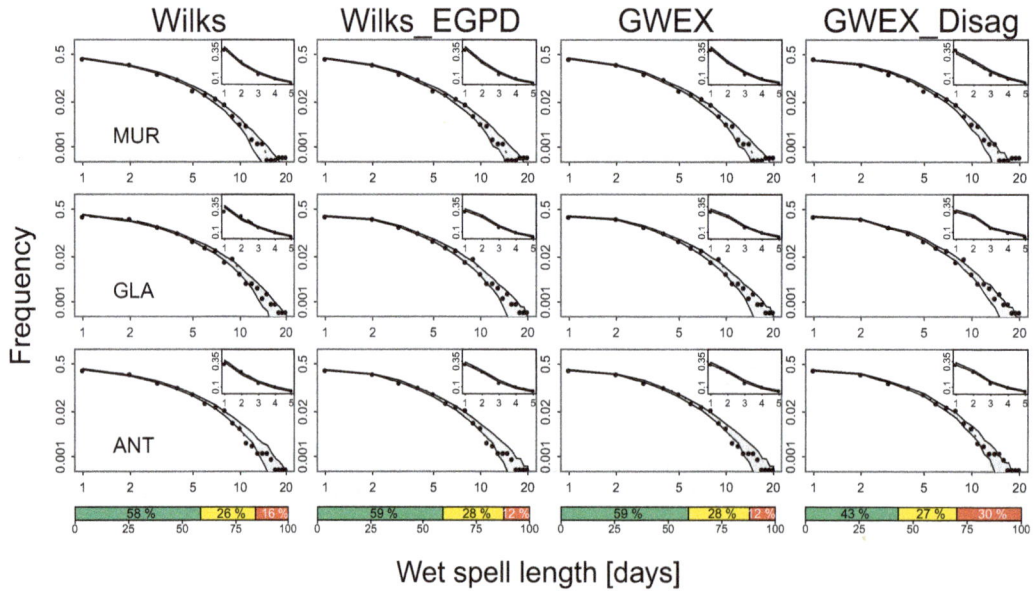

Figure 9. Distribution of wet spell lengths at the stations: the 90 % probability limits are shown. Overall performance is represented by the indicated percentages of good, fair, and poor performance for all sites. Inset plots provide a zoom for durations of 1 to 5 days.

50-year return periods, at daily and 3-day scales, respectively, for all spatial scales. The percentiles corresponding to these return periods are estimated empirically using the Gringorten formula (Gringorten, 1963). These figures provide an overview of model performance regarding extreme precipitation amounts.

At the daily scale (Fig. 12), there is no major difference in performance between the four models. For the 10-year and 50-year return periods, the number of poor perfor-

mance cases is below 20 % for all models. The relative differences are globally centered around zero, which means that the mixture of exponentials (Wilks model) and the E-GPD (Wilks_EGPD, GWEX and GWEX_Disag models) all produce a reasonable performance at this temporal scale. However, if we compare the 50-year return periods simulated by the Wilks and Wilks_EGPD models, we note an increase of 10 % of good performance cases (from 65 to 75 %), which

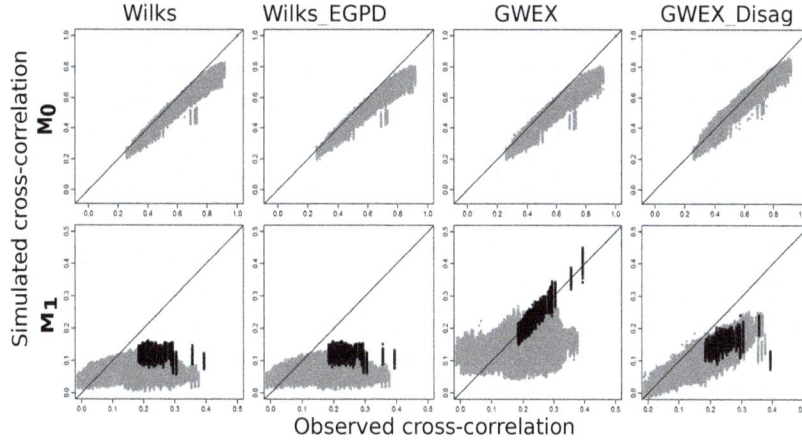

Figure 10. Comparison of unlagged inter-site correlations ($\mathbf{M}_0$) and lag-1 inter-site correlations ($\mathbf{M}_1$) in observed and simulated precipitation series, for the winter (DJF) and summer (JJA) seasons and for the different model versions considered. Black points indicate lag-1 serial autocorrelations at the stations.

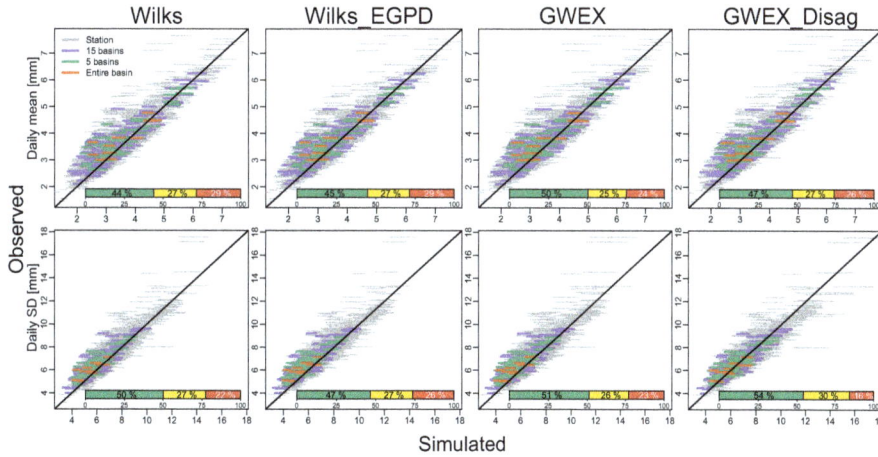

Figure 11. Daily amounts for all spatial scales and months: inter-annual mean (top) and standard deviation (SD, bottom). The 90 % probability limits are shown. Overall performance is represented by the indicated percentages of good, fair, and poor performance for all spatial scales and months.

can be explained by a slight underestimation of the largest maxima with Wilks, for some stations.

Comparing Wilks_EGPD and GWEX, the scores are almost identical, which suggests that the tail dependence introduced by the Student copula in GWEX does not produce a significant improvement for the reproduction of extremes. However, if we focus on the largest spatial scales (at the basins), and in particular on the entire Aare River basin (orange lines), it seems that the slight underestimation of the 50-year return periods obtained with Wilks_EGPD is reduced thanks to this tail dependence. GWEX_Disag also reproduces the largest precipitation amounts at all spatial scales adequately, even if a slight overestimation of the maxima at the largest spatial scales can be suspected. Nevertheless, this performance shows that the disaggregation process leads to an adequate reproduction of the daily maxima.

At the 3-day scale (Fig. 13), the underestimation of the maxima by Wilks and Wilks_EGPD is clear at all spatial scales. GWEX does not suffer from the same shortcomings, which means that the MAR(1) process (Eq. 10) improves the temporal structure of the largest 3-day precipitation amounts. As GWEX_Disag is fitted at a 3-day scale, this model logically leads to an adequate reproduction of extreme 3-day precipitation amounts. The strategy consisting of simulating 3-day precipitation amounts, which are then disaggregated at a daily scale, presents several advantages.

– As the model is fitted at a 3-day scale, 3-day maxima are adequately reproduced.

– As the method of fragments uses observed 3-day temporal structures to disaggregate 3-day amounts, the daily amounts resulting from a generated 3-day maxima are

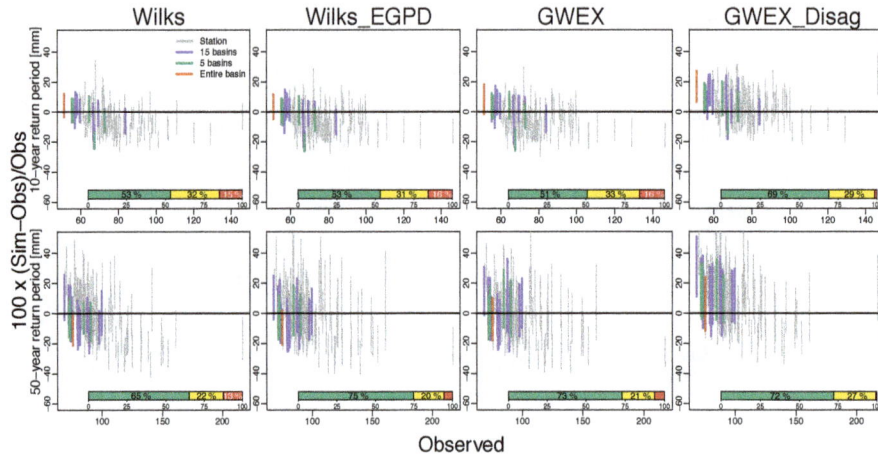

Figure 12. Daily annual maxima for all spatial scales: relative differences, expressed as a percentage, between observed and simulated 10-year (top plots) and 50-year (bottom plots) return periods. The 90 % probability limits are shown. Overall performance is represented by the indicated percentages of good, fair, and poor performance for all spatial scales.

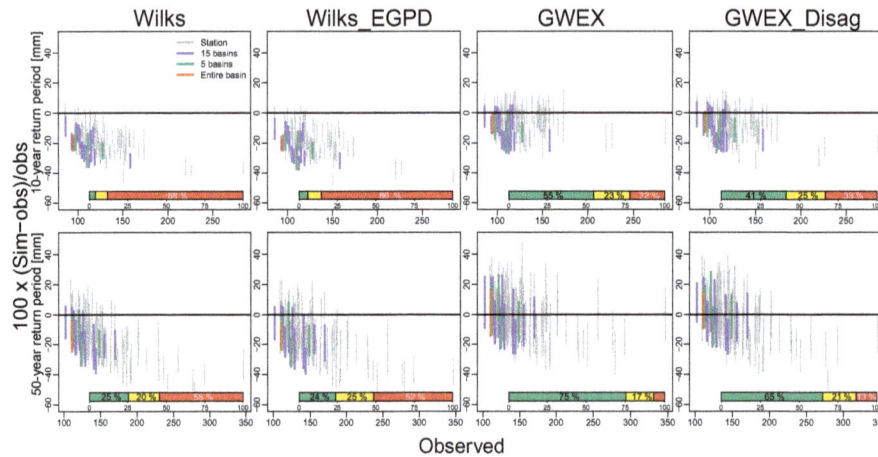

Figure 13. The 3-day annual maxima for all spatial scales: relative differences, expressed as a percentage, between observed and simulated 10-year (top plots) and 50-year (bottom plots) return periods. The 90 % probability limits are shown. Overall performance is represented by the indicated percentages of good, fair, and poor performance for all spatial scales.

physically plausible. In particular, the temporal and spatial structures of large and persistent observed precipitation events are used, which ensures consistency between the generated extreme events at the daily and 3-day scales.

GWEX and GWEX_Disag both adequately reproduce extreme precipitation amounts at daily and 3-day scales, as well as at all spatial scales. As indicated above, these models will be used to generate long precipitation scenarios, which will feed a hydrological model in order to produce flood scenarios. Ultimately, the reproduction of the flood properties using GWEX and GWEX_Disag will indicate which model is the most adequate. Since they correspond to the same model version fitted at daily and 3-day scales, respectively, we can

expect that resulting floods will have slightly different properties.

6 Conclusions and outlook

Precipitation models are usually developed for the purpose of risk assessment in relation to natural hazards (e.g., droughts, floods). Most existing precipitation models aim to reproduce a wide range of statistical properties of precipitation, at different scales, in order to be used as a general tool in different contexts. In this study, our main objective was to provide a precipitation generator that could be used together with a hydrological model for the evaluation of extreme flooding events in a region covering approximately half of Switzerland. As a consequence, we were especially interested in the reproduction of extreme precipitation amounts at medium to large spatial scales. As the daily and 3-day precipitation amounts are a major determinant of flood magnitudes in large Swiss basins (Froidevaux et al., 2015), an adequate reproduction of precipitation at these timescales was also required.

In this paper, we considered different multi-site precipitation models targeting the reproduction of extreme amounts at multiple temporal (daily, 3-day) and spatial scales. Different extended versions of the model introduced by Wilks (Wilks, 1998) have been proposed. A first direct extension, Wilks_EGPD, considers a Markov chain (of order 4 instead of order 1) for the at-site occurrence process. Furthermore, taking advantage of recent advances regarding extreme precipitation, a heavy-tailed distribution (instead of a mixture of exponential distributions), the E-GPD, is applied to the precipitation intensities at each station. Two important extensions of Wilks_EGPD, named GWEX and GWEX_Disag, are then considered. In the GWEX model, temporal and spatial dependencies of the occurrence and intensity process are introduced using the copula theory and a multivariate autoregressive process. A second version, GWEX_Disag, applies the same model, but at a 3-day scale. The 3-day simulated amounts are then disaggregated using an adaptation of the method of fragments (Wójcik and Buishand, 2003).

In this study, we support the use of a systematic evaluation framework. The CASE framework proposed by Bennett et al. (2017) provides a useful tool in this respect, making it possible to compare performance between precipitation models fairly. Regarding the reproduction of extreme precipitation, evaluations until now have usually been qualitative (e.g., interpretations based on one or two examples) and limited in terms of spatial scales (often only at the stations). The evaluation of extreme precipitation amounts proposed in this paper is multi-scale in time (daily and 3-day scale) and space (at the stations, for two different divisions of the study area into sub-basins, and for the entire Aare River basin).

The different multi-site precipitation models have been applied to 105 stations located in Switzerland. A multi-scale evaluation led to the following conclusions.

- A fourth-order Markov chain outperforms a first-order Markov chain for the transitions between dry and wet states, notably for the reproduction of dry spell lengths.

- At the scale of the stations, daily amounts (average, SD, and extremes) are reasonably well reproduced by all the models.

- With only three parameters, the E-GPD provides a parsimonious and flexible representation of the whole of precipitation amounts. Its GPD tail is in agreement with recent results, showing that extreme precipitation amounts must be modeled by heavy-tailed distributions (Papalexiou and Koutsoyiannis, 2013; Serinaldi and Kilsby, 2014a). Furthermore, robust estimates of

the parameter controlling the heaviness of the distribution tail are obtained using a regionalization method. In our study area, the E-GPD does not bring a significant improvement of the performance compared to the mixture of exponential distributions. However, the general framework proposed in this paper can be applied to very distinct precipitation regimes, and the possible heavy tail of the E-GPD might be valuable in other areas.

- At a 3-day scale, precipitation extremes are severely underestimated by Wilks and Wilks_EGPD. This underestimation can be explained by an incorrect representation of the persistence by these models.

- GWEX and GWEX_Disag adequately reproduce extreme precipitation amounts at daily and 3-day scales, and at all spatial scales. These models are deemed adequate for the evaluation of extreme flood events.

Future research will investigate whether the floods simulated by a hydrological model using the generated precipitation scenarios have statistical properties in agreement with observed floods. An extensive investigation is currently underway with a distributed version of the HBV hydrological model, applied to 87 sub-basins of the whole study area and using precipitation scenarios produced by GWEX as inputs. This hydrological evaluation of our weather scenarios will be presented in future publications.

Appendix A: Temporal disaggregation from a 3-day scale to a daily scale

For a 3-day period $\mathbf{D} = \{d, d+1, d+2\}$ starting on a day d, the observed and simulated precipitation amounts at a station k are denoted by $Y_{\mathbf{D}}(k)$ and $\widetilde{Y}_{\mathbf{D}}(k)$, respectively. We want to disaggregate the simulated 3-day amount for the period $\widetilde{\mathbf{D}} = \{\widetilde{d}, \widetilde{d}+1, \widetilde{d}+2\}$. This disaggregation is achieved in the following steps.

1. A set of observed 3-day sequences are retained as candidate periods $\mathbf{D}$ according to two criteria.

   - *Season.* Periods $\widetilde{\mathbf{D}}$ and $\mathbf{D}$ must belong to the same season, as defined in Sect. 3.3.

   - *Mean intensity.* Simulated and observed precipitation fields must have the same order of magnitude. Let $q_{0.5}$, $q_{0.75}$, $q_{0.9}$, and $q_{0.99}$ denote the quantiles of the mean observed precipitation intensities over all the stations associated with probabilities 0.5, 0.75, 0.9, and 0.99, respectively. Observed and simulated 3-day periods are classified into five groups according to their mean intensity $\overline{\mathbf{Y}} = \frac{1}{n} \sum_k Y_{\mathbf{D}}(k)$: dry periods $(\overline{\mathbf{Y}} < q_{0.5})$, moderately wet periods

$(q_{0.5} \leq \overline{Y} < q_{0.75})$, wet periods $(q_{0.75} \leq \overline{Y} < q_{0.9})$, very wet periods $(q_{0.9} \leq \overline{Y} < q_{0.99})$, and extremely wet periods $(q_{0.99} \geq \overline{Y})$.

This first selection of candidate periods aims to increase the chance of retaining periods corresponding to similar meteorological events.

2. For each observed 3-day candidate period **D**, we compute the following score:

$$\text{SCORE}(\widetilde{\mathbf{D}}, \mathbf{D}) = \sum_k \left| \frac{\widetilde{Y}_{\widetilde{d}-1}(k)}{\sum_k \widetilde{Y}_{\widetilde{d}-1}(k)} - \frac{Y_{d-1}(k)}{\sum_k Y_{d-1}(k)} \right|$$
$$+ \left| \frac{\widetilde{Y}_{\mathbf{D}}(k)}{\sum_k \widetilde{Y}_{\mathbf{D}}(k)} - \frac{Y_{\mathbf{D}}(k)}{\sum_k Y_{\mathbf{D}}(k)} \right|.$$

This score measures the similarity between the simulated spatial field for the period $\widetilde{Y}_{\mathbf{D}}(k)$ and the observed spatial field for the period $\widetilde{\mathbf{D}}$ and also takes into account the similarity between the spatial fields for the previous days $\widetilde{d} - 1$ and $d - 1$.

Absolute differences between relative precipitation intensities are computed (the lowest scores are therefore obtained for spatial fields with similar shapes) among the observed periods, corresponding to the same season and order of magnitude selected in the previous step.

3. For each simulated period $\widetilde{\mathbf{D}}$, the observed precipitation fields corresponding to the 10 lowest scores are retained. For each station k, if a positive precipitation amount has been simulated ($\widetilde{Y}_{\widetilde{\mathbf{D}}}(k) > 0$), we look at the corresponding observed amount $Y_{\mathbf{D}}(k)$. If $Y_{\mathbf{D}}(k) = 0$, this observed period cannot be used to disaggregate $\widetilde{Y}_{\widetilde{\mathbf{D}}}(k)$ and we look at the next best observed field among the 10 selected fields. If the observed field contains a positive precipitation amount at this station ($Y_{\mathbf{D}}(k) > 0$), then we obtain the simulated daily amount for day $\widetilde{d}$ as follows:

$$\widetilde{Y}_{\widetilde{d}}(k) = Y_d(k) \times \frac{\widetilde{Y}_{\widetilde{\mathbf{D}}}(k)}{Y_{\mathbf{D}}(k)}, \tag{A1}$$

with similar expressions for days $\widetilde{d} + 1$ and $\widetilde{d} + 2$. Simulated daily amounts correspond to the observed daily amounts, rescaled by the ratio between the simulated and observed 3-day amounts. The 3-day simulated amounts and observed temporal structures are thus preserved.

4. While the 3-day spatiotemporal consistency is generally conserved by applying the preceding steps, it can happen that the simulated 3-day amount is positive even though there is no positive precipitation among the 10 best 3-day observed fields. In this case, we seek similar observed amounts at this station only and randomly choose one 3-day period among the 10 best 3-day periods.

Competing interests. The authors declare that they have no conflict of interest.

Acknowledgements. We gratefully acknowledge financial support for this study provided by the Swiss Federal Office for Environment (FOEN), the Swiss Federal Nuclear Safety Inspectorate (ENSI), the Federal Office for Civil Protection (FOCP), and the Federal Office of Meteorology and Climatology, MeteoSwiss, through the project EXAR ("Evaluation of extreme Flooding Events within the Aare-Rhine hydrological system in Switzerland"). The authors would like to thank MeteoSwiss (the Swiss Federal Office of Meteorology and Climatology) for providing the meteorological data. We also thank the editor and two anonymous reviewers for their constructive comments, which helped us to improve the manuscript.

Edited by: Carlo De Michele

References

Ailliot, P., Allard, D., Monbet, V., and Naveau, P.: Stochastic weather generators: an overview of weather type models, Journal de la Société Franoaise de Statistique, 156, 101–113, 2015.

Allard, D. and Bourotte, M.: Disaggregating daily precipitations into hourly values with a transformed censored latent Gaussian process, Stoch. Environ. Res. Risk Assess., 29, 453–462, 2015.

Baxevani, A. and Lennartsson, J.: A spatiotemporal precipitation generator based on a censored latent Gaussian field, Water Resour. Res., 51, 4338–4358, 2015.

Bennett, B., Thyer, M., Leonard, M., Lambert, M., and Bates, B.: A comprehensive and systematic evaluation framework for a parsimonious daily rainfall field model, J. Hydrol., 556, 1123–1138, https://doi.org/10.1016/j.jhydrol.2016.12.043, 2018.

Buishand, T. A.: Extreme rainfall estimation by combining data from several sites, Hydrol. Sci. J., 36, 345–365, 1991.

Burn, D. H.: Evaluation of regional flood frequency analysis with a region of influence approach, Water Resour. Res., 26, 2257–2265, 1990.

Burton, A., Kilsby, C. G., Fowler, H. J., Cowpertwait, P. S. P., and O'Connell, P. E.: RainSim: A spatial-temporal stochastic rainfall modelling system, Environ. Modell. Softw., 23, 1356–1369, 2008.

Burton, A., Fowler, H. J., Kilsby, C. G., and O'Connell, P. E.: A stochastic model for the spatial-temporal simulation of nonhomogeneous rainfall occurrence and amounts, Water Resour. Res., 46, W11501, doi:10.1029/2009WR008884, 2010.

Bárdossy, A. and Pegram, G. G. S.: Copula based multisite model for daily precipitation simulation, Hydrol. Earth Syst. Sci., 13, 2299–2314, https://doi.org/10.5194/hess-13-2299-2009, 2009.

Cavanaugh, N. R., Gershunov, A., Panorska, A. K., and Kozubowski, T. J.: The probability distribution of intense daily precipitation, Geophys. Res. Lett., 42, 1560–1567, doi:10.1002/2015GL063238, 2015.

Chandler, R. E. and Wheater, H. S.: Analysis of rainfall variability using generalized linear models: A case study from the west of Ireland, Water Resour. Res., 38, 1192, doi:10.1029/2001WR000906, 2002.

Chen, J., Brissette, F. P., and Zhang, J. X.: A Multi-Site Stochastic Weather Generator for Daily Precipitation and Temperature, Trans. ASABE, 57, 1375–1391, 2014.

Dempster, A. P., Laird, N. M., and Rubin, D. B.: Maximum Likelihood from Incomplete Data via the EM Algorithm, J. Roy. Stat. Soc.-B, 39, 1–38, 1977.

Evin, G., Blanchet, J., Paquet, E., Garavaglia, F., and Penot, D.: A regional model for extreme rainfall based on weather patterns subsampling, J. Hydrol., 541, 1185–1198, 2016.

Froidevaux, P.: Meteorological characterisation of floods in Switzerland, Ph.D. thesis, Geographisches Institut, University of Bern, 2014.

Froidevaux, P., Schwanbeck, J., Weingartner, R., Chevalier, C., and Martius, O.: Flood triggering in Switzerland: the role of daily to monthly preceding precipitation, Hydrol. Earth Syst. Sci., 19, 3903–3924, https://doi.org/10.5194/hess-19-3903-2015, 2015.

Gringorten, I. I.: A plotting rule for extreme probability paper, J. Geophys. Res. , 68, 813–814, 1963.

Hundecha, Y., Pahlow, M., and Schumann, A.: Modeling of daily precipitation at multiple locations using a mixture of distributions to characterize the extremes, Water Resour. Res., 45, W12412, doi:10.1029/2008WR007453, 2009.

Keller, D. E., Fischer, A. M., Frei, C., Liniger, M. A., Appenzeller, C., and Knutti, R.: Implementation and validation of a Wilks-type multi-site daily precipitation generator over a typical Alpine river catchment, Hydrol. Earth Syst. Sci., 19, 2163–2177, https://doi.org/10.5194/hess-19-2163-2015, 2015.

Kleiber, W., Katz, R. W., and Rajagopalan, B.: Daily spatiotemporal precipitation simulation using latent and transformed Gaussian processes, Water Resour. Res., 48, W01523, doi:10.1029/2011WR011105, 2012.

Klemeš, V.: Tall Tales about Tails of Hydrological Distributions. I., J. Hydrol. Eng., 5, 227–231, 2000a.

Klemeš, V.: Tall Tales about Tails of Hydrological Distributions. II., J. Hydrol. Eng., 5, 232–239, 2000b.

Lafaysse, M., Hingray, B., Mezghani, A., Gailhard, J., and Terray, L.: Internal variability and model uncertainty components in future hydrometeorological projections: The Alpine Durance basin, Water Resour. Res., 50, 3317–3341, 2014.

Lamb, R., Faulkner, D., Wass, P., and Cameron, D.: Have applications of continuous rainfall-runoff simulation realized the vision for process-based flood frequency analysis?, Hydrol. Proc., 30, 2463–2481, 2016.

Leblois, E. and Creutin, J.-D.: Space-time simulation of intermittent rainfall with prescribed advection field: Adaptation of the turning band method, Water Resour. Res., 49, 3375–3387, 2013.

Leonard, M., Lambert, M. F., Metcalfe, A. V., and Cowpertwait, P. S. P.: A space-time Neyman-Scott rainfall model with defined storm extent, Water Resour. Res., 44, W09402, doi:10.1029/2007WR006110, 2008.

Maraun, D., Wetterhall, F., Ireson, A. M., Chandler, R. E., Kendon, E. J., Widmann, M., Brienen, S., Rust, H. W., Sauter, T., Themeßl, M., Venema, V. K. C., Chun, K. P., Goodess, C. M., Jones, R. G., Onof, C., Vrac, M., and Thiele-Eich, I.: Precipitation downscaling under climate change: Recent developments to bridge the gap between dynamical models and the end user, Rev. Geophys., 48, RG3003, doi:10.1029/2009RG000314, 2010.

McNeil, A. J., Frey, R., and Embrechts, P.: Quantitative Risk Management – Concepts, Techniques, and Tools, Princeton University Press, Princeton, N.J, 2005.

McRobie, F. H., Wang, L.-P., Onof, C., and Kenney, S.: A spatial-temporal rainfall generator for urban drainage design, Water Science and Technology: A J. Int. Assoc. Water Pollut. Res., 68, 240–249, 2013.

Mehrotra, R. and Sharma, A.: Preserving low-frequency variability in generated daily rainfall sequences, J. Hydrol., 345, 102–120, 2007a.

Mehrotra, R. and Sharma, A.: A semi-parametric model for stochastic generation of multi-site daily rainfall exhibiting low-frequency variability, J. Hydrol., 335, 180–193, 2007b.

Mehrotra, R. and Sharma, A.: Development and Application of a Multisite Rainfall Stochastic Downscaling Framework for Climate Change Impact Assessment, Water Resour. Res., 46, W07526, doi:10.1029/2009WR008423, 2010.

Mehrotra, R., Srikanthan, R., and Sharma, A.: A comparison of three stochastic multi-site precipitation occurrence generators, J. Hydrol., 331, 280–292, 2006.

Mezghani, A. and Hingray, B.: A combined downscaling-disaggregation weather generator for stochastic generation of multisite hourly weather variables over complex terrain: Development and multi-scale validation for the Upper Rhone River basin, J. Hydrol., 377, 245–260, 2009.

Naveau, P., Huser, R., Ribereau, P., and Hannart, A.: Modeling jointly low, moderate, and heavy rainfall intensities without a threshold selection, Water Resour. Res., 52, 2753–2769, 2016.

Papalexiou, S. M. and Koutsoyiannis, D.: Battle of extreme value distributions: A global survey on extreme daily rainfall, Water Resour. Res., 49, 187–201, 2013.

Papalexiou, S. M., Koutsoyiannis, D., and Makropoulos, C.: How extreme is extreme? An assessment of daily rainfall distribution tails, Hydrol. Earth Syst. Sci., 17, 851–862, https://doi.org/10.5194/hess-17-851-2013, 2013.

Papastathopoulos, I. and Tawn, J. A.: Extended generalised Pareto models for tail estimation, J. Stat. Plan. Infer., 143, 131–143, 2013.

Rasmussen, P. F.: Multisite precipitation generation using a latent autoregressive model, Water Resour. Res., 49, 1845–1857, 2013.

Rayner, D., Achberger, C., and Chen, D.: A multi-state weather generator for daily precipitation for the Torne River basin, northern Sweden/western Finland, Adv. Clim. Change Res., 7, 70–81, 2016.

Rebonato, R. and Jaeckel, P.: The Most General Methodology to Create a Valid Correlation Matrix for Risk Management and Option Pricing Purposes, SSRN Scholarly Paper ID 1969689, Social Science Research Network, Rochester, NY, 2011.

Rousseeuw, P. J. and Molenberghs, G.: Transformation of non positive semidefinite correlation matrices, Communications in Statistics – Theory and Methods, 22, 965–984, 1993.

Rudolph, J. V. and Friedrich, K.: Seasonality of Vertical Structure in Radar-Observed Precipitation over Southern Switzerland, J. Hydrometeorol., 14, 318–330, 2012.

Serinaldi, F. and Kilsby, C. G.: Rainfall extremes: Toward reconciliation after the battle of distributions, Water Resour. Res., 50, 336–352, 2014a.

Serinaldi, F. and Kilsby, C. G.: Simulating daily rainfall fields over large areas for collective risk estimation, J. Hydrol., 512, 285–302, 2014b.

Sharif, M. and Burn, D. H.: Improved K -Nearest Neighbor Weather Generating Model, J. Hydrol. Eng., 12, 2007.

Srikanthan, R. and Pegram, G. G. S.: A nested multisite daily rainfall stochastic generation model, J. Hydrol., 371, 142–153, 2009.

Thompson, C. S., Thomson, P. J., and Zheng, X.: Fitting a multisite daily rainfall model to New Zealand data, J. Hydrol., 340, 25–39, 2007.

Vischel, T., Lebel, T., Massuel, S., and Cappelaere, B.: Conditional simulation schemes of rain fields and their application to rainfall-runoff modeling studies in the Sahel, J. Hydrol., 375, 273–286, 2009.

Vrac, M. and Naveau, P.: Stochastic downscaling of precipitation: From dry events to heavy rainfalls, Water Resour. Res., 43, W07402, doi:10.1029/2006WR005308, 2007.

Wilks, D. S.: Multisite generalization of a daily stochastic precipitation generation model, J. Hydrol., 210, 178–191, 1998.

Wójcik, R. and Buishand, T.: Simulation of 6-hourly rainfall and temperature by two resampling schemes, J. Hydrol., 69–80, 2003.

Yates, D., Gangopadhyay, S., Rajagopalan, B., and Strzepek, K.: A technique for generating regional climate scenarios using a nearest-neighbor algorithm, Water Resour. Res., 39, 1199, doi:10.1029/2002WR001769, 2003.

6

Remapping annual precipitation in mountainous areas based on vegetation patterns: a case study in the Nu River basin

Xing Zhou, Guang-Heng Ni, Chen Shen, and Ting Sun

State Key Laboratory of Hydro-Science and Engineering, Department of Hydraulic Engineering, Tsinghua University, Beijing 100084, China

Correspondence to: Ting Sun (sunting@tsinghua.edu.cn)

Abstract. Accurate high-resolution estimates of precipitation are vital to improving the understanding of basin-scale hydrology in mountainous areas. The traditional interpolation methods or satellite-based remote sensing products are known to have limitations in capturing the spatial variability of precipitation in mountainous areas. In this study, we develop a fusion framework to improve the annual precipitation estimation in mountainous areas by jointly utilizing the satellite-based precipitation, gauge measured precipitation, and vegetation index. The development consists of vegetation data merging, vegetation response establishment, and precipitation remapping. The framework is then applied to the mountainous areas of the Nu River basin for precipitation estimation. The results demonstrate the reliability of the framework in reproducing the high-resolution precipitation regime and capturing its high spatial variability in the Nu River basin. In addition, the framework can significantly reduce the errors in precipitation estimates as compared with the inverse distance weighted (IDW) method and the TRMM (Tropical Rainfall Measuring Mission) precipitation product.

1 Introduction

Precipitation plays an important role in hydrological processes, land–atmospheric processes, and ecological dynamics. Accurate high-resolution precipitation is crucial for streamflow prediction, flood control, and water resources management in data-sparse regions such as mountainous areas (Song et al., 2016). However, it is a great challenge to obtain accurate precipitation in mountainous areas due to the sparse gauge network and the remarkable spatiotemporal variability of precipitation. Conventional gauge networks can provide accurate rainfall measurements at point scales, which can be interpolated within the region of interest to give estimates of precipitation in ungauged areas. However, such interpolated estimates might not be reliable in mountainous areas considering the very limited gauges there (Phillips et al., 1992; Mair and Fares, 2011; Jacquin and Soto-Sandoval, 2013; Wang et al., 2014; Borges et al., 2016).

Recently, remote-sensing-based precipitation (RSBP) products, such as the Global Precipitation Climatology Project (GPCP) (Schamm et al., 2014), the Tropical Rainfall Measuring Mission (TRMM) (Council, 2005), and the Climate Prediction Center Morphing Method (CMORPH) (Joyce et al., 2004), have been extensively used in ungauged or sparsely gauged areas to bridge the gap between the need for precipitation estimates and the scarcity in gauge observations (Akbari et al., 2012; Kneis et al., 2014; Li et al., 2015; Worqlul et al., 2015; Mourre et al., 2016; Wong et al., 2016). Also, data fusion across satellite and gauge observations is being conducted to further the application of RSBPs (Rozante et al., 2010; Woldemeskel et al., 2013; Arias-Hidalgo et al., 2013; Chen et al., 2016; Zhou et al., 2016). However, due to the relatively coarse spatial resolution (e.g., 0.25–5°) and uncertainties of RSBPs, their applications in mountainous basins, where the precipitation shows large spatial variability, are still very limited (Krakauer et al., 2013; Chen and Li, 2016).

Precipitation estimates can be influenced by a variety of ambient factors (e.g., topography, vegetation). In order to correct effects of topography on precipitation estimates, a digital elevation model (DEM) has been widely used in spatial interpolation of precipitation over mountainous areas (Marquínez et al., 2003; Lloyd, 2005). However, the relationship between elevation and precipitation is not clear. Meanwhile, strong correlations between the normalized difference vegetation index (NDVI) and precipitation have been found by several studies (Li et al., 2002; Kariyeva and Van Leeuwen, 2011; Li and Guo, 2012; Sun et al., 2013; Campo-Bescós et al., 2013). As such, establishing statistical models between the NDVI and precipitation so as to improve the spatial resolution of TRMM products in mountainous areas is becoming popular (Immerzeel et al., 2009; Jia et al., 2011; Duan and Bastiaanssen, 2013; Chen et al., 2014; Xu et al., 2015; Mahmud et al., 2015; Jing et al., 2016). For instance, Immerzeel et al. (2009) downscaled TRMM-3B43 to 1 km based on an exponential relationship between NDVI and TRMM precipitation on the Iberian Peninsula of Europe. Jia et al. (2011) established four multivariable linear regression models between TRMM-3B43 precipitation and two other factors (i.e., DEM and NDVI) of different resolutions (0.25, 0.5, 0.75, and 0.1°) to get 1 km estimates of precipitation in the Qaidam basin of China. Duan and Bastiaanssen (2013) used a nonlinear relationship between TRMM-3B43 and NDVI to downscale precipitation to 1 km in a humid area and a semi-arid area. Chen et al. (2014) established a spatially varying relationship between TRMM, NDVI, and DEM by using a local regression analysis approach known as geographically weighted regression (GWR) in South Korea. Xu et al. (2015) also used the GWR method to explore the spatial heterogeneity of the RSBP–NDVI and RSBP–DEM relationships over two mountainous areas in western China.

However, the present RSBP–NDVI-based schemes have several limitations: (1) significant errors can be introduced during the downscaling given the nonlinear relationship between RSBP and NDVI; (2) large uncertainties exist in the RSBP for mountainous areas; and (3) inter-comparison of existing NDVI datasets is missing in deriving the RSBP–NDVI relationships. In this study, we develop a fusion framework to obtain more accurate high-resolution estimates of precipitation in mountainous areas based on the relationship between precipitation and vegetation response. More specifically, in addition to RSBP, gauge measurements and different vegetation datasets will be used in this study to overcome the aforementioned limitations in current RSBP–NDVI-based schemes. The paper is organized as follows: Sect. 2 describes the development of the fusion framework; Sect. 3 documents the study area and related datasets; Sect. 4 presents the results of the fusion framework and discusses impacts of different determinants on the performance of the fusion framework; and Sect. 5 summarizes this work.

2 Framework development

The satellite–gauge–vegetation fusion framework (Fig. 1) involves three stages of development: (1) vegetation data merg-

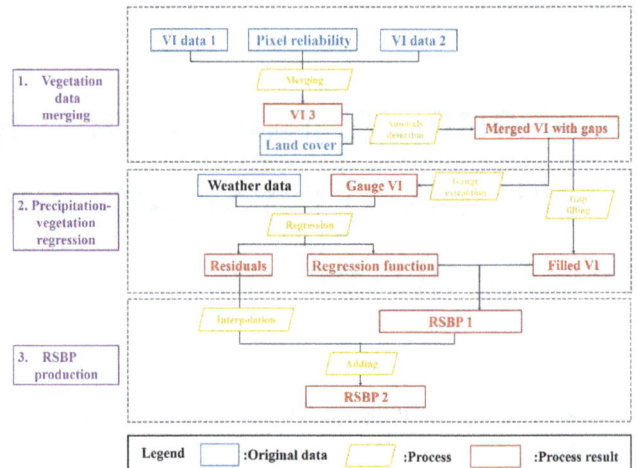

Figure 1. Flow chart of the satellite–gauge–vegetation fusion framework development.

ing, (2) precipitation–vegetation regression, and (3) RSBP product remapping, whose details are described in the following subsections.

2.1 Vegetation data merging

Vegetation closely interacts with soil moisture and is recognized as a good proxy of precipitation. The remote sensing technique provides us with various high-resolution vegetation products such as NDVI, EVI (enhanced vegetation index), and LAI (leaf area index). Among the vegetation indices, NDVI, an indicator of plant density and growth, is chosen as the proxy of precipitation in this study due to its wide availability. Considering the crucial role of NDVI in deriving precipitation estimates under our framework, we conduct an inter-comparison in data accuracy between two NDVI datasets (termed datasets A and B hereinafter) to reduce the error. First, the systematic errors of both datasets are eliminated by multiplying the reduction factor or using the simple regression model. After the correction, the final dataset is then obtained by selecting a better element between A and B if the quality criteria are satisfied, otherwise filling an anomaly value.

It should be noted that since the vegetation growth is suppressed or promoted on some land covers (e.g., rivers, lakes, snow and ice, and urban areas), the vegetation data of these land covers are excluded by filling anomaly values. Besides, due to the strong influence of farming activities (e.g., irrigation, fertilization, and harvest) on the crop growth, vegetation data of farmland are excluded as well. We note that although Moran's index (Li et al., 2007) is widely employed to detect anomalies in vegetation data (Jia et al., 2011; Duan and Bastiaanssen, 2013), it is not used in this study for its inapplicability in large areas with continuous anomaly pixels (e.g., farmland). As such, we identify anomaly pixels simply by land-use type: pixels categorized as water, wetland,

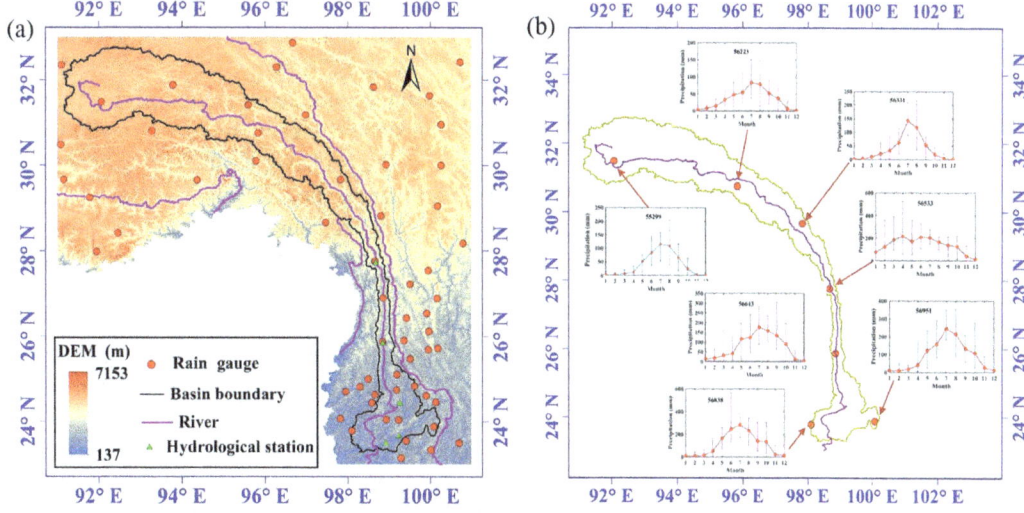

Figure 2. (a) Terrain map of the study area (the Nu-Salween basin and its adjacent areas). **(b)** The distribution of rainfall during the year across the Nu River.

urban, cropland, snow/ice, and barren will be identified as anomalies. The detected anomaly pixels are excluded from the original NDVI dataset and then filled with interpolated values using the IDW method so as to generate an optimized NDVI dataset.

Based on the optimized NDVI dataset, the NDVI data at the gauge locations are retrieved with the neighbor-average method (i.e., the value of a certain grid is determined as the average of all its eight neighboring grids) and will be used for the precipitation–vegetation regression.

2.2 Precipitation–vegetation regression

As far as we know, there is no widely accepted form of the precipitation–vegetation relationship. Therefore, the final regression form will be determined from several candidate relationships, including polynomial, exponential, logarithmic, and linear forms, according to the five metrics: correlation coefficient (R), coefficient of determination (R^2), root-mean-square error (E_{RMS}), mean relative error (E_{MR}), and mean absolute relative error (E_{MAR}), which are given as follows:

$$R = \frac{\sum_{i=1}^{n}(P_i - \overline{P})(O_i - \overline{O})}{\sqrt{\sum_{i=1}^{n}(P_i - \overline{P})^2}\sqrt{\sum_{i=1}^{n}(O_i - \overline{O})^2}}, \tag{1}$$

$$R^2 = \frac{\sum_{i=1}^{n}(P_i - O_i)^2}{\sqrt{\sum_{i=1}^{n}(O_i - \overline{O})^2}}, \tag{2}$$

$$E_{RMS} = \sqrt{\frac{\sum_{i=1}^{n}(P_i - O_i)^2}{n}}, \tag{3}$$

$$E_{MR} = \frac{1}{n}\sum_{i=1}^{n}(P_i - O_i), \tag{4}$$

$$E_{MAR} = \frac{1}{n}\sum_{i=1}^{n}\frac{|P_i - O_i|}{O_i}, \tag{5}$$

where $\overline{O}$ is the mean annual precipitation of all gauges, O_i the mean annual precipitation of gauge i, P_i the estimated precipitation at gauge i, and n the total number of gauges.

Also, considering the annual variability of precipitation, the regression model is further determined for two temporal scales: (1) the entire period covering all the study years and (2) the individual year of the entire study period. The regression models for the entire study period and for individual years are thus termed RME and RMI, respectively. RME can utilize the full knowledge of precipitation characteristics of the entire study period, whereas RMI implies the interannual variability. Besides, RME can reasonably reconstruct the precipitation series of the years when data gaps exist.

The calibration–validation procedure for each candidate model is conducted under three scenarios with different numbers of gauges and/or years:

Scenario a Fully random: a random number of gauges and a random number of years are independently used for calibration and validation;

Scenario b All gauges, partial period: all the gauges will be involved in both procedures, but only 2/3 of years will be randomly chosen for calibration, and the other years for validation;

Scenario c Partial gauges, entire period: all years will be used, but only 1/3 of gauges will be randomly chosen for calibration, and other gauges for validation.

For each scenario, the calibration–validation procedure will be performed for 100 samples determined based on the above criteria and the five evaluation metrics (i.e., R, R^2,

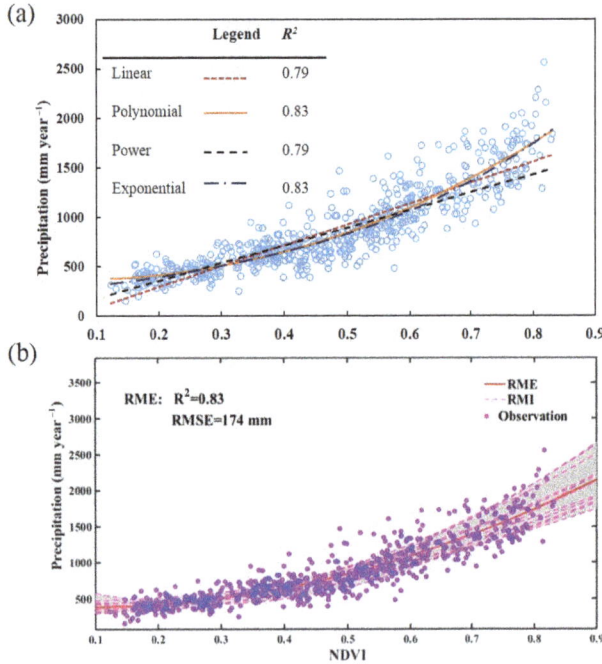

Figure 3. (a) Different regression form between annual precipitation and NDVI; **(b)** the NDVI–precipitation relationships for RME and RMI.

E_{RMS}, E_{MA}, and E_{MAR}) will be calculated for each sample accordingly. The best model is then determined based on the metrics.

2.3   RSBP product remapping

With the optimized vegetation dataset and the precipitation–vegetation regression model, the RSBP product is then remapped over the study region. Thanks to the finer resolution of the NDVI dataset than the RSBP product and the accurate estimate of precipitation by gauges, the remapped RSBP product is expected to provide more detailed spatial characteristics of precipitation over mountainous areas.

3   Study area and datasets for framework application

3.1   Study area

The Nu-Salween basin (Fig. 2a), where 6 million people live, is one of the largest river basins in South Asia and spreads across three countries with an area of $324\,000\,\mathrm{km}^2$. This study focuses on the Chinese part of the Nu-Salween basin (termed the Nu River basin hereafter), where the elevation ranges from 446 to 6134 m and the narrowest part is only 24 km. The annual precipitation of the Nu River basin ranges from 400 to 2000 mm with an average of 900 mm, and the mean annual runoff is $69\,\mathrm{km}^3$. The precipitation of the Nu River basin generally decreases from southwest to northeast and demonstrates high variability due to mountain

weather systems (e.g., the difference in annual precipitation between the mountaintop and valley of Gongshan is larger than 1000 mm). Annual rainfall varies significantly across this region. Figure 2b shows the annual rainfall distributions of seven stations located in the upstream, middle, and downstream of the Nu River basin. The upstream and downstream have similar rainfall distributions, with larger rainfall occurring in summer compared to winter, while the middle part observes relatively large rainfall in winter and spring. Thanks to the adequate rainfall and minimal human perturbation, the Nu River basin has an extensive vegetation coverage, with the dominant types grassland in the Qinghai–Tibetan Plateau (upper basin) and mixed forest in Yunnan Province (lower basin). However, the dense vegetation cover increases the difficulty in conducting precipitation observations and only 13 gauges are very unevenly distributed over the whole basin of $142\,479\,\mathrm{km}^2$, which makes it highly challenging to obtain the accurate spatial precipitation characteristics with traditional interpolation approaches. Although the RSBP products are available for this area, they are too coarse (usually with a spatial resolution of $\sim 50\,\mathrm{km}$) to capture the high spatial variability of precipitation.

Considering the limited number of gauges (i.e., 13) in the Nu River basin, an enlarged area covering 23–33° N and 91–101° E is chosen for the application of the fusion framework, where 59 gauges are available and the climatic and topographic conditions are similar: both regions are characterized as mountainous areas under the subtropical climate influenced by the southeast and southwest monsoons. Besides, given no rain gauges are available outside of China in this study region, the non-Chinese region is excluded from the study area.

3.2   Datasets

3.2.1   Vegetation data

In this study, we use two MODIS (MODerate resolution Imaging Spectoradiometer) vegetation products, MOD13A3 (termed MOD hereafter) and MYD13A3 (termed MYD hereafter), in the application of the fusion framework. Both the MOD and MYD datasets contain 10 sub-datasets consisting of NDVI, EVI, and pixel reliability. The temporal and spatial resolutions of the MOD13A3 and MYD13A3 products are 1 month and 1 km, respectively. The pixel reliability is an accuracy metric of the data quality pixel and has four valid values: 0 for good accuracy, 1 for marginal accuracy, 2 for snow/ice, and 3 for cloud. Based on the pixel reliability information, the NDVI values are either selected for corresponding pixel reliability levels of 0 and 1, or discarded as anomalies otherwise.

The MOD dataset is used as a benchmark while MYD is taken as the alternative for occasions when MOD data are missing or have large uncertainties. Since both the MOD and MYD datasets are extracted from different satellites at different transit times, systematic errors may exist in the difference between the two datasets. As such, we construct two regressions to remove their systematic errors: one is based on a subset with both MOD and MYD of good reliability ($= 0$), and the other on a subset with MOD of marginal reliability ($= 1$) and MOD of good reliability ($= 0$). After the removal of systematic errors, a merged dataset of MOD and MYD (termed MMD hereafter) is generated under the criteria given as follows:

$$
\mathrm{MMD} = \begin{cases} \mathrm{MOD} & (\mathrm{MOD} == 0), \\ \mathrm{MYD} & (\mathrm{MOD} > 1 \,\&\, \mathrm{MYD} == 0), \\ \mathrm{MOD} & (\mathrm{MOD} == 1 \,\&\, \mathrm{MYD} == 1), \\ \mathrm{NULL} & (\mathrm{MOD} > 1 \,\&\, \mathrm{MYD} > 0). \end{cases} \quad (6)
$$

The annual MMD dataset is then calculated by averaging the 12 monthly images.

3.2.2 Land-use data

The MCD12Q1 Version 51 (MODIS/Terra+Aqua Land Cover Type Yearly L3 Global 500 m SIN Grid V051) land-use dataset in the period of 2001–2013 is used to identify the outliers of MMD, while the IGBP (International Geosphere Biosphere Programme) classification is adopted for its wide applications. Due to mismatch in spatial resolutions between the MMD and MCD12Q1 datasets, the MCD12Q1 dataset is upscaled to 1 km as MMD for outlier identification. It should be noted that for any of the four 500 m pixels in MCD12Q1 classified as water, urban, snow or ice and cropland, the upscaled 1 km pixel will be assigned with a missing value (i.e., -9999) and the corresponding NDVI pixel will be identified as an outlier.

3.2.3 Weather data

Datasets consisting of daily precipitation and air temperature collected at the 59 gauges in the study area are obtained via the China Meteorological Data Sharing Service system (http://data.cma.cn/data/detail/dataCode/SURF_CLI_CHN_MUL_DAY_V3.0/keywords/v3.0.html). The air temperature measurements will be used for dependence analysis later in Sect. 4.5. The streamflow data provided by Yunnan University will be used for calculating sub-basin-scale precipitation based on water balance. The five hydrological stations are Gongshan, Liuku, Jiucheng, Gulaohe, and Dawanjiang, with drainage areas of 101146, 106681, 6308, 4185, and 7986 km^2, respectively. MODIS evapotranspiration (ET) product MOD16 (http://www.ntsg.umt.edu/project/mod16) with the spatiotemporal resolution of 1 km / 1 weekly will also be used in calculating precipitation based on water balance.

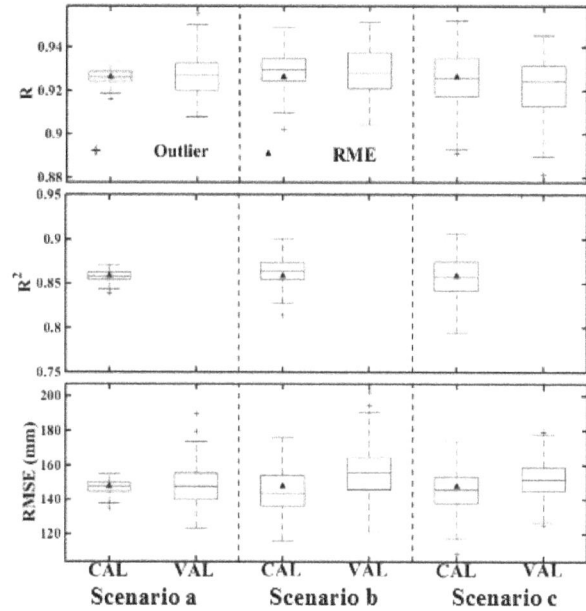

Figure 4. Box plots of R, R^2, and E_{RMS} of the RME model under three scenarios: **(a)** fully random; **(b)** all gauges, partial period; and **(c)** partial gauges, entire period. Details of the three scenarios refer to Sect. 2.2. The triangle marker corresponds to the value (R, R^2, RMSE) of the RME model. Plus signs represent the outlier of the sample used to draw the box diagram whose value is out of the range from (Q1 $-$ 1.5IQR) to (Q3 $+$ 1.5IQR). Q1 and Q3 represent the lower and upper quartiles, IQR $=$ Q3–Q1.

4 Results and discussion

4.1 Model calibration and validation

Based on the results of six evaluation metrics for different regression form candidates (Fig. 3a), the second-order polynomial is chosen as the regression model form in this study:

$$
p = a\,\mathrm{NDVI}^2 + b\,\mathrm{NDVI} + c, \quad (7)
$$

where p denotes the precipitation amount in millimeters, and a, b, and c are regression coefficients. The results of regression coefficients and evaluation metrics are given in Table 1, and the NDVI–precipitation relationships for the study period are demonstrated in Fig. 3b.

The best performance of the regression model is found within $0.2 < \mathrm{NDVI} < 0.7$ and $400\,\mathrm{mm\,yr}^{-1} < p < 1500\,\mathrm{mm\,yr}^{-1}$. Larger errors are found at pixels with NDVI larger than 0.7 or annual rainfall higher than 1500 mm, implying the water supply is no longer a determinant of vegetation growth as annual rainfall exceeds a certain threshold.

In general, the RMIs demonstrate better performance than RME, which can be attributable to the lower variability of precipitation in a single year than the whole study period. It is also noted that the R^2 values of RMIs for drier years (2003, 2009, and 2011) are less than wetter years, indicating

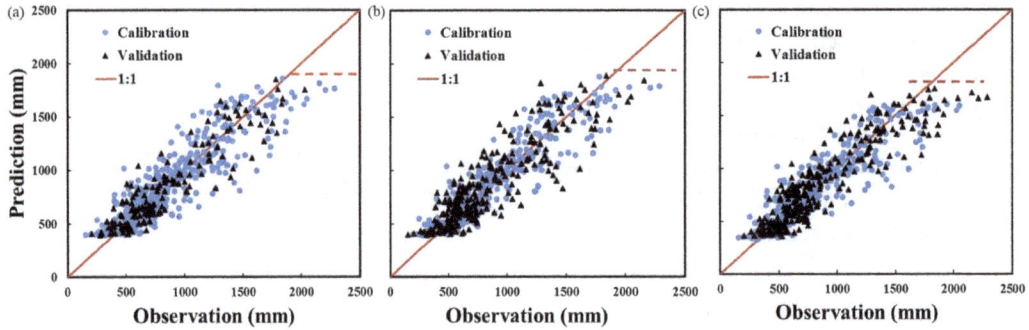

Figure 5. Comparison in annual precipitation between the gauged measurements and predictions by the regression model for scenarios **(a)** fully random; **(b)** all gauges, partial period; and **(c)** partial gauges, entire period. Details of the three scenarios refer to Sect. 2.2.

Table 1. Regression model performance and regression coefficients.

Year	Mean (mm)	R^2	E_{RMS} (mm)	E_{MAR} (%)	a	b	c
2001	961	0.91	138	10.6	3038.1	−345.3	359.8
2002	887	0.90	119	10.2	1354.7	687.5	212.0
2003	828	0.75	155	14.0	1700.2	−115.5	472.7
2004	1018	0.89	171	12.4	3784.3	−1047.7	517.4
2005	810	0.93	97	9.5	2465.4	−265.0	363.2
2006	737	0.88	122	11.4	2065.2	−112.2	287.5
2007	928	0.84	184	14.6	2306.9	53.5	286.4
2008	960	0.91	121	9.4	2504.0	−258.1	433.5
2009	726	0.89	119	13.2	2091.3	−168.0	294.5
2010	937	0.94	124	9.1	4094.8	−1293.3	512.6
2011	824	0.84	167	14.2	4697.8	−2613.7	792.7
2012	791	0.89	114	10.6	1966.4	3.5	308.1
RME	848	0.83	174	15.2	2670.4	−471.2	409.2

Table 2. Statistics of regression models for validation and calibration under three scenarios.

Scenario	Statistics	Calibration				Validation		
		R	R^2	E_{RMS} (mm)	E_{MAR} (%)	R	E_{RMS} (mm)	E_{MAR} (%)
a	mean	0.91	0.83	175	16.6	0.91	173.9	16.8
	max	0.92	0.85	186.2	17.8	0.94	211.8	19.9
	min	0.9	0.81	161.1	15.7	0.88	141	13.2
b	mean	0.92	0.84	166.6	15.8	0.91	186.1	17.8
	max	0.94	0.89	207	19.7	0.95	229.7	23.3
	min	0.89	0.8	126.2	12.8	0.89	148.6	12.9
c	mean	0.91	0.82	172.7	16.5	0.91	180.8	17.3
	max	0.95	0.91	207.9	19.1	0.94	204.8	24.4
	min	0.85	0.73	144.6	13.9	0.85	143.4	13.9

the weaker coupling effect between vegetation growth and precipitation.

The performance of regression models is assessed under three scenarios as described in Sect. 2.2. A total of 300 tests are conducted and performance metrics (i.e., R, R^2, E_{RMS}, and E_{MAR}) are calculated accordingly (Fig. 4 and Table 2). The high R values (> 0.85) indicate a strong correlation between NDVI and precipitation independent of sampling method. Also, the regression models demonstrate good performance, with R^2 larger than 0.75 and E_{MAR} less than 20 %.

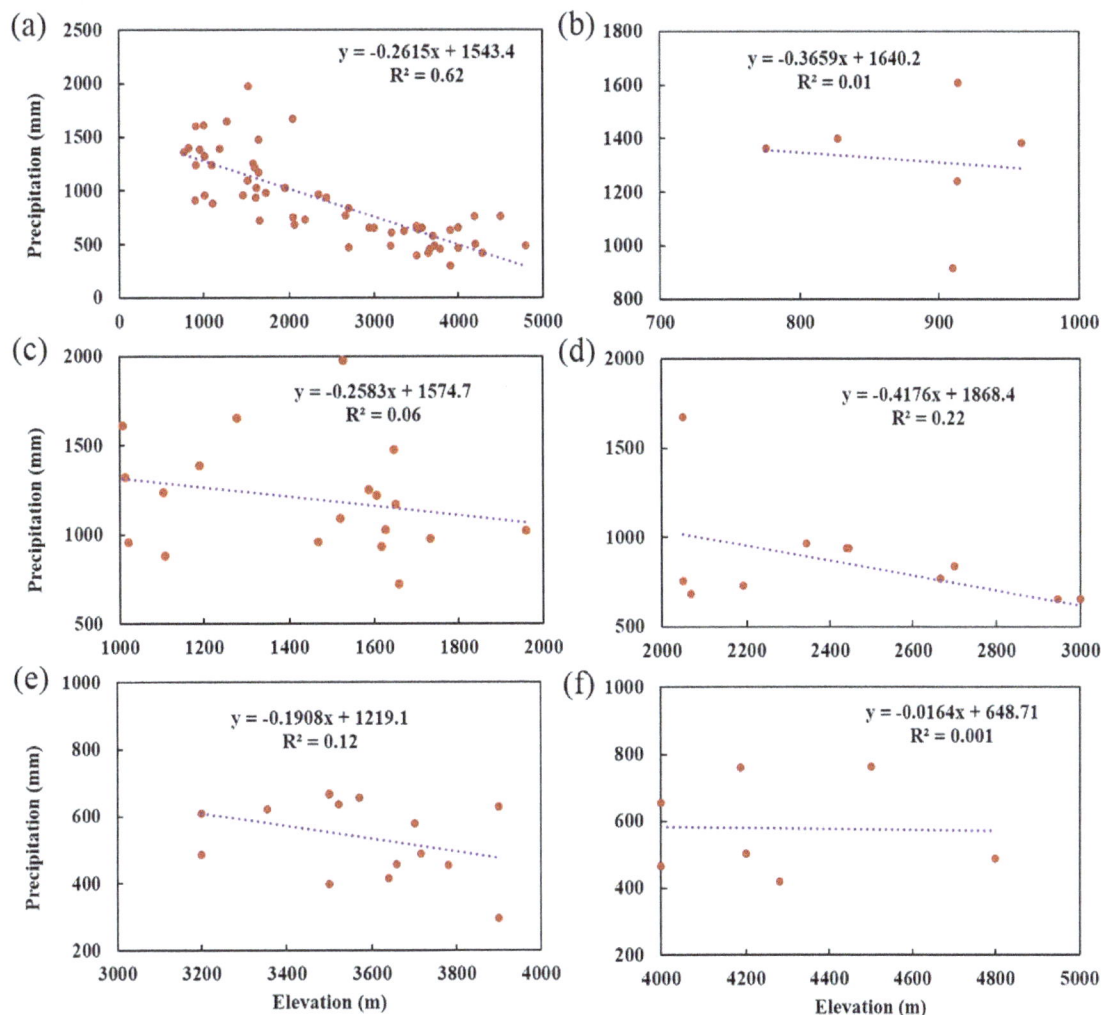

Figure 6. The relationship between mean annual precipitation and elevation at different elevation bands: **(a)** whole elevation bands; **(b)** elevation band: <1000 m; **(c)** band: 1000–2000 m; **(d)** band: 2000–3000 m; **(e)** band: 3000–4000 m; **(f)** band: >4000 m.

In addition, the metrics of regression models fluctuate around that of the RME, with narrow inter-quartile ranges, indicating the regression models have remarkable consistency with the RME model.

Scenario a is designed to examine inter-annual stability in the performance of regression models, where the good performance indicates the acceptable ability of the RME model in estimating precipitation during periods when precipitation measurements are not available. Scenarios b and c investigate the impacts of spatial and temporal coverages of measurements, respectively. It is noteworthy that under Scenario b better performance in regression models is observed as compared with Scenario c, implying the greater importance of spatial coverage of measurements in conducting the regressions. In addition, the results of calibration are better than validation, as revealed by all metrics criteria, as expected. However, the differences between calibration and validation

are not significant, implying the consistent performance of regression models under various scenarios.

The performance of RME is further assessed by comparing the estimates against observations (Fig. 5), and good agreement between estimates and observations is observed. It should be noted that the RME shows difficulty in estimating precipitation higher than 2000 mm (cf. the dashed line in Fig. 5), implying the limitation of the fusion framework inherited from the oversaturation effect of the vegetation index.

Elevation effect on the relationship between precipitation and NDVI is a concern to appreciate. An overall negative relationship is found between precipitation and elevation for the whole elevation range (i.e., 0–5000 m) with the R^2 value of 0.62 (Fig. 6a), whereas there is only an unapparent/weak relationship at different elevation bands (Fig. 6b–f). Given the spatial heterogeneity of orographic effects on precipitation (Brunsdon et al., 2001; Daly et al., 2008) and the insufficient data of this study, a more thorough investigation of

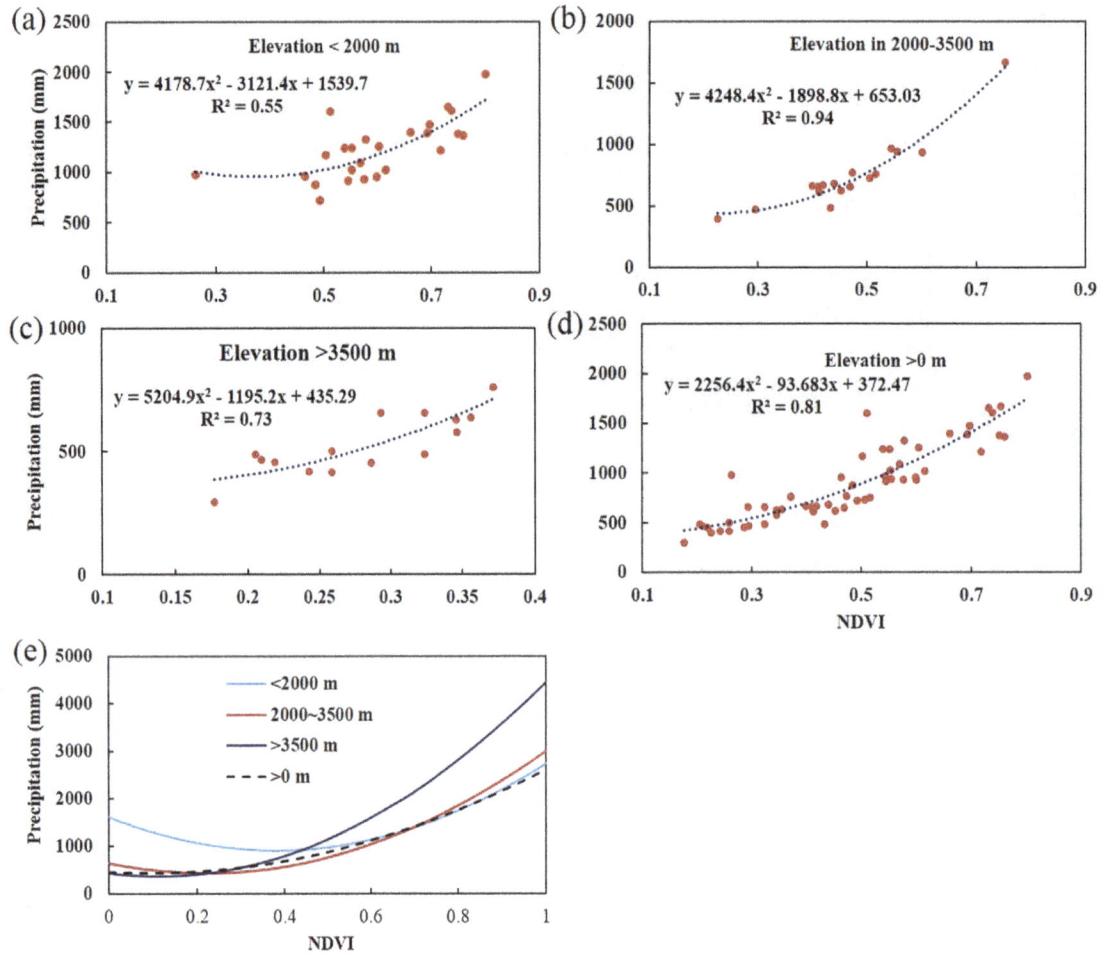

Figure 7. The relationship between mean annual precipitation and NDVI at different elevation bands: **(a)** elevation band: < 200 m; **(b)** band: 2000–3500 m; **(c)** band: > 3500 m; **(d)** whole bands; **(e)** comparison of the precipitation–NDVI relationship for different bands.

the relationship between precipitation and elevation needs to be conducted with more information that might be available in the future. Positive precipitation–NDVI relationships are found at different elevation bands (Fig. 7), with the best and worst fitness observed at elevation band 2000–3500 m with an R^2 value of 0.94 and at elevation band 0–2000 m with an R^2 value of 0.62, respectively. By comparing the three regressions at different bands with the global regression, we notice that more significant overestimates of precipitation are observed with the range of lower NDVI values (< 0.4) at band 0–2000 m than the other three regressions, whereas regression at band > 3500 m has a significant overestimation of precipitation than the other three regressions for higher NDVI values (> 0.5).

4.2 Spatial characteristics of precipitation

The spatial characteristics of the precipitation of the study area are investigated with RME for the whole study period (Fig. 8). Annual precipitation in the Nu River is ob-

served to decrease from south to north and from west to east with prominent spatial variability. Two "hot-spot" regions, whose annual precipitation exceeds 1500 mm, can be identified in the study areas: one near the southern border and the other close to the southwestern mountain border. The eastern part of the Nu River basin featuring a dry and warm climate receives an average annual precipitation of 800 mm with large inter-annual variability. A precipitation product (DEMP) based on a precipitation–elevation relationship is used to compare with RME. There is no obvious distribution pattern of precipitation (Fig. 9a) and a smaller spatial variability compared to RME in the DEMP product, indicating the advantage of RME in representing the spatial variability of annual precipitation. And the overall underestimation of precipitation is observed in the DEMP product across the whole study area (Fig. 9b). In addition, the pixels in Fig. 8 with a value out of the valid range (i.e., 400 mm yr^{-1} < P < 1500 mm yr^{-1}) may have a relatively large error as discussed in Sect. 4.1. As there is no justifiable method for such a correction and given the limited fraction of invalid

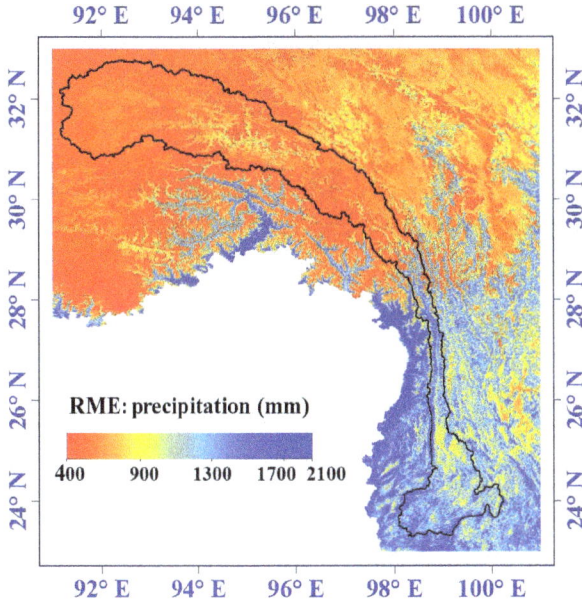

Figure 8. Average annual precipitation distribution of 2003–2012 from RME.

Table 3. Performance comparison between IDW, RME, and TRMM.

Method	Statistics	E_{RMS} (mm)	E_{MR}	E_{MAR}
IDW	max	273	0.1	0.26
	min	249	0.08	0.23
	mean	223	0.05	0.21
TRMM	max	220	0.17	0.24
	min	213	0.16	0.23
	mean	203	0.15	0.22
RME	max	183	0.07	0.18
	min	177	0.05	0.17
	mean	168	0.04	0.16
RME–IDW (%)	max	−32.9	−33	−30.5
	min	−26.3	−9.8	−21.4
	mean	−20.4	−1.2	−18.9
RME–TRMM (%)	max	−16.8	−59.5	−23.8
	min	−16.6	−66	−25.9
	mean	−17.4	−71.5	−28.3

pixels (10 % in the whole study area and 7 % in the Nu River basin), the figure can be used to demonstrate a full picture of the spatial precipitation pattern in the study area, but we note those pixels are of large uncertainties and should be interpreted with caution.

4.3　Model performance comparison

The performance between the IDW approach, the TRMM product and the fusion framework is compared in this section. IDW is one of the most popular methods for spatial interpolation of rainfall due to its easy implementation and flexibility in incorporating other auxiliary information (e.g., elevation). In general, the IDW approach is unable to demonstrate the high spatial variability, though it can capture the general spatial distribution of the whole basin (Fig. 10a), as TRMM (Fig. 10b). Due to the coarse spatial resolution, TRMM cannot capture the high variability in the river valley, where the elevation varies significantly. Although large rainfall (> 1800 mm) is observed in both our and TRMM products in the southwest of the study area region, our product gives lower rainfall compared to TRMM. As discussed above, the regression model tends to underestimate rainfall as the annual rainfall exceeds a certain threshold because the water supply is no longer a determinant of vegetation growth.

To demonstrate the advantage of the fusion framework, a cross-validation is conducted against the randomly sampled gauge observations by varying the number of samples (1–40). The cross-validation shows a higher E_{RMS} for the IDW approach, followed by TMMM and RME (Fig. 11a). A higher mean E_{MR} of 15 % is observed for TRMM than for IDW (8 %) and RME (5 %), while the differences in E_{MAR}

are minimal between TRMM and IDW. The results indicate an overestimated precipitation by TRMM as compared to gauge observations. Table 3 summarizes the maximum, minimum, and mean values of each method and shows the relative difference between RME and the other two methods. On average, the E_{RMS} of RME is smaller than that of IDW and TRMM by 20.4 and 17.4 %, respectively. In general, the fusion framework demonstrates better performance than the other approaches.

To further evaluate the performance of RME, the annual averages of precipitation of five hydrological stations (Fig. 12a) and the whole basin estimated by the three approaches (IDW, RME, and TRMM) are compared. At the whole basin scale, the estimate by RME is 5.2 % higher than that of IDW but 7.9 % lower than TRMM. Although the difference between the three approaches is minimal at the basin scale, the difference at the sub-basin scale is remarkable. In the upstream region (i.e., the Gongshan sub-basin) located on the Tibetan Plateau, TRMM overestimates precipitation by 13.2 %, while IDW underestimates it by 7.6 % as compared with RME. In the other four downstream sub-basins, estimates by RME are larger than those by IDW and TRMM. In general, in the midstream and downstream regions with large variability in terrain height, RME gives larger estimates of precipitation than IDW and TRMM.

To validate the accuracy of different precipitation estimates, we utilize the monthly MODIS (MOD16) global ET (evapotranspiration) product with 1 km spatial resolution (Mu et al., 2011) (i.e., ET + R) and to compare it with five products, including RME, BandP (rainfall based on the precipitation–NDVI relationship with the consideration ele-

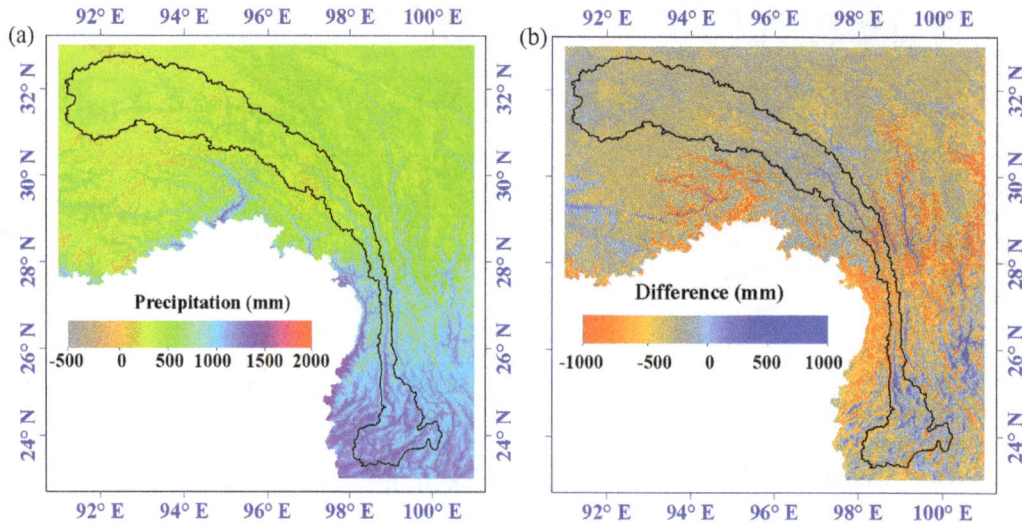

Figure 9. (a) The map of precipitation estimates of DEMP; **(b)** difference in precipitation estimates between RME and DEMP.

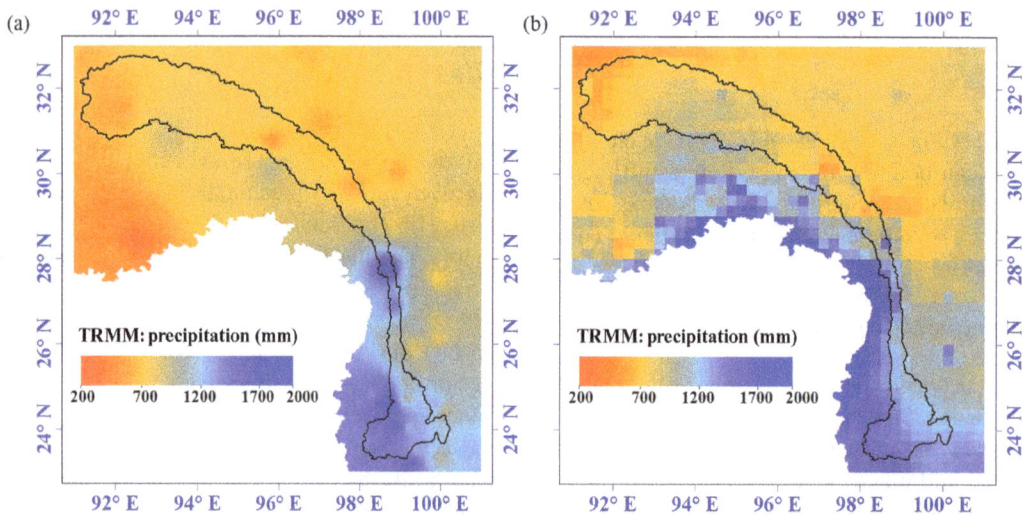

Figure 10. Spatial distribution of mean annual precipitation of 2003–2012 estimated by **(a)** IDW and **(b)** TRMM.

Table 4. Regression model performance and coefficients of regression.

	R^2	E_{RMS} (mm)	E_{MAR} (%)	a	b	c
NDVI	0.83	174.7	14.8	2670.4	−471.2	409.2
EVI	0.87	143.8	12.4	5129.6	702.5	254.7

vation band), DEMP, TRMM, and IDW (Fig. 12b). Although all five products underestimate the sub-basin-scale precipitation, RME and BandP give the closest estimates to the water-budget-based precipitation, indicating the effectiveness of the precipitation–NDVI relationship in precipitation remapping.

We also compared our products with the Multi-Source Weighted-Ensemble Precipitation (MSWEP) product. The dataset takes advantage of a wide range of data sources, including gauges, satellites, and atmospheric reanalysis models, to obtain the best possible precipitation estimates at the global scale with a high 3-hourly temporal and 0.25° spatial resolution (Beck et al., 2016). Comparison in the annual mean precipitation between the gauge measurements and predictions by the MSWEP and TRMM products (Fig. 13) shows acceptable performance of both MSWEP and TRMM in predicting the precipitation with an overall overestimation. The RMSE values for MSWEP, TRMM, and RME are 241, 196, and 174 mm, respectively, indicating that RME gives the best prediction among the three products. The possible reason why MSWEP shows no superiority over TRMM in

Table 5. Results of two regression models established with extra independent variables: RME + T for temperature, RME + H for elevation.

Model	R^2	E_{RMS} (mm)	E_{MAR} (%)	a	b	c	Extra b
RME	0.83	174.7	15	2670.4	−471.2	409.2	–
RME + T	0.84	172.6	15	2728.8	−496	407.3	−0.2
RME + H	0.84	172.6	15	2838.4	−638.7	492.9	−0.02

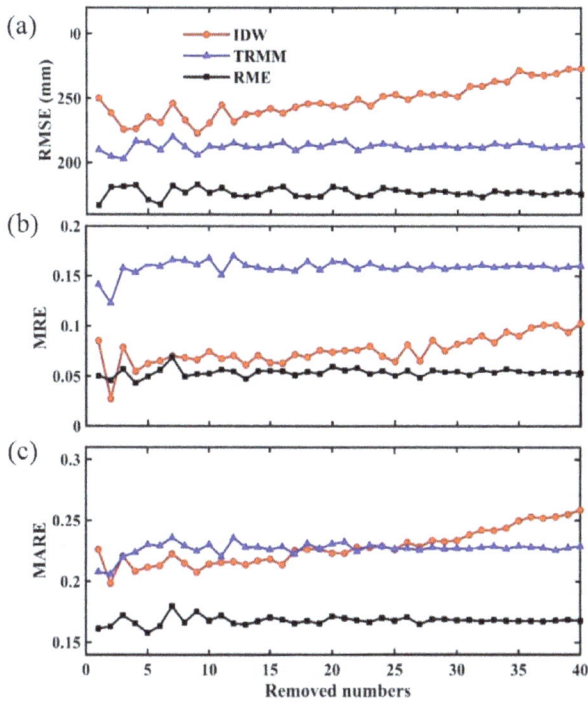

Figure 11. Performance of E_{RMS}, E_{MR}, and E_{MAR} for three methods in different removed numbers.

predicting annual precipitation is that very few gauges are available in this region that might limit the applicability of the MSWEP methodology. However, the MSWEP methodology does provide insights into the production of high temporal resolution (3-hourly) rainfall, which we believe will be helpful to our future work.

4.4 Influence of different vegetation indices

Considering the possible degradation in model performance caused by oversaturation of NDVI in high biomass areas, another vegetation indicator, the enhanced vegetation index (EVI), is suggested as an alternative for estimating vegetation growth (Matsushita et al., 2007; Liao et al., 2015). As such, we also test the fusion framework with EVI in addition to NDVI and the results are assessed against the gauge observations.

Based on the chosen metrics, EVI is found to outperform NDVI with better regression quality (Table 4): the EVI-based regression model gives higher R^2, and smaller E_{RMS} and E_{MAR} compared to the NDVI-based model. Also, a remarkable difference is observed in the precipitation estimates based on the two vegetation indices (Fig. 14). It is noted that the curvature of the EVI-based model is larger than the NDVI-based model, suggesting higher sensitivity of the EVI-based model in a humid environment. Although the EVI-based model demonstrates better performance than the NDVI-based one, it should be noted that NDVI is the most popular vegetation index used in operational applications among the available vegetation index products. Besides, NDVI has a relative longer temporal coverage compared to other vegetation index products. For instance, the AVHRR (Advanced Very High Resolution Radiometer) NDVI data have been available since 1982 with a global coverage. As such, under scenarios when EVI is unavailable, NDVI is a satisfactory index that can be used in the fusion framework.

4.5 Influence of other ambient determinants

One major assumption of the proposed framework is that precipitation is the only determinant of vegetation growth, and thus NDVI is regarded as a proxy for precipitation. However, other ambient factors, such as soil properties, solar radiation, air temperature, and elevation, may significantly influence the vegetation growth as well as NDVI values. Considering the data availability of various ambient factors, air temperature and elevation, in addition to NDVI, are adopted as extra determinants to establish the regression models, which are thus termed RME + T and RME + H for air temperature and elevation, respectively. We note that, for simplicity, the extra determinants are assumed to have a linear relationship with precipitation.

The differences in R^2, E_{RMS}, and E_{MAR} between the three models are minimal, and the regression coefficients of the three models are very close to each other (Table 5). The negative regression coefficient of temperature in RME + T indicates inconsistent trends between precipitation and temperature. Since the temperature decreases with the increase in elevation, RME + T and RME + H essentially provide consistent estimates of precipitation which are also clearly shown in Fig. 15. It is also noted that the information added by extra determinants (i.e., air temperature and elevation) is in fact minimal. Overall there is little difference between RME and the other two products. As such, we consider the RME-only-

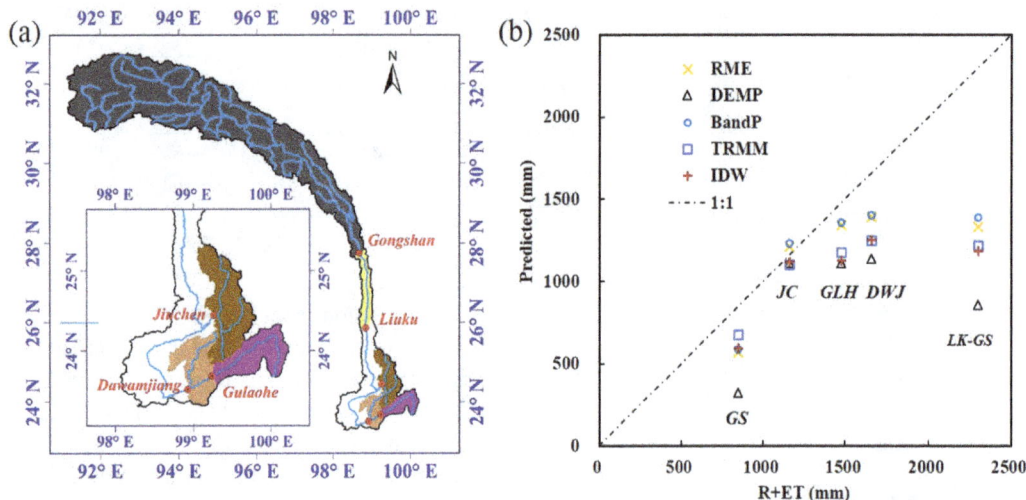

Figure 12. (a) Sub-basins based on hydrological stations. **(b)** Comparison between precipitations based on basin water balance ($R + \mathrm{ET}$) and different annual rainfall products: DEMP (P elevation relationship), BandP (P–NDVI relationship with consideration elevation band), RME, TRMM, and IDW. GS, JC, GLH, DWJ, and LK-GS are the abbreviations for Gongshan, Jiuchen, Gulaohe, Dawanjing, and Liuku-Gongshan, respectively.

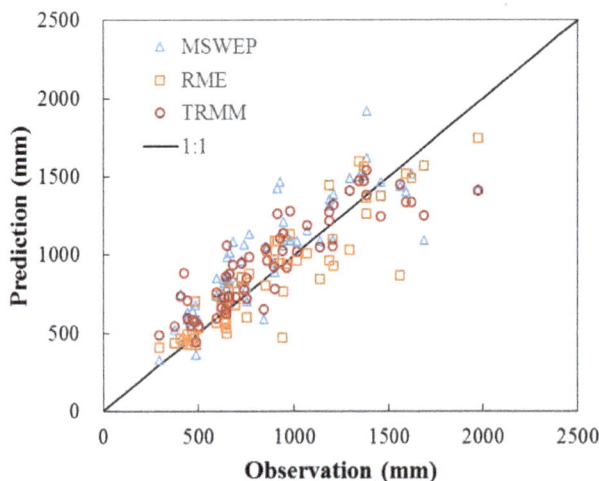

Figure 13. Comparison in mean annual precipitation between the gauged measurements and predictions by the MSWEP, RMM and RME.

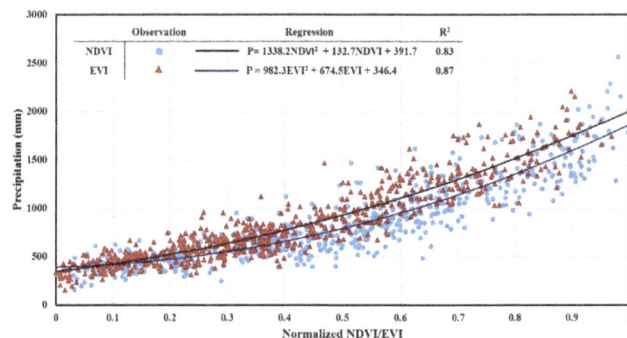

Figure 14. Regression relationship between annual precipitation and normalized NDVI/EVI.

based vegetation index to be a simple and efficient model for precipitation estimation.

5 Conclusion

In this study, a satellite–gauge–vegetation fusion framework has been developed for estimating the precipitation in mountainous areas by establishing a regression relationship between gauge-based precipitation observations and a satellite-based vegetation dataset. The fusion framework was then ap-

plied in the Nu River basin of Southwest China for estimating precipitation between 2001 and 2012.

The fusion framework for the Nu River basin adopted a second-order polynomial form and demonstrated promising ability in capturing the high spatial variability of precipitation in the river valley. Five evaluation metrics, including R, R^2, E_{RMS}, E_{MR}, and E_{MAR}, indicated good performance of the fusion framework in precipitation estimation. The performance of the fusion framework was also compared with the IDW approach and TRMM product and the comparison results indicated that the fusion framework generally outperformed other approaches in estimating precipitation in mountainous areas. On average, the E_{RMS} of the fusion framework is 20.4 %, 17.4 % smaller than that of IDW and TRMM, respectively. The E_{MR} of the fusion framework is 1.2 %, 71.5 % smaller than that of IDW and TRMM. The E_{MAR} of

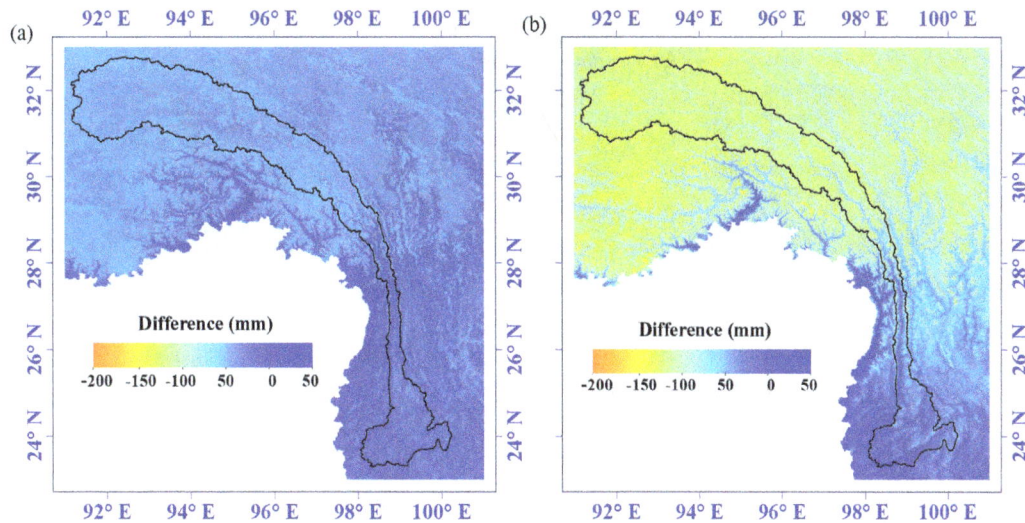

Figure 15. Spatial precipitation difference between RME and **(a)** RME + *H*; **(b)** RME + *T* **(b)**.

the fusion framework is 18.9 %, 28.3 % smaller than that of IDW and TRMM.

The success of application of the fusion framework in the Nu River sheds light on the precipitation estimation in mountainous areas by using multi-source datasets. However, this framework does have certain limitations that are important to appreciate. First, the framework is applied only in the Nu River basin. More mountainous areas under different climates need to be examined to further test the robustness of this framework. In addition, although the RME model can utilize the full knowledge of precipitation in the entire study period compared with RMI models, the difference in the coefficients suggests apparent inter-annual variability of precipitation that should be considered when applying these models. Given the duration of study period and purpose, we suggest the RME model be used for long-term climatology identification while RMI models for inter-annual variability examination. Also, to fully verify the theoretical basis of this framework that vegetation actively interacts with precipitation in mountainous areas, future work is required to re-fine the spatiotemporal resolution of this study to enable better scrutiny into vegetation–precipitation interactions at submonthly scales across more detailed vegetation species.

Appendix A: Merging of NDVI datasets

The merging of NDVI datasets improves the accuracy as expected (Fig. A1); the monthly error rates (i.e., the ratio of the pixel whose quality value is over 1) of MOD and MMD are generally reduced with an average of 5 % and over 20 % in several months. Figure A2 shows that the accuracy of MMD is significantly improved in a ridge area covering $23°10'$–$23°40'$ N and $98°30'$–$99°0'$ E. Figure A2b shows that the NDVI value near the right and left boundaries is underestimated by MOD. Figure A2c shows that the NDVI value in the middle boundary is underestimated by MYD. The underestimates in both products near the boundary of MOD and MYD are amended (Fig. A2a). Figure A3 shows the three NDVI series for one rain gauge. Comparing with MOD series, the improved accuracy in MMD is mainly observed in the wet season (from May to October), when the NDVI values could be often underestimated due to the overcasts.

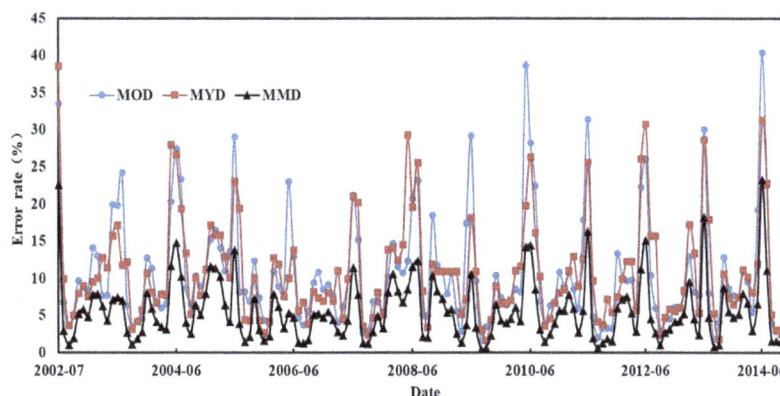

Figure A1. Monthly error rate of MOD, MYD, and MMD.

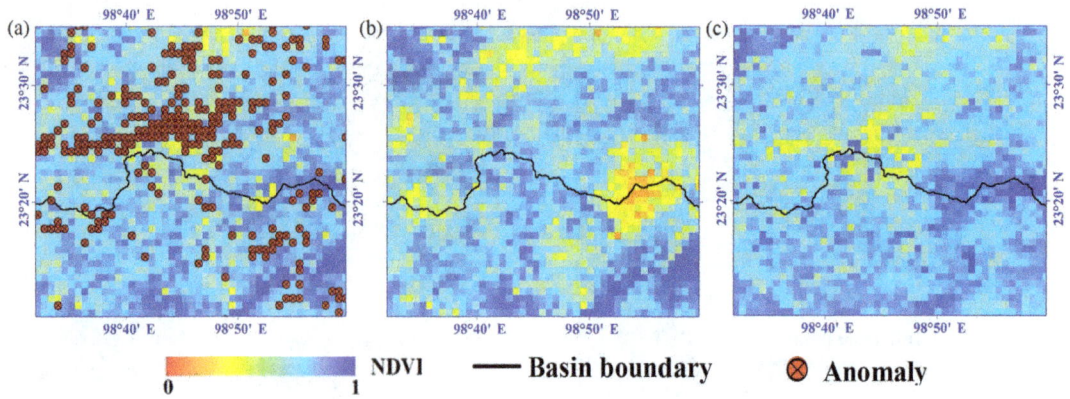

Figure A2. Comparison of three NDVI products over a ridge area on June 2006, **(a)** for MMD, **(b)** for MOD, and **(c)** for MYD.

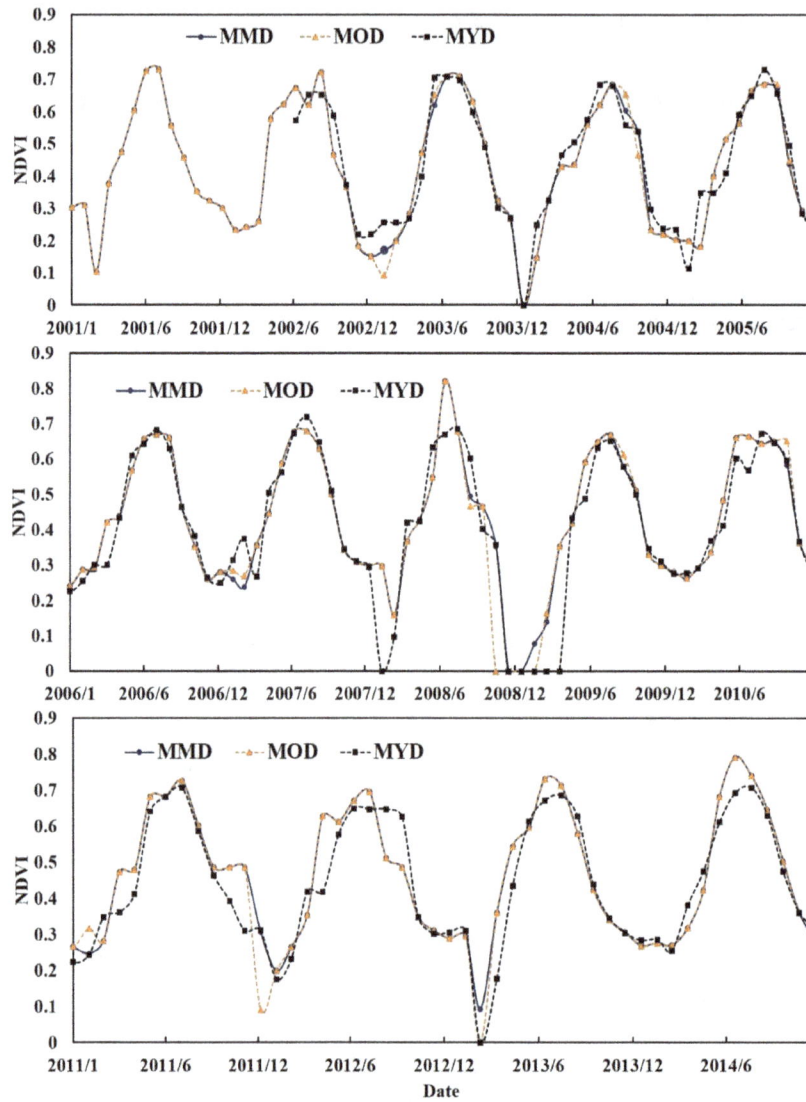

Figure A3. Comparison of three NDVI monthly time series over one gauge.

Competing interests. The authors declare that they have no conflict of interest.

Acknowledgements. The study is supported by the NSFC under grants U1202231, 51679119, and 91647107, the National Key Technology Support Program under grant 2011BAC09B07-3, and by the China Postdoctoral Science Foundation under grant 2015T80093. The authors thank the China Meteorological Administration, Yunnan University, MODIS NDVI, the Tropical Rainfall Measuring Mission (TRMM), and the Shuttle Radar Topography Mission (SRTM) for providing the data used in this study.

Edited by: L. Wang

References

Akbari, A., Abu Samah, A., and Othman, F.: Integration of SRTM and TRMM date into the GIS-based hydrological model for the purpose of flood modelling, Hydrol. Earth Syst. Sci. Discuss., 9, 4747–4775, doi:10.5194/hessd-9-4747-2012, 2012.

Arias-Hidalgo, M., Bhattacharya, B., Mynett, A. E., and van Griensven, A.: Experiences in using the TMPA-3B42R satellite data to complement rain gauge measurements in the Ecuadorian coastal foothills, Hydrol. Earth Syst. Sci., 17, 2905–2915, doi:10.5194/hess-17-2905-2013, 2013.

Beck, H. E., van Dijk, A. I. J. M., Levizzani, V., Schellekens, J., Miralles, D. G., Martens, B., and de Roo, A.: MSWEP: 3-hourly 0.25° global gridded precipitation (1979–2015) by merging gauge, satellite, and reanalysis data, Hydrol. Earth Syst. Sci. Discuss., doi:10.5194/hess-2016-236, in review, 2016.

Brunsdon, C., McClatchey, J., and Unwin, D. J.: Spatial variations in the average rainfall-altitude relationship in Great Britain: an approach using geographically weighted regression, Int. J. Climatol., 21, 455–466, doi:10.1002/joc.614, 2001.

Campo-Bescós, M. A., Muñoz-Carpena, R., Southworth, J., Zhu, L., Waylen, P. R., and Bunting, E.: Combined Spatial and Temporal Effects of Environmental Controls on Long-Term Monthly NDVI in the Southern Africa Savanna, Remote Sensing, 5, 6513–6538, doi:10.3390/rs5126513, 2013.

Chen, F. and Li, X.: Evaluation of IMERG and TRMM 3B43 Monthly Precipitation Products over Mainland China, Remote Sensing, 8, 472, doi:10.3390/rs8060472, 2016.

Chen, F., Liu, Y., Liu, Q., and Li, X.: Spatial downscaling of TRMM 3B43 precipitation considering spatial heterogeneity, Int. J. Remote Sens., 35, 3074–3093, doi:10.1080/01431161.2014.902550, 2014.

Chen, J., Yong, B., Ren, L., Wang, W., Chen, B., Lin, J., Yu, Z., and Li, N.: Using a Kalman Filter to Assimilate TRMM-Based Real-Time Satellite Precipitation Estimates over Jinghe Basin, China, Remote Sensing, 8, 899, doi:10.3390/rs8110899, 2016.

Council, N. R.: Assessment of the Benefits of Extending the Tropical Rainfall Measuring Mission: A Perspective from the Research and Operations Communities, Interim Report, available from: https://www.nap.edu/catalog/11195/assessment-of-the-benefits (last access: 18 November 2016), 2005.

Daly, C., Halbleib, M., Smith, J. I., Gibson, W. P., Doggett, M. K., Taylor, G. H., Curtis, J., and Pasteris, P. P.: Physiographically sensitive mapping of climatological temperature and precipitation across the conterminous United States, Int. J. Climatol., 28, 2031–2064, doi:10.1002/joc.1688, 2008.

Borges, P. A., Franke, J., da Anunciação, Y. M. T., Weiss, H., and Bernhofer, C.: Comparison of spatial interpolation methods for the estimation of precipitation distribution in Distrito Federal, Brazil, Theor. Appl. Climatol., 123, 335–348, doi:10.1007/s00704-014-1359-9, 2016.

Duan, Z. and Bastiaanssen, W. G. M.: First results from Version 7 TRMM 3B43 precipitation product in combination with a new downscaling–calibration procedure, Remote Sens. Environ., 131, 1–13, doi:10.1016/j.rse.2012.12.002, 2013.

Immerzeel, W. W., Rutten, M. M., and Droogers, P.: Spatial downscaling of TRMM precipitation using vegetative response on the Iberian Peninsula, Remote Sens. Environ., 113, 362–370, doi:10.1016/j.rse.2008.10.004, 2009.

Jacquin, A. P. and Soto-Sandoval, J. C.: Interpolation of monthly precipitation amounts in mountainous catchments with sparse precipitation networks, Chil. J. Agr. Res., 73, 406–413, doi:10.4067/S0718-58392013000400012, 2013.

Jia, S., Zhu, W., Lü, A., and Yan, T.: A statistical spatial downscaling algorithm of TRMM precipitation based on NDVI and DEM in the Qaidam Basin of China, Remote Sens. Environ., 115, 3069–3079, doi:10.1016/j.rse.2011.06.009, 2011.

Jing, W., Yang, Y., Yue, X., and Zhao, X.: A Spatial Downscaling Algorithm for Satellite-Based Precipitation over the Tibetan Plateau Based on NDVI, DEM, and Land Surface Temperature, Remote Sensing, 8, 655, doi:10.3390/rs8080655, 2016.

Joyce, R. J., Janowiak, J. E., Arkin, P. A., and Xie, P.: CMORPH: A Method that Produces Global Precipitation Estimates from Passive Microwave and Infrared Data at High Spatial and Temporal Resolution, J. Hydrometeorol., 5, 487–503, doi:10.1175/1525-7541(2004)005<0487:CAMTPG>2.0.CO;2, 2004.

Kariyeva, J. and Van Leeuwen, W. J. D.: Environmental Drivers of NDVI-Based Vegetation Phenology in Central Asia, Remote Sensing, 3, 203–246, doi:10.3390/rs3020203, 2011.

Kneis, D., Chatterjee, C., and Singh, R.: Evaluation of TRMM rainfall estimates over a large Indian river basin (Mahanadi), Hydrol. Earth Syst. Sci., 18, 2493–2502, doi:10.5194/hess-18-2493-2014, 2014.

Krakauer, N. Y., Pradhanang, S. M., Lakhankar, T., and Jha, A. K.: Evaluating Satellite Products for Precipitation Estimation in Mountain Regions: A Case Study for Nepal, Remote Sensing, 5, 4107–4123, doi:10.3390/rs5084107, 2013.

Li, B., Tao, S., and Dawson, R. W.: Relations between AVHRR NDVI and ecoclimatic parameters in China, Int. J. Remote Sens., 23, 989–999, doi:10.1080/014311602753474192, 2002.

Li, D., Ding, X., and Wu, J.: Simulating the regional water balance through hydrological model based on TRMM satellite rainfall data, Hydrol. Earth Syst. Sci. Discuss., 12, 2497–2525, doi:10.5194/hessd-12-2497-2015, 2015.

Li, H., Calder, C. A., and Cressie, N.: Beyond Moran's I: Testing for Spatial Dependence Based on the Spatial Autoregressive Model, Geogr. Anal., 39, 357–375, doi:10.1111/j.1538-4632.2007.00708.x, 2007.

Li, Z. and Guo, X.: Detecting Climate Effects on Vegetation in Northern Mixed Prairie Using NOAA AVHRR 1-

km Time-Series NDVI Data, Remote Sensing, 4, 120–134, doi:10.3390/rs4010120, 2012.

Liao, Z., He, B., and Quan, X.: Modified enhanced vegetation index for reducing topographic effects, J. Appl. Remote Sens., 9, 096068–096068, doi:10.1117/1.JRS.9.096068, 2015.

Lloyd, C. D.: Assessing the effect of integrating elevation data into the estimation of monthly precipitation in Great Britain, J. Hydrol., 308, 128–150, doi:10.1016/j.jhydrol.2004.10.026, 2005.

Mahmud, M. R., Numata, S., Matsuyama, H., Hosaka, T., and Hashim, M.: Assessment of Effective Seasonal Downscaling of TRMM Precipitation Data in Peninsular Malaysia, Remote Sensing, 7, 4092–4111, doi:10.3390/rs70404092, 2015.

Mair, A. and Fares, A.: Comparison of Rainfall Interpolation Methods in a Mountainous Region of a Tropical Island, J. Hydrol. Eng., 16, 371–383, doi:10.1061/(ASCE)HE.1943-5584.0000330, 2011.

Marquínez, J., Lastra, J., and García, P.: Estimation models for precipitation in mountainous regions: the use of GIS and multivariate analysis, J. Hydrol., 270, 1–11, doi:10.1016/S0022-1694(02)00110-5, 2003.

Matsushita, B., Yang, W., Chen, J., Onda, Y., and Qiu, G.: Sensitivity of the Enhanced Vegetation Index (EVI) and Normalized Difference Vegetation Index (NDVI) to Topographic Effects: A Case Study in High-density Cypress Forest, Sensors, 7, 2636–2651, doi:10.3390/s7112636, 2007.

Mourre, L., Condom, T., Junquas, C., Lebel, T. E., Sicart, J., Figueroa, R., and Cochachin, A.: Spatio-temporal assessment of WRF, TRMM and in situ precipitation data in a tropical mountain environment (Cordillera Blanca, Peru), Hydrol. Earth Syst. Sci., 20, 125–141, doi:10.5194/hess-20-125-2016, 2016.

Phillips, D. L., Dolph, J., and Marks, D.: A comparison of geostatistical procedures for spatial analysis of precipitation in mountainous terrain, Agr. Forest Meteorol., 58, 119–141, doi:10.1016/0168-1923(92)90114-J, 1992.

Rozante, J. R., Moreira, D. S., de Goncalves, L. G. G., and Vila, D. A.: Combining TRMM and Surface Observations of Precipitation: Technique and Validation over South America, Weather Forecast., 25, 885–894, doi:10.1175/2010WAF2222325.1, 2010.

Schamm, K., Ziese, M., Becker, A., Finger, P., Meyer-Christoffer, A., Schneider, U., Schröder, M., and Stender, P.: Global gridded precipitation over land: a description of the new GPCC First Guess Daily product, Earth Syst. Sci. Data, 6, 49–60, doi:10.5194/essd-6-49-2014, 2014.

Song, J., Xia, J., Zhang, L., Wang, Z.-H., Wan, H., and She, D.: Streamflow prediction in ungauged basins by regressive regionalization: a case study in Huai River Basin, China, Hydrol. Res., 47, 1053–1068, doi:10.2166/nh.2015.155, 2016.

Sun, J., Cheng, G., Li, W., Sha, Y., and Yang, Y.: On the Variation of NDVI with the Principal Climatic Elements in the Tibetan Plateau, Remote Sensing, 5, 1894–1911, doi:10.3390/rs5041894, 2013.

Wang, S., Huang, G. H., Lin, Q. G., Li, Z., Zhang, H., and Fan, Y. R.: Comparison of interpolation methods for estimating spatial distribution of precipitation in Ontario, Canada, Int. J. Climatol., 14, 3745–3751, doi:10.1002/joc.3941, 2014.

Woldemeskel, F. M., Sivakumar, B., and Sharma, A.: Merging gauge and satellite rainfall with specification of associated uncertainty across Australia, J. Hydrol., 499, 167–176, doi:10.1016/j.jhydrol.2013.06.039, 2013.

Wong, J. S., Razavi, S., Bonsal, B. R., Wheater, H. S., and Asong, Z. E.: Evaluation of various daily precipitation products for large-scale hydro-climatic applications over Canada, Hydrol. Earth Syst. Sci. Discuss., doi:10.5194/hess-2016-511, in review, 2016.

Worqlul, A. W., Collick, A. S., Tilahun, S. A., Langan, S., Rientjes, T. H. M., and Steenhuis, T. S.: Comparing TRMM 3B42, CFSR and ground-based rainfall estimates as input for hydrological models, in data scarce regions: the Upper Blue Nile Basin, Ethiopia, Hydrol. Earth Syst. Sci. Discuss., 12, 2081–2112, doi:10.5194/hessd-12-2081-2015, 2015.

Xu, S., Wu, C., Wang, L., Gonsamo, A., Shen, Y., and Niu, Z.: A new satellite-based monthly precipitation downscaling algorithm with non-stationary relationship between precipitation and land surface characteristics, Remote Sens. Environ., 162, 119–140, doi:10.1016/j.rse.2015.02.024, 2015.

Zhou, L., Chen, Y., Liang, N., and Ni, Y.: Daily rainfall model to merge TRMM and ground based observations for rainfall estimations, in 2016 IEEE International Geoscience and Remote Sensing Symposium (IGARSS), 601–604, 2016.

7

Rainfall-runoff modelling using river-stage time series in the absence of reliable discharge information: a case study in the semi-arid Mara River basin

Petra Hulsman, Thom A. Bogaard, and Hubert H. G. Savenije

Water Resources Section, Faculty of Civil Engineering and Geosciences, Delft University of Technology, Stevinweg 1, 2628 CN Delft, the Netherlands

Correspondence: Petra Hulsman (p.hulsman@tudelft.nl)

Abstract. Hydrological models play an important role in water resources management. These models generally rely on discharge data for calibration. Discharge time series are normally derived from observed water levels by using a rating curve. However, this method suffers from many uncertainties due to insufficient observations, inadequate rating curve fitting procedures, rating curve extrapolation, and temporal changes in the river geometry. Unfortunately, this problem is prominent in many African river basins. In this study, an alternative calibration method is presented using water-level time series instead of discharge, applied to a semi-distributed rainfall-runoff model for the semi-arid and poorly gauged Mara River basin in Kenya. The modelled discharges were converted into water levels using the Strickler–Manning formula. This method produces an additional model output; this is a "geometric rating curve equation" that relates the modelled discharge to the observed water level using the Strickler–Manning formula and a calibrated slope-roughness parameter. This procedure resulted in good and consistent model results during calibration and validation. The hydrological model was able to reproduce the water levels for the entire basin as well as for the Nyangores sub-catchment in the north. The newly derived geometric rating curves were subsequently compared to the existing rating curves. At the catchment outlet of the Mara, these differed significantly, most likely due to uncertainties in the recorded discharge time series. However, at the "Nyangores" sub-catchment, the geometric and recorded discharge were almost identical. In conclusion, the results obtained for the Mara River basin illustrate that with the proposed calibration method, the water-level time series can be simulated well, and that the discharge-water-level relation can also be derived, even in catchments with uncertain or lacking rating curve information.

1 Introduction to rating curve uncertainties

Hydrological models play an important role in water resources management. In hydrological modelling, discharge time series are of crucial importance. For example, discharge is used when estimating flood peaks (Di Baldassarre et al., 2012; Kuczera, 1996), calibrating models (Domeneghetti et al., 2012; McMillan et al., 2010) or determining the model structure (McMillan and Westerberg, 2015; Bulygina and Gupta, 2011). Discharge is commonly measured indirectly through the interpolation of velocity measurements over the cross-section (WMO, 2008; Di Baldassarre and Montanari, 2009). However, to obtain frequent or continuous discharge data, this method is time consuming and cost-inefficient. Moreover, in African river catchments, the quantity and quality of the available discharge measurements are often unfortunately inadequate for the reliable calibration of hydrological models (Shahin, 2002; Hrachowitz et al., 2013).

There are several sources of uncertainty in discharge data when using rating curves that cannot be neglected. First, measurement errors in the individual discharge measurements affect the estimated continuous discharge data, for example in the velocity-area method, uncertainties in the cross-section and velocity can arise due to poor sampling (Pelletier, 1988; Sikorska et al., 2013). Second, these measurements are usually conducted during normal flows. However during floods, the rating curve needs to be extrapolated. Therefore, the uncertainty increases for discharges under extreme conditions (Di Baldassarre and Claps, 2011; Domeneghetti et al., 2012). Thirdly, the fitting procedure does not always account well for irregularities in the profile, particularly when banks are overtopped. Finally, the river is a dynamic, nonstationary system which influences the rating curve, for example changes in the cross-section due to sedimentation or erosion, backwater effects or hysteresis (Petersen-Øverleir, 2006). The lack of incorporating such temporal changes in the rating curve increases the uncertainty in discharge data (Guerrero et al., 2012; Jalbert et al., 2011; Morlot et al., 2014). As a result, the rating curve should be regularly updated to take such changes into account. The timing of adjusting the rating curve relative to the changes in the river affects the number of rating curves and the uncertainty (Tomkins, 2014). Previous studies focused on assessing the uncertainty of rating curves (Di Baldassarre and Montanari, 2009; Clarke, 1999) and their effect on model predictions (Karamuz et al., 2016; Sellami et al., 2013; Thyer et al., 2011).

In the absence of reliable rating curves, remotely sensed river characteristics related to the discharge such as river width and water level can provide valuable information on the flow dynamics for model calibration and validation. For instance, previous studies derived the discharge from remotely sensed river width (Revilla-Romero et al., 2015; Yan et al., 2015; Sun et al., 2015) or river water levels measured with radar altimetry (Pereira-Cardenal et al., 2011; Michailovsky et al., 2012; Ričko et al., 2012; Schwatke et al., 2015; Tourian et al., 2017; Sun et al., 2012). In previous studies, hydrological models were calibrated on river width or surface water extent (Sun et al., 2015; Revilla-Romero et al., 2015). Also radar altimetry observations of river water levels have been used to calibrate or validate hydrological models by using empirical equations transforming discharge to the water level without using cross-section information (Sun et al., 2012; Getirana, 2010), for instance conceptual hydrological models (Sun et al., 2012; Pereira-Cardenal et al., 2011) or process-based models (Getirana, 2010; Paiva et al., 2013).

Besides remotely sensed river characteristics, locally measured river water-level time series are also valuable for model calibration and validation (van Meerveld et al., 2017). In general, water-level time series are more reliable than discharge data or remotely sensed river characteristics, as these are direct measurements and not processed data. In previous studies, hydrological models have been calibrated on river water-level time series using the Spearman rank correlation coefficient (Jian et al., 2017; Seibert and Vis, 2016) or by including an inverse rating curve with three new calibration parameters to convert the modelled discharge to water level (Jian et al., 2017). When using the Spearman rank correlation function, the focus is on correlating the ranks instead of the magnitudes, which as a result, introduces biases in the model results. Alternatively, rainfall-runoff models can

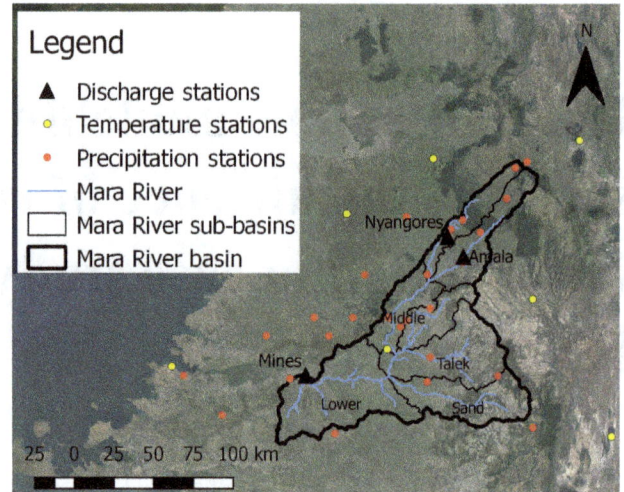

Figure 1. Map of the Mara River basin and the hydrometeorological stations for which data are available.

be calibrated on water-level time series combined with a hydraulic equation introducing only one new calibration parameter. Data-driven models have also been calibrated successfully on water-level time series; for example artificial neural network or fuzzy logic approaches were applied (Liu and Chung, 2014; Panda et al., 2010; Alvisi et al., 2006).

The goal of this study is to illustrate the potential of water-level time series for hydrological model calibration by incorporating a hydraulic equation describing the rating curve within the model. This calibration method is applied to the semi-arid and poorly gauged Mara River basin in Kenya. For three gauging stations within this basin, the quality of the recorded rating curves have been analysed and compared to the model results. For this purpose, a semi-distributed rainfall-runoff model has been developed on a daily timescale applying the FLEX-Topo modelling concept (Savenije, 2010).

2   Site description of the Mara River basin and data availability

The Mara River originates in Kenya in the Mau Escarpment and flows through the Maasai Mara National Reserve in Kenya into Lake Victoria in Tanzania. The main tributaries are the Nyangores and Amala rivers in the upper reach and the Lemek, Talak and Sand in the middle reach (Fig. 1). The first two tributaries are perennial, while the remaining tributaries are ephemeral, which generally dry out during dry periods. In total, the river is 395 km long (Dessu et al., 2014) and its catchment covers an area of about 11 500 km^2 (McClain et al., 2013), of which 65 % is located in Kenya (Mati et al., 2008).

Within the Mara River basin, there are two wet seasons linked to the annual oscillations of the ITCZ (Intertropical

Table 1. Hydro-meteorological data availability in the Mara River basin. The temporal coverage for water level and discharge can be different due to poor administration.

	Precipitation	Temperature	Water level, discharge		
Number of stations	28	7	3		
Station ID	–	–	1LA03	1LB02	5H2
Station location	–	–	Nyangores at Bomet	Amala at Kapkimolwa	Mara at Mines
Time range	1959–2011	1957–2014	1963–2009	1955–2015	1969–2013
Duration [years]	0–43	3–57	46	60	44
Coverage	8 %–100 %	30 %–100 %	Discharge: 85 % Water level: 85 %	Discharge: 72 % Water level: 70 %	Discharge: 53 % Water level: 61 %

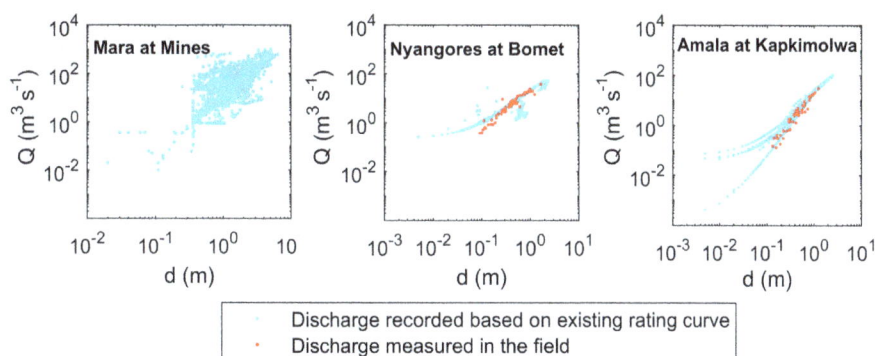

Figure 2. Discharge–water depth graphs for the three main river gauging stations in the Mara River basin; these are the Mara at Mines, Nyangores at Bomet and Amala at Kapkimolwa. For each location, the following are visualised: (1) recorded discharge and water-level time series between 1960 and 2010 (light blue), (2) discharge field measurements from the Nile Decision Support Tool (NDST) for the time period 1963–1989 (Nyangores) and 1965–1992 (Amala); no data were available for Mines (red).

Convergence Zone). The first wet season is from March to May and the second from October to December (McClain et al., 2013). The precipitation varies spatially over the catchment following the local topography. The largest annual rainfall can be found in the upstream area of the catchment, which is between 1000 and 1750 mm yr^{-1}. In the middle and downstream areas, the annual rainfall is between 900 and 1000 mm yr^{-1} and between 300 and 850 mm yr^{-1}, respectively (Dessu et al., 2014).

The elevation of the river basin varies between 3000 m a.s.l. (metres above sea level) at the Mau Escarpment, 1480 m at the border to Tanzania and 1130 m at Lake Victoria (McClain et al., 2013). In the Mara River basin, the main land cover types are agriculture, grass, shrubs and forests. The main forest in the catchment is the Mau forest, which is located in the north. Croplands are mainly found in the north and in the south, whereas the middle part is dominated by grasslands.

2.1 Data availability

2.1.1 In situ monitoring data

In the Mara River basin, long-term daily water level and discharge time series are available for 44–60 years be-

tween 1955 and 2015 at the downstream station near Mines and in the two main tributaries, the Nyangores and Amala. In addition, precipitation and air temperature is measured at 27 and 7 stations, respectively (Fig. 1 and Table 1). However, the temporal coverage of these data are poor, as there are many gaps.

There are many uncertainties in the discharge and precipitation data in the Mara River basin. Discharge data analyses indicated that the time series were unreliable due to various inconsistencies in the data, especially at Mines and Amala. At Mines, a high scatter in the discharge-water-level graph was observed (Fig. 2) and back-calculated cross-section average flow velocities were below 1 m s^{-1} (Fig. S1 in the Supplement), whereas in 2012 the measured velocity was 2.13 m s^{-1} and the discharge 529.3 m^3 s^{-1} (GLOWS-FIU, 2012). At Amala, the rating curves were adjusted multiple times, affecting mostly the low flows. Only the rating curve at Nyangores was stable and consistent with field measurements. The precipitation data analysis showed a high spatial variability between the limited number of rainfall stations available. More information can be found in "S1 Data quality" in the Supplement.

During field trips, point discharge measurements were done in September and October 2014 at Emarti Bridge, Ser-

Table 2. Discharge measured in the field using an Acoustic Doppler Profiler (SonTek RiverSurveyor M9) mounted on a portable raft that is also equipped with a Power Communications Module and a DGPS (Rey et al., 2015).

Station name	Date	Mean discharge	Standard deviation
Emarti Bridge	13 Sep 2014	$19.2\,\mathrm{m^3\,s^{-1}}$	$0.7\,\mathrm{m^3\,s^{-1}}$
	4 Oct 2014	$13.4\,\mathrm{m^3\,s^{-1}}$	$0.6\,\mathrm{m^3\,s^{-1}}$
Serena Pump House	9 Oct 2014	$16.6\,\mathrm{m^3\,s^{-1}}$	$0.4\,\mathrm{m^3\,s^{-1}}$
New Mara Bridge	19 Sep 2014	$19.6\,\mathrm{m^3\,s^{-1}}$	$0.6\,\mathrm{m^3\,s^{-1}}$
	6 Oct 2014	$21.9\,\mathrm{m^3\,s^{-1}}$	$0.4\,\mathrm{m^3\,s^{-1}}$

ena Pump House and New Mara Bridge (see Table 2 and Fig. 3). At each location, the discharge was derived using an Acoustic Doppler Profiler (SonTek RiverSurveyor M9) mounted on a portable raft that is also equipped with a Power Communications Module and a Differential Global Positioning System (DGPS) antenna (Rey et al., 2015).

2.1.2 Remotely sensed data

Besides ground observations, remotely sensed data were also used for setting up the rainfall-runoff model. Catchment classification was based on topography and land cover. For the topography, a digital elevation map (SRTM) with a resolution of 90 m and vertical accuracy of 16 m was used (US Geological Survey, 2014). The land cover was based on Africover, a land cover database based on ground truth and satellite images (FAO-UN, 2002). For the climate, remotely sensed precipitation was used from the Famine Early Warning Systems Network (FEWS NET) on a daily timescale from 2001 to 2010 and monthly actual evaporation from USGS from 2001 to 2013. Moreover, normalised difference vegetation index (NDVI) maps derived from Landsat images were used to define parameter constraints.

3 Hydrological model setup for the Mara River basin

3.1 Catchment classification based on landscape and land use

For this study, the modelling concept of FLEX-Topo has been used (Savenije, 2010). It is a semi-distributed rainfall-runoff modelling framework that distinguishes hydrological response units (HRUs) based on landscape features. The landscape classes were identified based on the topographical indices HAND (Height Above Nearest Drain) and slope using a digital elevation map. Hillslopes are defined by a strong slope and high HAND, wetlands by a low HAND, and terraces by a high HAND and mild slope. The threshold for the slope (21.9 %) was based on a sensitivity analyses within the Mara basin, which revealed that the area of hillslopes changed asymptotically with the threshold. Therefore, the

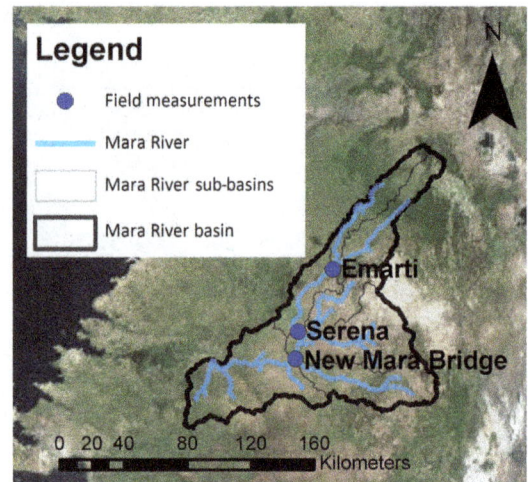

Figure 3. Map of discharge measurement locations during field trips in September and October 2014.

slope threshold was chosen at the point where changes in the sloped area become insignificant. As the wetland area was insignificant based on field observations, the HAND threshold was set to zero. In the Mara River basin, there are mainly terraces and hillslopes.

To further delimit these two main landscape units, the land cover is taken into account as well. In the upper sub-catchments, there are mainly croplands and forests, whereas further south, the land use is dominated by grasslands. In the lower sub-catchment, there are mostly croplands and grasslands. This resulted in four HRUs within the sub-basin of the Mara River basin, namely forested hillslopes, shrubs on hillslopes, agriculture and grassland (Figs. 4 and 5 and Table 3).

3.2 Hydrological model structure

Each HRU is represented by a lumped conceptual model; the model structure is based on the dominant flow processes observed during field trips or deducted from interviews with local people. For example, in forests and shrub lands, shallow subsurface flow (SSF) was seen to be the dominating flow mechanism; rainwater infiltrates into the soil and flows through preferential flow paths to the river. In contrast, grassland and cropland generate overland flow. The observed soil compaction, due to cattle trampling and ploughing, reduces the preferential infiltration capacity resulting in overland flow during heavy rainfall. Consequently, the Hortonian overland flow (HOF) occurs at high rainfall intensities exceeding the maximum infiltration capacity. The perception of the dominant flow mechanisms (Fig. 5) was then used to develop the model structure (Fig. 6). This approach of translating a perceptual model into a model concept (Beven, 2012) was applied successfully in previous FLEX-Topo applications (Gao et al., 2014a; Gharari et al., 2014).

Table 3. Classification results, namely the area percentage of each hydrological response unit per sub-catchment in the Mara River basin.

Sub-catchment	Agriculture	Shrubs on hillslopes	Grassland	Forested hillslopes
Amala	67 %	0 %	0 %	33 %
Nyangores	61 %	0 %	0 %	39 %
Middle	19 %	16 %	65 %	0 %
Lemek	10 %	39 %	51 %	0 %
Talek	0 %	21 %	79 %	0 %
Sand	0 %	42 %	58 %	0 %
Lower	26 %	23 %	52 %	0 %

The model structure contains multiple storage components schematised as reservoirs (Fig. 6). For each reservoir, the inflow, outflow and storage are defined by water balance equations, see Table 4. Process equations determine the fluxes between these reservoirs as a function of input drivers and their storage. HRUs function in parallel and independently from each other. However, they are connected to the groundwater system and the drainage network. To find the total runoff at the sub-catchment outlet $Q_{m,sub}$, the outflow $Q_{m,i}$ of each HRU is multiplied by its relative area and then added up together with the groundwater discharge Q_s. The relative area is the area of a specific HRU divided by the entire sub-catchment area. Subsequently, the modelled discharge at the catchment outlet is obtained by using a simple river routing technique, where a delay from sub-catchment outlet to catchment outlet was added assuming an average river flow velocity of $0.5\,\mathrm{m\,s^{-1}}$. In the Sand sub-catchment, it is schematised that runoff can percolate to the groundwater from the riverbed and that moisture can evaporate from the groundwater through deep rooting or riparian vegetation.

3.3 Model constraints

Parameters and process constraints were applied to eliminate unrealistic parameter combinations and constrain the flow volume. Parameter constraints were applied to the maximum interception, reservoir coefficients, the storage capacity in the root zone or on the surface, and the slope-roughness parameter, Table 5. Process constraints were applied to the runoff coefficient, groundwater recharge, interception and infiltration, Table 6. The effect of including these parameter and process constraints is illustrated in Fig. S5. For instance, the maximum storage in the unsaturated zone $S_{u,max}$ equals the root zone storage capacity and was estimated using the method of Gao et al. (2014b) based on remotely sensed precipitation and evaporation (Gao et al., 2014b; Wang-Erlandsson et al., 2016). The dry season evaporation has been derived from the actual evaporation using the NDVI.

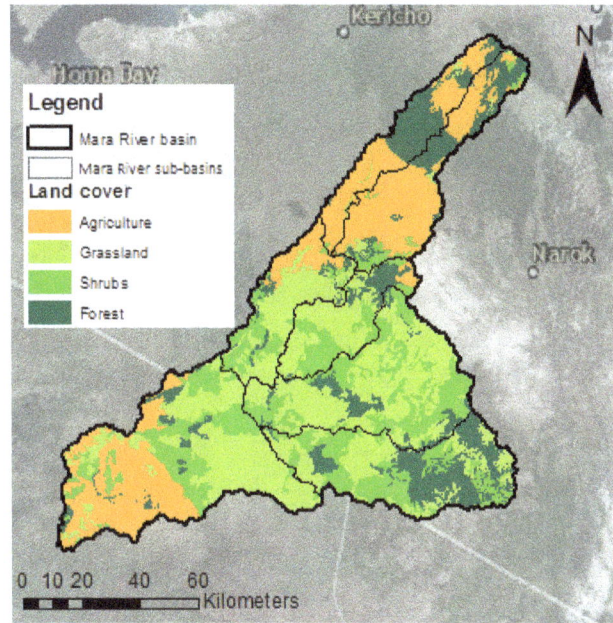

Figure 4. Classification of the Mara River basin into four hydrological response units for each sub-catchment based on land use and landscape.

3.4 Model calibration method using water levels

The hydrological model was calibrated on a daily timescale applying the MOSCEM-UA algorithm (Vrugt et al., 2003), with parameter ranges and values as indicated in Tables S1 and S2 in the Supplement. For the calibration, the Nash–Sutcliffe coefficient (Nash and Sutcliffe, 1970) was applied to the water-level duration curve (Eq. 1 linear, and Eq. 2 logscale). This frequently used objective function is advantageous, as it is sensitive not only to high flows, but also to low flows when using logarithmic values (Krause et al., 2005; McCuen Richard et al., 2006; Pushpalatha et al., 2012). By calibrating on the duration curve, the focus is on the flow statistics and not on the timing of individual flow peaks. This information is also in the time series. This is justified, since there were high uncertainties in the timings of floods events due to the limited number of available rainfall stations to capture the spatial variability of the rainfall input well. Therefore, duration curves were considered as a good signature for calibrating this model; this was also concluded in previous studies (Westerberg et al., 2011; Yadav et al., 2007). This signature was incorporated in the objective functions with the following equations:

$$\mathrm{NS_d} = 1 - \frac{\Sigma\left(h_{\mathrm{mod,sorted}} - h_{\mathrm{obs,sorted}}\right)^2}{\Sigma\left(h_{\mathrm{obs,sorted}} - h_{\mathrm{obs,avg}}\right)^2}, \tag{1}$$

$$\mathrm{NS_{log(d)}} = 1 - \frac{\Sigma\left(\log\left(h_{\mathrm{mod,sorted}}\right) - \log\left(h_{\mathrm{obs,sorted}}\right)\right)^2}{\Sigma\left(\log\left(h_{\mathrm{obs,sorted}}\right) - \log\left(h_{\mathrm{obs,avg}}\right)\right)^2}. \tag{2}$$

Table 4. Equations applied in the hydrological model. The formulas for the unsaturated zone are written for the hydrological response units, namely *forested hillslopes* and *shrubs on hillslopes*; for grass and agriculture, the inflow P_e changes to Q_F. The modelling time step is $\Delta t = 1$ day. Note that at a time daily step, the transfer of interception storage between consecutive days is assumed to be negligible.

Reservoir system	Water balance equation	Process functions
Interception	$\frac{\Delta S_i}{\Delta t} = P - P_e - E_i \approx 0$	$E_i = \min\left(E_p, \min\left(P, I_{\max}\right)\right)$
Surface	$\frac{\Delta S_o}{\Delta t} = P_e - Q_F - Q_{HOF} - E_o$	$Q_F = \min\left(\frac{S_o}{\Delta t}, F_{\max}\right)$ $Q_{HOF} = \max\frac{(0, S_o - S_{\max})}{\Delta t}$ $E_o = \max\left(0, \min\left(E_p - E_i, \frac{S_o}{\Delta t}\right)\right)$
Unsaturated zone	$\frac{\Delta S_u}{\Delta t} = (1 - C) \cdot P_e - E$	$C = 1 - \left(1 - \frac{S_u}{S_{u,\max}}\right)^{\beta}$ $E = \min\left(\left(E_p - E_i\right)\min\left(\frac{S_u}{\Delta t}, \left(E_p - E_i\right) \cdot \frac{S_u}{S_{u,\max}} \cdot \frac{1}{C_e}\right)\right)$
Groundwater recharge		$R_s = W \cdot C \cdot P_e$
Fast runoff	$\frac{\Delta S_f}{\Delta t} = R_{fl} - Q_f$	$R_{fl} = T_{lag}\left(C \cdot P_e - R_s\right) \rightarrow$ in a linear delay function T_{lag} $Q_f = \frac{S_f}{K_f}$
Groundwater	$\frac{\Delta S_s}{\Delta t} = R_{s,tot} - Q_s - E_s + Q_{inf}$	$R_{s,tot} = \sum_{i=1}^{i=4} R_s; \text{HRU}_i$ $Q_s = \frac{S_s}{K_s}$ $E_s = 0$ and $Q_{inf} = 0$ for all sub-basins except Sand $Q_{inf} = \min\left(\frac{S_{s,\max} - S_s}{\Delta t}, Q_f\right)$ for Sand sub-basin $E_s = \max\left(0, \min\left(E_p - E_i - E_o - E, \frac{S_s}{\Delta t}\right)\right)$ for Sand sub-basin
Total runoff		$Q_m = Q_s + \sum_{i=1}^{i=4} Q_f; \text{HRU}_i$

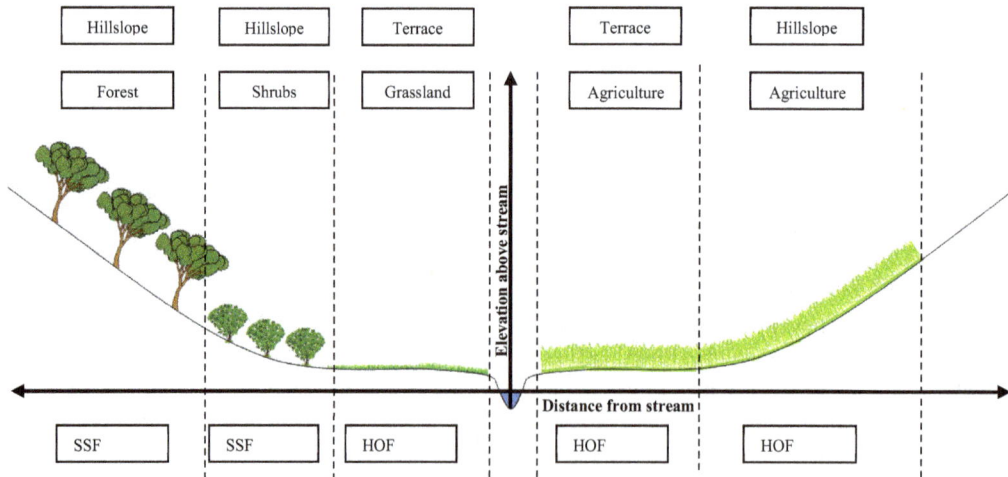

Hillslope	Hillslope	Terrace	Terrace	Hillslope
Forest	Shrubs	Grassland	Agriculture	Agriculture

Elevation above stream

Distance from stream

SSF	SSF	HOF	HOF	HOF

Figure 5. Schematisation of the landscape and land-use-based classification.

For the water-level-based calibration, the modelled discharge needs to be converted to the modelled water level. This calculation was done with the Strickler–Manning formula, in which the discharge is a function of the water level (Eq. 3) and where R is the hydraulic radius (Eq. 6), A the cross-sectional area (Eq. 5), i the slope, k the roughness and c the slope-roughness parameter (Eq. 4). The hydraulic radius and cross-section are a function of the water depth d, which is the water level subtracted h by the reference level h_0 (Eq. 7). The cross-sections were simplified as a trapezium with river

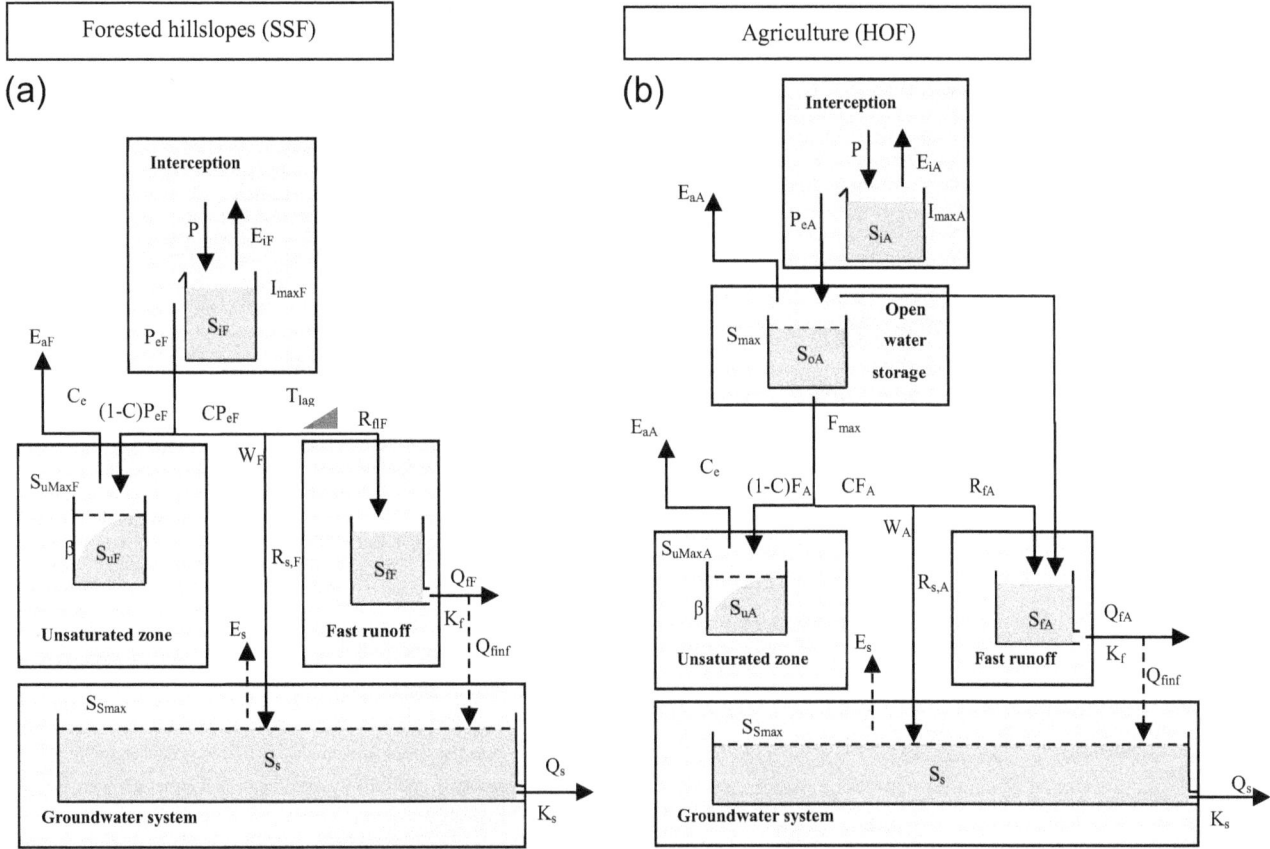

Figure 6. Model structure of the HRUs for *forested hillslopes* (**a**) and *agriculture* (**b**). The structure for *shrubs on hillslopes* is similar to the left one, replacing the indices F with S. The structure for *grassland* is similar to the right one, replacing the indices A with G. Parameters are marked in red and storages and fluxes in black. In terms of the symbol explanation for *fluxes*, precipitation is denoted by (P), evaporation of the interception zone by (E_i), actual evaporation by (E_a), evaporation from groundwater only applied in the sub-catchment Sand by (E_s), effective precipitation by (P_e), infiltration into the unsaturated zone by (F), discharge from unsaturated zone to the fast runoff zone by (R_f), groundwater recharge by (R_s), discharge from the fast runoff by (Q_f), infiltration into groundwater system only applied in the sub-catchment Sand by ($Q_{f,inf}$) and discharge from the slow runoff by (Q_s). For *storages*, storage in the interception zone is denoted by (S_i), open water storage by (S_o), storage in the root zone by (S_u), storage for the slow runoff by (S_s), storage for the fast runoff by (S_f). For the *remaining symbols*, splitter is denoted by (W) and (C), the soil moisture distribution coefficient by (β), the transpiration coefficient by ($C_e = 0.5$), and the reservoir coefficient by (K); indices f and s indicate the fast and slow runoff. Units used are for fluxes [mm day^{-1}], storages [mm], reservoir coefficient [day] and remaining parameters [–].

width B and two different riverbank slopes i_1 and i_2; these coefficients (Table 7) were estimated based on the available cross-section information (Figs. S6–S8). Since the slope and roughness are unknown, the slope-roughness parameter c was calibrated. The following equations were applied for these calculations:

$$Q = k \cdot i^{\frac{1}{2}} \cdot A \cdot R^{\frac{2}{3}} = c \cdot A \cdot R^{\frac{2}{3}}, \tag{3}$$

$$c = k \cdot i^{\frac{1}{2}}, \tag{4}$$

$$A = B \cdot d + \frac{1}{2} \cdot d \cdot (i_1 + i_2) \cdot d, \tag{5}$$

$$R = \frac{A}{B + d \cdot \left(\left(1 + i_1^2\right)^{\frac{1}{2}} + \left(1 + i_2^2\right)^{\frac{1}{2}} \right)}, \tag{6}$$

$$d = h - h_0. \tag{7}$$

This model calibration method, illustrated graphically in Fig. 7, was applied to three basins individually, namely to the entire river basin using the station Mines and for the sub-catchments Nyangores and Amala. At each location, the model was calibrated and validated for time periods indicated in Table 8; at Mines, two time periods were used for validation to maximise the use of the available ground measurements.

3.5 Rating curve analysis

After calibration, the modelled water levels and discharges were analysed. For the model calibration and validation, the modelled and recorded water levels were compared at the

Table 5. Overview of all parameter constraints applied in the hydrological model for the Mara River basin.

Parameter	Symbol	Formula	Comment
Interception	I_{max}	$I_{max,forest} > I_{max,grass}, I_{max,shrubs}, I_{max,cropland}$ $I_{max,shrubs} > I_{max,grass}, I_{max,cropland}$	Based on perception
Reservoir coefficient	K_s, K_f	$K_s > K_f$	Based on perception
Storage capacity in unsaturated zone	$S_{u,max}$	$S_{R,y_i} = \int P_e - E_d dt$ with: $\frac{E_d}{E_a} = \frac{NDVI_D}{NDVI_A}$ thus: $E_d = E_a \cdot \frac{NDVI_D}{NDVI_A}$	Based on NDVI, equivalent to the root zone storage capacity (Gao et al., 2014b) S_{R,y_i}: required storage for year i P_e: effective rainfall over dry season E_d: annual mean dry season evaporation, calculated assuming a linear relation between the evaporation and the NDVI E_a: actual mean annual evaporation $NDVI_D$: annual mean dry season NDVI $NDVI_A$: annual mean actual NDVI Through a statistical analysis of S_R using the Gumbel distribution, the storage capacity $S_{u,max}$ with a return period of 20 years is calculated.
Reservoir coefficient for groundwater system	K_s	$Q_s = Q_{t=0} \cdot \exp\left(-\frac{t}{K_s}\right)$	Based on hydrograph recession analysis Q_s: groundwater discharge
Maximum surface water storage	S_{max}	–	Based on DEM, assuming S_{max} is equal to the sink volumes
Slope-roughness parameter	c	$Q = c \cdot A \cdot R^{\frac{2}{3}} = u \cdot A$ $u = c \cdot R^{\frac{2}{3}} \rightarrow c_{calculated} = \frac{u}{R^{\frac{2}{3}}}$ $c_{calculated, -25\% error} < c < c_{calculated, +25\% error}$	Based on Strickler formula, cross-section data and a single discharge and velocity measurement at Mines allowing a wide error margin of 25 %

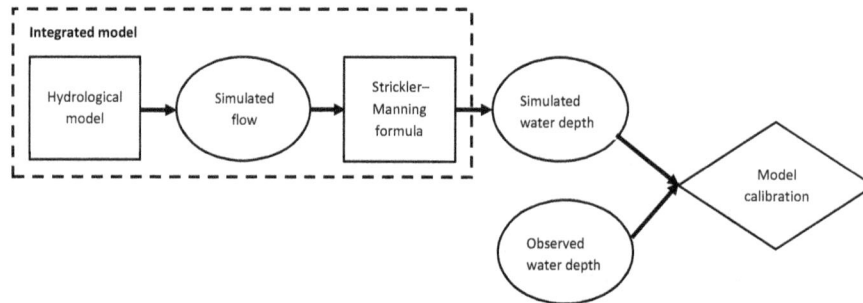

Figure 7. Flow chart of the proposed calibration method.

basin level, focusing on the time series and the duration curves. Hereafter, water-level–discharge relations were analysed taking two rating curves into consideration:

- The "recorded rating curve" relates Q_{rec} to h_{obs}.

- The "geometric rating curve" relates $Q_{Strickler}$ to h_{obs}.

The geometric rating curve relates the modelled discharge $Q_{Strickler}$ to the observed water level h_{obs}. This discharge $Q_{Strickler}$ was calculated with the Strickler–Manning formula using the calibrated slope-roughness parameter c,

cross-section data, and the observed water level h_{obs}. Therefore, the equation behind the geometric rating curve is basically the Strickler–Manning formula (Eq. 3) instead of the traditional rating curve equation (Eq. 8). The advantage of the Strickler–Manning formula is that only one parameter is unknown (riverbed slope and roughness c, Eq. 4), instead of two (fitting parameters a and b). However, the Strickler–Manning rating curve approach requires additional information on the cross-section. This is represented by

$$Q = a \cdot (h - h_0)^b. \tag{8}$$

Table 6. Overview of all process constraints applied in the hydrological model for the Mara River basin.

Process	Symbol	Formula	Comment
Average annual runoff coefficient	C	$C = 1 - \frac{E}{P} = e^{-\frac{E_p}{P}}$	Based on the Budyko curve using the 95 % percentile, hence the modelled average annual runoff coefficient should be below the 95-percentile of the observations
Groundwater recharge	R_s	$R_{s,F} > R_{s,C}, R_{s,G}$	Based on the assumption that deeper rooting vegetation creates preferential drainage patterns
Annual interception	E_i	$E_{i,F} > E_{i,G}, E_{i,S}$	Based on the assumption that the interception is higher in forests than in grassland and shrublands
Fast runoff infiltration	–	$f_{Q_{river}} < 3\,\mathrm{yr}^{-1}$	Frequency of river runoff. Based on interviews, locals seldom observed runoff more than 3 times a year.

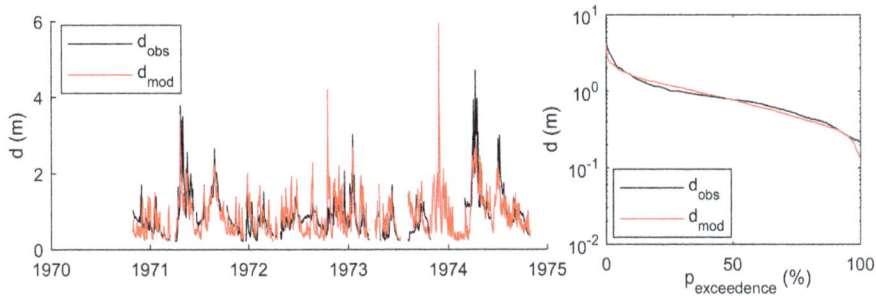

Figure 8. Model results at Mines during calibration for water depth time series and water depth exceedance.

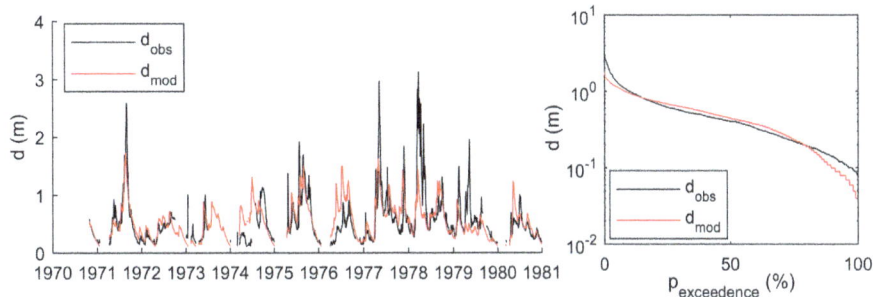

Figure 9. Model results at Nyangores during calibration for water depth time series and water depth exceedance.

4 Results and discussion

4.1 Water-level time series and duration curve

Model results were analysed graphically (Figs. 8 to 10 and Figs. S9 to S19) and numerically based on the Nash–Sutcliffe values for the objective functions (Table 9). The results of the objective functions indicate that at Nyangores and Mines, the calibration and validation results were consistent. At Mines, the modelled water level was simulated well, particularly with regard to the duration curve (Fig. 8). At individual events, there were substantial differences. In some years, for example in 1974, the observed data were very well represented by the model outcome. However, in other years this was not the case. In general, the model captured the dynamics in the water level well. This was the case during both calibration and validation (see Figs. S12 and S13).

At Nyangores the observed and modelled water levels were also similar during calibration and validation, extremely high flows excluded (Fig. 9). However, at Amala,

Table 7. Coefficients used for the river cross-section.

	Riverbank width B [m]	Riverbank slope i_1 [–]	Riverbank slope i_2 [–]	Reference level h_0 [m]
Amala	10.0	3.50	1.83	0
Nyangores	19.05	2.65	5.56	0
Mines	43.81	3.53	3.66	10

Table 8. Time periods used for the calibration and validation at the three basins of Mines, Nyangores and Amala.

	Mines	Nyangores	Amala
Calibration time period	1970–1974	1970–1980	1991–1992
Validation time period	1980–1981 1982–1983	1981–1992	1985–1986

the observed and modelled water levels differed significantly during calibration (Fig. 10) and validation (Fig. S15). The model missed several discharge events completely, likely related to missing rainfall events in the input data due to the high heterogeneity in precipitation.

4.2 Discharge at sub-catchment level

At Mines, the discharge originates from seven different sub-catchments, each with a different contribution. Based on field observations, the mountainous upstream sub-catchments from the north should have the largest contribution, whereas the contribution from the relatively drier and flatter Lemek and Talek tributaries from the eastern part of the catchment should be relatively low. The contribution of each sub-catchment to the total modelled discharge was assessed on a monthly timescale and compared with observations.

As shown in Fig. 11, the contribution varied throughout the year. In the summer (July–September), the modelled discharge mainly originates from the northern sub-catchments, Nyangores and Amala. However, in the winter (November–April), the modelled discharge mainly originates from the Sand and Lower sub-catchments. The eastern Middle, Talek and Lemek sub-catchments have the lowest discharge throughout the entire year, as has been similarly observed.

In previous studies, it has been shown that only a few discharge measurements can contain sufficient information to constrain model predictive uncertainties effectively (Seibert and Beven, 2009). To evaluate the model at the sub-catchment level, model results were compared with discharge measurements done during field trips in September and October 2014 at the Emarti Bridge, Serena Pump House and New Mara Bridge. At all three locations, the point measurements fitted well within the range of the modelled discharge (see Fig. 12).

4.3 Rating curve analysis

In this study, the recorded and geometric (Strickler–Manning) rating curves were compared (Fig. 13). At Mines, these two rating curves differed significantly. On the one hand, for medium to high flows, both the recorded and geometric rating curves run parallel, indicating similar cross-sectional properties; only the offset differed through changing riverbed levels. On the other hand, the simulated cross-section average flow velocities were realistic compared to the point measurements at Mines indicating that velocities are greater than $2 \, \mathrm{m \, s^{-1}}$ during high flows (see Fig. 13). At Nyangores, the recorded and geometric rating curves were almost identical, while there were significant differences at the Amala gauging station, especially in the low flows. Interestingly, these observations also hold for the validation period for all three stations.

The difference between the recorded and geometric rating curves at Mines probably resulted from uncertainties in the available recorded discharge data. In the complete discharge–water-level graphs for all available data (Fig. S2), large scatter was found. This could be the result of natural variability in the reference water level h_0 in the rating curve equation, which was not taken into account. A sensitivity analysis of the recorded rating curve equation at Mines showed that a deviation of 0.1 m in the reference water level could alter the discharge with 4 % for high flows and 46 % for low flows. However, a deviation of 0.5 m resulted in a 19 %–325 % change in the discharge. Therefore, unnoticed variations in the riverbed level strongly affect the uncertainty in the recorded rating curve at Mara Mines, which is located in a morphologically dynamic section of the river (Stoop, 2017).

At Amala, the difference between both rating curves could be related to the effect of missing rain events in the input data as result of the short time series for calibration and validation. This resulted in absent discharge peaks and hence an underestimation of the flow; these were the most extreme at Amala. During model calibration, this was compensated by increasing the parameter c in the Strickler–Manning formula (Eq. 4). As a result, discharge values not only increased during missed events, but also for all other days. The compensation effect was limited though, since the model was calibrated on the duration curves instead of the time series. As parameter c is linearly related to the geometric rating curve (Eq. 3), the latter was overestimated as well. Therefore, missing rain events in the input data resulted in the overestimation of the geometric rating curve.

In short, at the two stations with inconsistent rating curves, Amala and Mines, the geometric rating curve deviated significantly from the recordings. Strikingly, the deviations were observed at the same flow magnitudes where large inconsistencies were found in the observations, for instance in the low flows at Amala. However, at the gauging station with a reliable rating curve, Nyangores, the geometric and recorded discharge–water-level relations were almost identical.

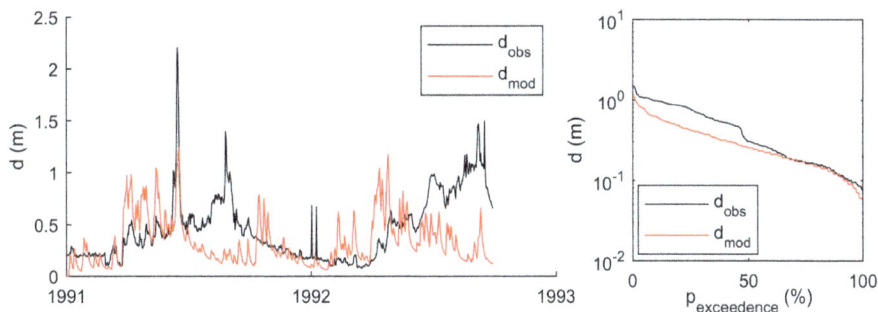

Figure 10. Model results at Amala during calibration for water depth time series and water depth exceedance.

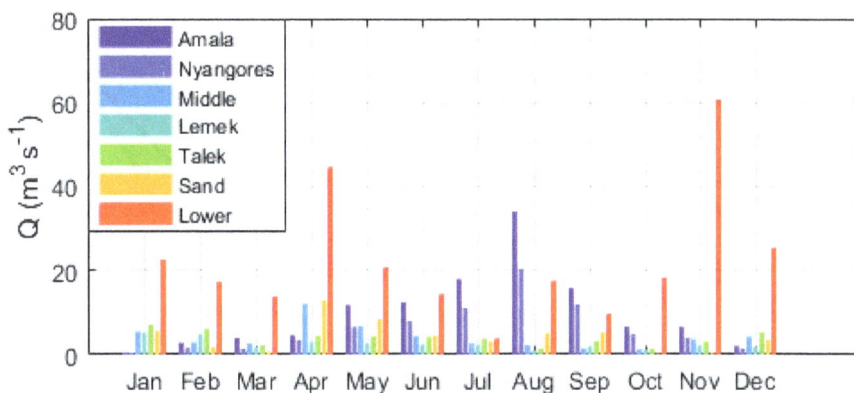

Figure 11. Monthly averaged modelled discharge for each sub-catchment.

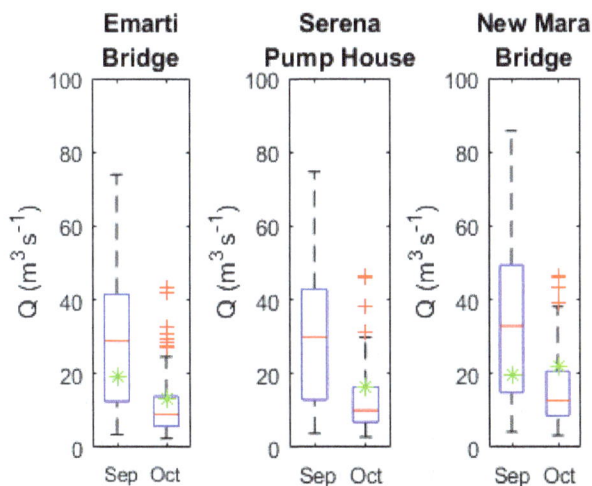

Figure 12. Box plot of the modelled discharge at three locations; the green asterisk represents the measured discharge in September and October 2014.

4.4 Limitations

This paper illustrated that the proposed water-level calibration method simulated the discharge-water-level relation well for the gauging station where consistent rating curve information was available. However, there are several limitations to this method. First, the slope-roughness parameter compensates for non-closure effects in the water balance, for instance due to errors in the precipitation, which is extremely heterogeneous in the semi-arid Mara basin. Unfortunately, this heterogeneity is poorly described in our study area with the available rain gauges (see Sect. S7.2 on the precipitation data analysis), influencing the modelling results. Therefore, this parameter should be constrained to minimise this compensation as much as possible. Second, the cross-section was assumed to be constant during the modelling time period. Data analyses indicated that expected changes in the river width or slope cannot affect the rating curve significantly. However, if this is not the case, then this cross-section change should be included during the model calibration.

In previous studies, river water-level time series were used for model calibration by using the Spearman rank correlation function (Seibert and Vis, 2016) or an inverse rating curve to convert the modelled discharge to water level (Jian et al., 2017). Compared to these approaches, the calibration method proposed in this paper has the following advantages: (1) Water-level time series are direct measurements and are therefore more reliable compared to processed data such as satellite based measurements. (2) Merely one new

Table 9. Overview of the values of the objective functions for each model simulation. Calibration was done based on the water level using $NS_{\log(h)}$ and NS_h; for comparison, objective functions using the discharge were added here as well.

	Nyangores		Amala		Mines		
	Calibration	Validation	Calibration	Validation	Calibration	Validation 1	Validation 2
$NS_{\log(d)}$	0.92	0.75	0.92	−0.23	0.97	0.81	0.93
NS_d	0.80	0.69	0.26	0.37	0.97	0.92	0.89
$NS_{\log(Q)}$	0.92	0.69	0.57	0.63	0.97	0.81	0.93
NS_Q	0.55	0.37	0.08	−1.67	0.90	0.76	0.77

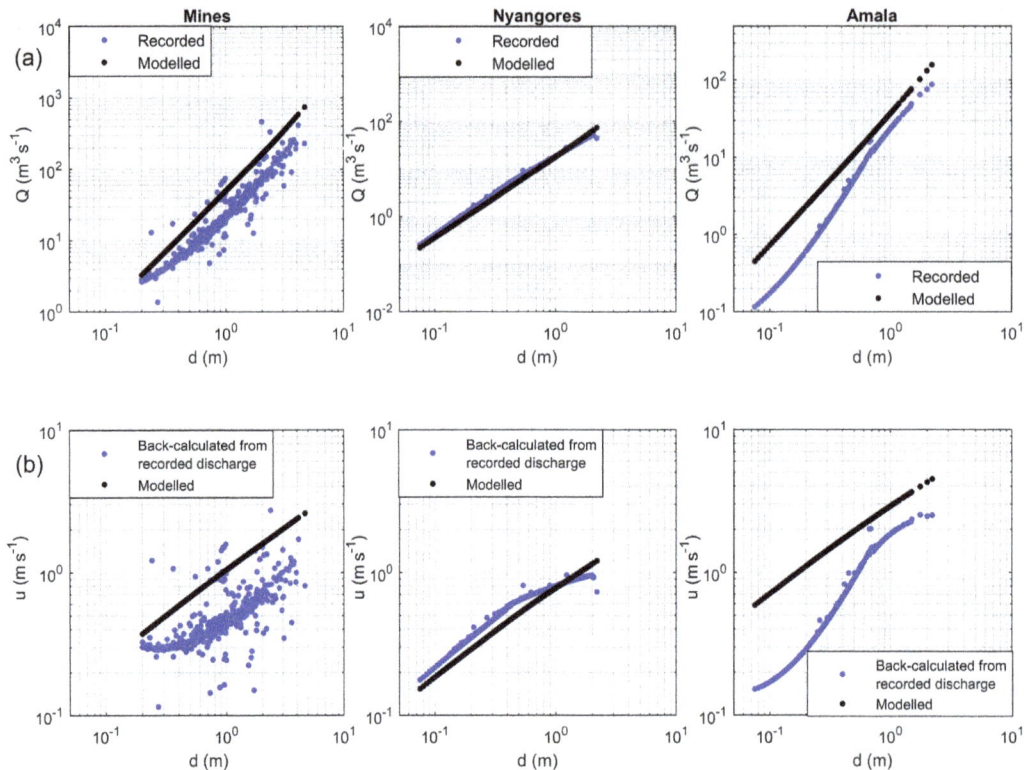

Figure 13. Model calibration results at Mines, Nyangores and Amala for discharge–water depth graphs **(a)** and velocity–water depth graphs **(b)**.

calibration parameter (the slope-roughness parameter) is introduced instead of three when using an inverse rating curve. (3) The model is calibrated on water-level magnitudes instead of only the ranks, which would introduce biases. However, this method also has several disadvantages; (1) cross-section information is needed and is assumed to be constant over the time period for which it is applied and (2) the newly introduced slope-roughness parameter compensates for non-closure effects in the water balance when not constrained well.

5 Summary and conclusion

The goal of this paper was to illustrate a new calibration method using water-level time series and the Strickler–Manning formula instead of discharge in a semi-arid and poorly gauged basin. This method offers a potential alternative for calibration on discharge data, as is also common practice in poorly gauged catchments. The semi-distributed rainfall-runoff modelling framework FLEX-Topo was applied. The catchment was divided into four hydrological response units (HRUs) and seven sub-catchments based on the river tributaries. For each HRU, a unique model structure was defined based on the observed dominant flow processes. By constraining the parameters and processes, unrealistic parameter sets were excluded from the calibration parameter set and the flow volume was constrained. This model was calibrated based on water levels to capture the flow dynamics. For this purpose, the modelled discharge was converted to water levels using the Strickler–Manning formula. The unknown slope-roughness parameter was calibrated.

An important output of this calibration approach is the "geometric rating curve equation", which relates the discharge to the water level using the Strickler–Manning formula. The geometric and recorded rating curves were significantly different at two gauging stations, namely Mines, the catchment outlet, and Amala, a sub-catchment outlet. At both locations, the deviations were at the same flow magnitudes where large inconsistencies were found in the observations. However, at the gauging station with a reliable rating curve, Nyangores, the recorded and geometric discharge-water-level relations were almost identical. In conclusion, this calibration method allows reliable simulations of the discharge–water-level relation, even in a data-poor region.

In addition, this paper analysed the current status of the hydro-meteorological network in the Mara River basin, focusing on the data availability and quality. Moreover, a hydrological model and an improved geometric rating curve equation were developed for this river. All three aspects contribute to improving the assessment of the water resources availability in the Mara River basin.

For future studies, it would be interesting to apply this calibration method to other studies of river basins with different climatic conditions and better data availability. Furthermore, it is recommended to assess the effect of rainfall uncertainties on this calibration method. Moreover, the hydrological model was calibrated on two signatures and two objective functions only. However, whether these signatures and objective functions provide sufficient information for calibration has not been analysed. Therefore, the procedures for water-level-based calibration should be analysed in more detail.

Author contributions. This paper has been co-authored by TAB and HHGS. Their contribution and support to this research has been valued very much.

Competing interests. The authors declare that they have no conflict of interest.

Acknowledgements. This research was part of the MaMaSe project (Mau Mara Serengeti) led by IHE Delft.

Edited by: Elena Toth

References

Alvisi, S., Mascellani, G., Franchini, M., and Bárdossy, A.: Water level forecasting through fuzzy logic and artificial neural network approaches, Hydrol. Earth Syst. Sci., 10, 1–17, https://doi.org/10.5194/hess-10-1-2006, 2006.

Beven, K. J.: Rainfall-runoff modelling: the primer, John Wiley & Sons, Chichester, England, https://doi.org/10.1002/9781119951001, 2012.

Bulygina, N. and Gupta, H.: Correcting the mathematical structure of a hydrological model via Bayesian data assimilation, Water Resour. Res., 47, https://doi.org/10.1029/2010WR009614, 2011.

Clarke, R. T.: Uncertainty in the estimation of mean annual flood due to rating-curve indefinition, J. Hydrol., 222, 185–190, https://doi.org/10.1016/S0022-1694(99)00097-9, 1999.

Dessu, S. B., Melesse, A. M., Bhat, M. G., and McClain, M. E.: Assessment of water resources availability and demand in the Mara River Basin, Catena, 115, 104–114, https://doi.org/10.1016/j.catena.2013.11.017, 2014.

Di Baldassarre, G. and Claps, P.: A hydraulic study on the applicability of flood rating curves, Hydrol. Res., 42, 10–19, https://doi.org/10.2166/nh.2010.098, 2011.

Di Baldassarre, G. and Montanari, A.: Uncertainty in river discharge observations: a quantitative analysis, Hydrol. Earth Syst. Sci., 13, 913–921, https://doi.org/10.5194/hess-13-913-2009, 2009.

Di Baldassarre, G., Laio, F., and Montanari, A.: Effect of observation errors on the uncertainty of design floods, Phys. Chem. Earth Pt. A/B/C, 42–44, 85–90, https://doi.org/10.1016/j.pce.2011.05.001, 2012.

Domeneghetti, A., Castellarin, A., and Brath, A.: Assessing rating-curve uncertainty and its effects on hydraulic model calibration, Hydrol. Earth Syst. Sci., 16, 1191–1202, https://doi.org/10.5194/hess-16-1191-2012, 2012.

FAO-UN: Multipurpose Landcover Database for Kenya – Africover, available at: http://www.fao.org/geonetwork/srv/en/metadata.show?id=38098&currTab=simple (last access: 27 September 2018), 2002.

Gao, H., Hrachowitz, M., Fenicia, F., Gharari, S., and Savenije, H. H. G.: Testing the realism of a topography-driven model (FLEX-Topo) in the nested catchments of the Upper Heihe, China, Hydrol. Earth Syst. Sci., 18, 1895–1915, https://doi.org/10.5194/hess-18-1895-2014, 2014a.

Gao, H., Hrachowitz, M., Schymanski, S. J., Fenicia, F., Sriwongsitanon, N., and Savenije, H. H. G.: Climate controls how ecosystems size the root zone storage capacity at catchment scale, Geophys. Res. Lett., 41, https://doi.org/10.1002/2014GL061668, 2014b.

Getirana, A. C. V.: Integrating spatial altimetry data into the automatic calibration of hydrological models, J. Hydrol., 387, 244–255, https://doi.org/10.1016/j.jhydrol.2010.04.013, 2010.

Gharari, S., Hrachowitz, M., Fenicia, F., Gao, H., and Savenije, H. H. G.: Using expert knowledge to increase realism in environmental system models can dramatically reduce the need for calibration, Hydrol. Earth Syst. Sci., 18, 4839–4859, https://doi.org/10.5194/hess-18-4839-2014, 2014.

GLOWS-FIU: Environmental Flow Recommendation for the Mara River, Kenya and Tanzania, in: Global Water for Sustainability program (GLOWS), Miami, FL, 2012.

Guerrero, J. L., Westerberg, I. K., Halldin, S., Xu, C. Y., and Lundin, L. C.: Temporal variability in stage-discharge relationships, J. Hydrol., 446–447, 90–102, https://doi.org/10.1016/j.jhydrol.2012.04.031, 2012.

Hrachowitz, M., Savenije, H. H. G., Blöschl, G., McDonnell, J. J., Sivapalan, M., Pomeroy, J. W., Arheimer, B., Blume, T., Clark, M. P., Ehret, U., Fenicia, F., Freer, J. E., Gelfan, A., Gupta, H. V., Hughes, D. A., Hut, R. W., Montanari, A., Pande, S., Tetzlaff, D., Troch, P. A., Uhlenbrook, S., Wagener, T., Winsemius, H. C., Woods, R. A., Zehe, E., and Cudennec, C.: A decade of Predictions in Ungauged Basins (PUB) – a review, Hydrolog. Sci. J., 58, 1198–1255, https://doi.org/10.1080/02626667.2013.803183, 2013.

Jalbert, J., Mathevet, T., and Favre, A. C.: Temporal uncertainty estimation of discharges from rating curves using a variographic analysis, J. Hydrol., 397, 83–92, https://doi.org/10.1016/j.jhydrol.2010.11.031, 2011.

Jian, J., Ryu, D., Costelloe, J. F., and Su, C.-H.: Towards hydrological model calibration using river level measurements, J. Hydrol.: Reg. Stud., 10, 95–109, https://doi.org/10.1016/j.ejrh.2016.12.085, 2017.

Karamuz, E., Osuch, M., and Romanowicz, R. J.: The influence of rating curve uncertainty on flow conditions in the River Vistula in Warsaw, in: Hydrodynamic and Mass Transport at Freshwater Aquatic Interfaces. GeoPlanet: Earth and Planetary Sciences, edited by: Rowiński P. and Marion A., Springer, Cham, 153–166, 2016.

Krause, P., Boyle, D. P., and Bäse, F.: Comparison of different efficiency criteria for hydrological model assessment, Adv. Geosci., 5, 89–97, https://doi.org/10.5194/adgeo-5-89-2005, 2005.

Kuczera, G.: Correlated rating curve error in flood frequency inference, Water Resour. Res., 32, 2119–2127, https://doi.org/10.1029/96WR00804, 1996.

Liu, W.-C. and Chung, C.-E.: Enhancing the Predicting Accuracy of the Water Stage Using a Physical-Based Model and an Artificial Neural Network-Genetic Algorithm in a River System, Water, 6, https://doi.org/10.3390/w6061642, 2014.

Mati, B. M., Mutie, S., Gadain, H., Home, P., and Mtalo, F.: Impacts of land-use/cover changes on the hydrology of the transboundary Mara River, Kenya/Tanzania, Lakes Reserv.: Res. Manage., 13, 169–177, 2008.

McClain, M. E., Subalusky, A. L., Anderson, E. P., Dessu, S. B., Melesse, A. M., Ndomba, P. M., Mtamba, J. O. D., Tamatamah, R. A., and Mligo, C.: Comparing flow regime, channel hydraulics and biological communities to infer flow-ecology relationships in the Mara River of Kenya and Tanzania, Hydrolog. Sci. J., 59, 1–19, https://doi.org/10.1080/02626667.2013.853121, 2013.

McCuen Richard, H., Knight, Z., and Cutter, A. G.: Evaluation of the Nash–Sutcliffe Efficiency Index, J. Hydrol. Eng., 11, 597–602, https://doi.org/10.1061/(ASCE)1084-0699(2006)11:6(597), 2006.

McMillan, H., Freer, J., Pappenberger, F., Krueger, T., and Clark, M.: Impacts of uncertain river flow data on rainfall-runoff model calibration and discharge predictions, Hydrol. Process., 24, 1270–1284, https://doi.org/10.1002/hyp.7587, 2010.

McMillan, H. K. and Westerberg, I. K.: Rating curve estimation under epistemic uncertainty, Hydrol. Process., 29, 1873–1882, https://doi.org/10.1002/hyp.10419, 2015.

Menne, M. J., Durre, I., Korzeniewski, B., McNeal, S., Thomas, K., Yin, X., Anthony, S., Ray, R., Vose, R. S., Gleason, B. E., and Houston, T. G.: Global Historical Climatology Network – Daily (GHCN-Daily), Version 3.12, https://doi.org/10.7289/V5D21VHZ, 2012.

Michailovsky, C. I., McEnnis, S., Berry, P. A. M., Smith, R., and Bauer-Gottwein, P.: River monitoring from satellite radar altimetry in the Zambezi River basin, Hydrol. Earth Syst. Sci., 16, 2181–2192, https://doi.org/10.5194/hess-16-2181-2012, 2012.

Morlot, T., Perret, C., Favre, A. C., and Jalbert, J.: Dynamic rating curve assessment for hydrometric stations and computation of the associated uncertainties: Quality and station management indicators, J. Hydrol., 517, 173–186, https://doi.org/10.1016/j.jhydrol.2014.05.007, 2014.

Nash, J. E. and Sutcliffe, J. V.: River flow forecasting through conceptual models part I – A discussion of principles, J. Hydrol., 10, 282–290, https://doi.org/10.1016/0022-1694(70)90255-6, 1970.

Paiva, R. C. D., Collischonn, W., Bonnet, M. P., de Gonçalves, L. G. G., Calmant, S., Getirana, A., and Santos da Silva, J.: Assimilating in situ and radar altimetry data into a large-scale hydrologic-hydrodynamic model for streamflow forecast in the Amazon, Hydrol. Earth Syst. Sci., 17, 2929–2946, https://doi.org/10.5194/hess-17-2929-2013, 2013.

Panda, R. K., Pramanik, N., and Bala, B.: Simulation of river stage using artificial neural network and MIKE 11 hydrodynamic model, Comput. Geosci., 36, 735–745, https://doi.org/10.1016/j.cageo.2009.07.012, 2010.

Pelletier, P. M.: Uncertainties in the single determination of river discharge: a literature review, Can. J. Civ. Eng., 15, 834–850, 1988.

Pereira-Cardenal, S. J., Riegels, N. D., Berry, P. A. M., Smith, R. G., Yakovlev, A., Siegfried, T. U., and Bauer-Gottwein, P.: Real-time remote sensing driven river basin modeling using radar altimetry, Hydrol. Earth Syst. Sci., 15, 241–254, https://doi.org/10.5194/hess-15-241-2011, 2011.

Petersen-Øverleir, A.: Modelling stage-discharge relationships affected by hysteresis using the Jones formula and nonlinear regression, Hydrolog. Sci. J., 51, 365–388, https://doi.org/10.1623/hysj.51.3.365, 2006.

Pushpalatha, R., Perrin, C., Moine, N. L., and Andréassian, V.: A review of efficiency criteria suitable for evaluating low-flow simulations, J. Hydrol., 420–421, 171–182, https://doi.org/10.1016/j.jhydrol.2011.11.055, 2012.

Revilla-Romero, B., Beck, H. E., Burek, P., Salamon, P., de Roo, A., and Thielen, J.: Filling the gaps: Calibrating a rainfall-runoff model using satellite-derived surface water extent, Remote Sens. Environ., 171, 118–131, https://doi.org/10.1016/j.rse.2015.10.022, 2015.

Rey, A., de Koning, D., Rongen, G., Merks, J., van der Meijs, R., and de Vries, S.: Water in the Mara Basin: Pioneer project for the MaMaSe project, unpublished MSc project report, Delft University of Technology, Delft, the Netherlands, 2015.

Ričko, M., Birkett, C. M., Carton, J. A., and Crétaux, J. F.: Intercomparison and validation of continental water level products derived from satellite radar altimetry, J. Appl. Remote Sens., 6, https://doi.org/10.1117/1.JRS.6.061710, 2012.

Savenije, H. H. G.: HESS Opinions "Topography driven conceptual modelling (FLEX-Topo)", Hydrol. Earth Syst. Sci., 14, 2681–2692, https://doi.org/10.5194/hess-14-2681-2010, 2010.

Schwatke, C., Dettmering, D., Bosch, W., and Seitz, F.: DAHITI – an innovative approach for estimating water level time series over inland waters using multi-mission satellite altimetry, Hydrol. Earth Syst. Sci., 19, 4345–4364, https://doi.org/10.5194/hess-19-4345-2015, 2015.

Seibert, J. and Beven, K. J.: Gauging the ungauged basin: how many discharge measurements are needed?, Hydrol. Earth Syst. Sci., 13, 883–892, https://doi.org/10.5194/hess-13-883-2009, 2009.

analysis of the SWAT model for two small Mediterranean catchments, Hydrolog. Sci. J., 58, 1635–1657, https://doi.org/10.1080/02626667.2013.837222, 2013.

Shahin, M.: Hydrology and Water Resources of Africa, Water Science and Technology Library, Springer, the Netherlands, 2002.

Sikorska, A. E., Scheidegger, A., Banasik, K., and Rieckermann, J.: Considering rating curve uncertainty in water level predictions, Hydrol. Earth Syst. Sci., 17, 4415–4427, https://doi.org/10.5194/hess-17-4415-2013, 2013.

Stoop, B. M.: Morphology of the Mara River: Assessment of the long term morphology and the effect on the riverine physical habitat, Delft University of Technology, Delft, the Netherlands, 2017.

Sun, W., Ishidaira, H., and Bastola, S.: Calibration of hydrological models in ungauged basins based on satellite radar altimetry observations of river water level, Hydrol. Process., 26, 3524–3537, https://doi.org/10.1002/hyp.8429, 2012.

Sun, W., Ishidaira, H., Bastola, S., and Yu, J.: Estimating daily time series of streamflow using hydrological model calibrated based on satellite observations of river water surface width: Toward real world applications, Environ. Res., 139, 36–45, https://doi.org/10.1016/j.envres.2015.01.002, 2015.

Thyer, M., Renard, B., Kavetski, D., Kuczera, G., and Clark, M.: Improving hydrological model predictions by incorporating rating curve uncertainty, in: Engineers Australia, Australia, Proceedings of the 34th IAHR World Congress, Brisbane, Australia, 26 June–1 July 2011, 1546–1553, 2011.

Tomkins, K. M.: Uncertainty in streamflow rating curves: Methods, controls and consequences, Hydrol. Process., 28, 464–481, https://doi.org/10.1002/hyp.9567, 2014.

Tourian, M. J., Schwatke, C., and Sneeuw, N.: River discharge estimation at daily resolution from satellite altimetry over an entire river basin, J. Hydrol., 546, 230–247, https://doi.org/10.1016/j.jhydrol.2017.01.009, 2017.

US Geological Survey: Digital Elevation Map: https://earthexplorer.usgs.gov/ (last access: 25 September 2018), 2014.

van Meerveld, H. J. I., Vis, M. J. P., and Seibert, J.: Information content of stream level class data for hydrological model calibration, Hydrol. Earth Syst. Sci., 21, 4895–4905, https://doi.org/10.5194/hess-21-4895-2017, 2017.

Vrugt, J. A., Gupta, H. V., Bastidas, L. A., Bouten, W., and Sorooshian, S.: Effective and efficient algorithm for multiobjective optimization of hydrologic models, Water Resour. Res., 39, https://doi.org/10.1029/2002WR001746, 2003.

Wang-Erlandsson, L., Bastiaanssen, W. G. M., Gao, H., Jägermeyr, J., Senay, G. B., van Dijk, A. I. J. M., Guerschman, J. P., Keys, P. W., Gordon, L. J., and Savenije, H. H. G.: Global root zone storage capacity from satellite-based evaporation, Hydrol. Earth Syst. Sci., 20, 1459–1481, https://doi.org/10.5194/hess-20-1459-2016, 2016.

Westerberg, I. K., Guerrero, J. L., Younger, P. M., Beven, K. J., Seibert, J., Halldin, S., Freer, J. E., and Xu, C. Y.: Calibration of hydrological models using flow-duration curves, Hydrol. Earth Syst. Sci., 15, 2205–2227, https://doi.org/10.5194/hess-15-2205-2011, 2011.

WMO: World Meteorological Organization: Guide to Hydrological Practices, Volume I, Hydrology – From Measurement to Hydrological Information, No. 168, Sixth edition, Geneva, Switzerland, 2008.

Yadav, M., Wagener, T., and Gupta, H.: Regionalization of constraints on expected watershed response behavior for improved predictions in ungauged basins, Adv. Water Resour., 30, 1756–1774, https://doi.org/10.1016/j.advwatres.2007.01.005, 2007.

Yan, K., Di Baldassarre, G., Solomatine, D. P., and Schumann, G. J. P.: A review of low-cost space-borne data for flood modelling: topography, flood extent and water level, Hydrol. Process., 29, 3368–3387, https://doi.org/10.1002/hyp.10449, 2015.

8

Tree-, stand- and site-specific controls on landscape-scale patterns of transpiration

Sibylle Kathrin Hassler[1,2]**, Markus Weiler**[3]**, and Theresa Blume**[2]

[1]Karlsruhe Institute of Technology (KIT), Institute of Water and River Basin Management, Chair of Hydrology, Karlsruhe, Germany

[2]Helmholtz Centre Potsdam, GFZ German Research Centre for Geosciences, Section Hydrology, Potsdam, Germany

[3]Hydrology, Faculty of Environment and Natural Resources, University of Freiburg, Freiburg, Germany

Correspondence: Sibylle Kathrin Hassler (sibylle.hassler@kit.edu)

Abstract. Transpiration is a key process in the hydrological cycle, and a sound understanding and quantification of transpiration and its spatial variability is essential for management decisions as well as for improving the parameterisation and evaluation of hydrological and soil–vegetation–atmosphere transfer models. For individual trees, transpiration is commonly estimated by measuring sap flow. Besides evaporative demand and water availability, tree-specific characteristics such as species, size or social status control sap flow amounts of individual trees. Within forest stands, properties such as species composition, basal area or stand density additionally affect sap flow, for example via competition mechanisms. Finally, sap flow patterns might also be influenced by landscape-scale characteristics such as geology and soils, slope position or aspect because they affect water and energy availability; however, little is known about the dynamic interplay of these controls.

We studied the relative importance of various tree-, stand- and site-specific characteristics with multiple linear regression models to explain the variability of sap velocity measurements in 61 beech and oak trees, located at 24 sites across a 290 km^2 catchment in Luxembourg. For each of 132 consecutive days of the growing season of 2014 we modelled the daily sap velocity and derived sap flow patterns of these 61 trees, and we determined the importance of the different controls.

Results indicate that a combination of mainly tree- and site-specific factors controls sap velocity patterns in the landscape, namely tree species, tree diameter, geology and aspect. For sap flow we included only the stand- and site-specific predictors in the models to ensure variable independence. Of those, geology and aspect were most important. Compared to these predictors, spatial variability of atmospheric demand and soil moisture explains only a small fraction of the variability in the daily datasets. However, the temporal dynamics of the explanatory power of the tree-specific characteristics, especially species, are correlated to the temporal dynamics of potential evaporation. We conclude that transpiration estimates on the landscape scale would benefit from not only consideration of hydro-meteorological drivers, but also tree, stand and site characteristics in order to improve the spatial and temporal representation of transpiration for hydrological and soil–vegetation–atmosphere transfer models.

1 Introduction

Transpiration makes up 65 % of total terrestrial evapotranspiration and it is a key process in the hydrological cycle, but knowledge about transpiration fluxes in landscapes is still poor (Jasechko et al., 2013). While the main atmospheric drivers for transpiration are radiation and vapour pressure deficit, the most important terrestrial controls of this water flux are plant physiological properties and soil characteristics. The magnitude and dynamics of transpiration in turn affect the system's energy balance, soil water storage, groundwater recharge and stream flow (Barnard et al., 2010; Bond et al., 2002; Fahle and Dietrich, 2014; Moore et al., 2011; Pielke Sr., 2005). Spatial patterns of transpiration affect hydrological processes and feedbacks within the catchment and are therefore important to consider in distributed hydrological modelling. While most of these models rely on estimates of evapotranspiration gained from meteorological measurements, for example using the Penman–Monteith equation, a better representation of spatio-temporal transpiration dynamics can inform model setups (Fenicia et al., 2016), serve for multi-response evaluation of models (Loritz et al., 2017; Scudeler et al., 2016) and improve model performance (Seibert et al., 2017). However, studies on the influences on spatial patterns of transpiration in landscapes are still scarce.

Methods to measure transpiration span a wide range of scales, from water and CO_2-exchange measurements on individual leaves to characterisation of the convective boundary layer which integrates transpiration on a landscape scale. On the plot and stand-level scale, eddy-covariance techniques are applied, whereas on a tree scale, measuring xylem sap velocity and deriving sap flow by including an estimate of the sapwood area is a common method. Determining transpiration of stands using sap flow entails the challenges of reliably estimating whole-tree water use and applying appropriate empirical relationships when upscaling to stands (Köstner et al., 1998). However, for the investigation of the main controls for individual trees' water use, sap flow measurements are a suitable tool.

Atmospheric conditions and water availability are the main temporally variable abiotic controls for sap flow, influencing hourly, daily and yearly dynamics (Bovard et al., 2005; Clausnitzer et al., 2011; Ghimire et al., 2014; Granier et al., 2000; Oren et al., 1996; Schume et al., 2004). However, tree-, stand- or site-specific characteristics can also govern the magnitude of sap flow. Under similar external conditions, different tree species show contrasts in sap flow due to their different hydraulic architectures and mechanisms for coping with water stress (Bovard et al., 2005; Gebauer et al., 2012; Oren and Pataki, 2001; Traver et al., 2010). Tree diameter and thus tree size and crown area affects not only sap flow rates, but also radial sap velocity patterns (Bosch et al., 2014; Hölscher et al., 2005; Lüttschwager and Remus, 2007; Vertessy et al., 1995). Within stands, variation in sap velocity can occur because of competition for light and water resources, depending on the species composition (Cienciala et al., 2002; Dalsgaard et al., 2011; Gebauer et al., 2012; Oren and Pataki, 2001; Vincke et al., 2005).

On the landscape scale, site-specific characteristics such as geology, soil type, soil depth or depth to groundwater, elevation, slope position and aspect could potentially control spatial sap flow patterns because of their influence on water and energy availability. Many of these characteristics can be derived from maps and digital elevation models, and quantifying their importance is thus especially interesting for modelling purposes requiring landscape-scale transpiration. For instance, the geological setting and associated soil types determine soil water holding capacities, the location of the tree within the landscape's topography can influence its access to

groundwater resources and the stand's microclimatic conditions, and differences of aspect also entail variation in energy input (Čermák and Prax, 2001; Vilhar et al., 2005). However, few studies have focused on the relative strength and possible temporal dynamics of these controls. While the impact of differences in accessible soil volume and groundwater depth on sap flow dynamics has been well described (Angstmann et al., 2013; Čermák and Prax, 2001; Tromp-van Meerveld and McDonnell, 2006), there have been few attempts to empirically use geological or soil units as large-scale proxies for water availability or potentially also for rooting depth limitations (Boer-Euser et al., 2016).

Slope position and elevation have been investigated at a few sites as site-specific controls of sap flow that possibly influence soil characteristics and microclimate. Bond et al. (2002) report no significant differences in sap flow with slope position for red alders and Douglas fir in Oregon, whereas Kumagai et al. (2007) found larger sap flux density values for cedars in a downslope stand compared to upslope trees; however, this effect was confounded by differences in tree size and stand structure, so that transpiration for the stands did not differ between the two slope positions. Similarly, in a drought-prone eucalypt forest in Australia, Mitchell et al. (2012) attribute lower sap flow values at their upslope plot compared to downslope positions to the differences in stand structure (lower basal area (BA) and sapwood area) and lower LAI. Otieno et al. (2014) compared two stands of subtropical evergreen forest in China at two different elevations and highlighted the structural differences of the two stands, but did not find differences in stand transpiration. However, differences were found among individual trees and were attributed to tree size as well as social position of the crown. Jung et al. (2014) studied the elevation effect in deciduous forests on a mountain slope in South Korea at three different elevations, at 450 m, 650 m and 950 a.m.s.l, and found a decrease of total annual canopy transpiration with elevation as a consequence of decreasing length of the growing season, and hence of differences in local climate. Maximum sap flux density of individual trees during clear-sky days, however, did not vary significantly due to these effects. Using a geostatistical approach, Adelman et al. (2008) studied a suspected influence on transpiration due to differences in water availability on a slope inducing contrasts in species composition; however, they did not see this effect in the data, possibly because of overall seasonal dryness during the study period. Another study on controls of patterns of spatial autocorrelation in extensive sap flow dataset found the clear species influence on transpiration patterns; however, the effect of a slope-related moisture gradient could not be confirmed (Loranty et al., 2008), adding to the contrasting findings about the influence of slope position on transpiration we see in the literature. Even though hillslope aspect at least partially controls radiation input, sap flow studies on the influence of aspect are scarce. In a simulation study, Holst et al. (2010) examined water balances for two beech stands on opposite slopes in south-western Germany and found higher transpiration values for the south-west slope compared to the north-east slope, which the authors explained with the higher evaporative demand and higher precipitation input on that slope. Focusing on limits of atmospheric exchange, Renner et al. (2016) found that stand composition compensated for differences in sensitivities of sap velocity to evaporative demand on the south- and north-facing slopes of a valley transect, which led to overall similar transpiration rates on both slopes.

To summarise, the reported studies have shown that in addition to the obvious atmospheric and tree-scale physiological controls, site-specific characteristics can influence sap flow patterns in a landscape. So far this influence has mainly been studied as individual plot comparisons or on a seasonal basis. However, this approach does not provide information on the possible short-term, day-to-day changes in the importance of the different controls as a consequence of varying hydro-meteorological conditions. Yet estimating the dynamics of the various controls of sap flow is essential for understanding and predicting spatial patterns of transpiration on a landscape scale.

In this study we aim to explore daily spatial patterns of sap velocity and derived sap flow on the landscape scale, by applying multiple linear regression models and identifying the influence of tree-, stand- and site-specific characteristics that could be gained from maps or surveys and hence would be available for modelling purposes. We also examine the temporal dynamics of these influences and to what extent this can be linked to hydro-meteorological conditions. Our analysis is based on an extensive sap velocity dataset, measured on 61 beech and oak trees on 132 consecutive days in the growing season of 2014, spread over 24 locations in a 290 km^2 catchment in Luxembourg.

2 Methods

2.1 Study site

The study site is located in the Attert catchment in western Luxembourg. The catchment covers three geological units (Fig. 1), predominantly Devonian schists of the Ardennes massif in the north-west, Triassic sandy marls, and a small area underlain by Luxembourg sandstone (Jurassic) on the southern catchment border (Martínez-Carreras et al., 2012). These different geological units gave rise to soils with different water retention properties. The soils on schists developed to haplic Cambisols, the soils on marls can be classified as different types of Stagnosols depending on their clay content of 20–60 % and the sandy textures on the Luxembourg sandstone gave rise to Arenosols. The soils were classified according to the WRB classification system (IUSS Working Group WRB, 2006) and described further by Sprenger et al. (2016). Plant available water was determined from

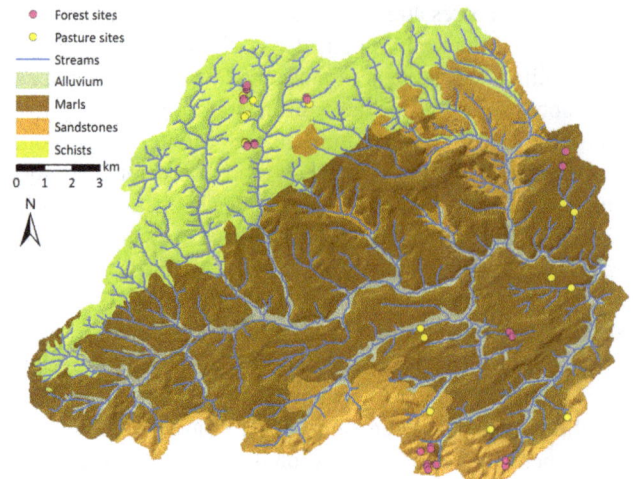

Figure 1. Map of the study site, the Attert catchment in Luxembourg.

mean water retention curves (using water tensions at 60 hPa and 10$^{4.2}$ hPa) based on 120 soil samples (Jackisch, 2015), amounting to 0.30 m^3 m^{-3} for Cambisols and Stagnosols and 0.25 m^3 m^{-3} for the Arenosols. However, the access to water is not only determined by the soil type. For example the Cambisols in the schist are very shallow and of high rock content. There are cracks filled with soil material in the underlying schist which could provide water for tree roots. The Stagnosols in the marl area are very clayey in the subsurface, probably limiting plant-available water resources and root penetration in these layers. We observed maximum rooting depths, averaged for each soil type, of 68 cm for the Cambisols, 90 cm for the Stagnosols and 98 cm for the Arenosols (Sprenger et al., 2016). Mean annual precipitation of the study area is approximately 850 mm (Pfister et al., 2000). Land use varies from mainly pasture and agriculture in the marl area and mainly forests in the sandstone to a mixture of agriculture and pasture on the plateaus and forests on the steep slopes of the schist area.

The catchment is the focus area of the CAOS (Catchments As Organised Systems) research unit which investigates landscape-scale structures, patterns and interactions in hydrological processes for model development (Zehe et al., 2014). A monitoring network of 45 sensor clusters was installed in 2012–2013, covering the different geological units, the land use types deciduous forest and pasture, and different slope positions and aspects. Measurements at the individual sites include meteorological parameters such as air temperature and humidity (Campbell CS215) and solar radiation (Apogee Pyranometer SP110) as well as soil moisture (Decagon 5TE) at three depths and three locations at each site. Sap flow is monitored with East 30 Sap Flow Sensors at all 29 forest sites.

The forests covered by the monitoring network mainly consist of mixed deciduous stands with European beech (*Fa-*

gus sylvatica L.), pedunculate and sessile oak (Quercus robur L. and Q. petraea (Matt.) Liebl., common hornbeam (Carpinus betulus L.), and a few maples (Acer pseudoplatanus L.) and alders (Alnus glutinosa (L.) Gaertn.). However, in this study only the most common species, beech and oaks, are considered. Identification to species level was not possible for the two oak species as they both occur in the study area, show morphologically intermediate forms and are possibly hybridised (Elsner, 1993; Zanetto et al., 1994). Our term "oak" refers to the whole group and we are aware that the oak species might differ somewhat in their transpiration characteristics, but the physiological contrast to beech trees should by far surpass these differences.

2.2 Sap velocity measurements and calculation of sap flow

The sites in the forest were characterised with a forest inventory on $20 \times 20\,\text{m}^2$ plots, recording stem numbers, diameter at breast height (DBH) and BA for all trees with a circumference of more than 4 cm, and tree height for a representative subset of the trees in the stand. Heights were gauged roughly as the canopy tops were not always clearly visible and we were interested in the social status of the trees rather than the precise height. Four trees per site were selected for sap flow sensor installation. The tree species and diameter were chosen to roughly represent the stand structure at the site but also allow a comparison to other sites where possible. Sap flow sensors were installed at breast height on the north-facing side of the stem and protected with a reflective cover to minimise the effects of radiation-induced changes in stem temperatures. After removing the bark, holes for the sensors were drilled using a drilling guide to ensure parallel installation of the sensor needles. The sensors, manufactured by East 30 Sensors in Washington, US, use the heat ratio method with a central heater needle and a thermistor needle upstream and downstream of the heater. Each thermistor needle contains three thermistors, at 5, 18 and 30 mm depth in the wood. Sap velocities (V_{sap} in m s^{-1}) at each of these locations are calculated from the temperatures measured at the corresponding thermistor pairs according to Eq. (1) (equations after Campbell et al., 1991):

$$V_{\text{sap}} = \frac{2k}{C_w(r_u + r_d)} \ln\left(\frac{\Delta T_u}{\Delta T_d}\right), \tag{1}$$

where k is the thermal conductivity of the sapwood, set to $0.5\,\text{W m}^{-1}\,\text{K}^{-1}$, C_w is the specific heat capacity of water ($\text{J m}^{-3}\,\text{K}^{-1}$), r is the distance (m) from the heater needle to the thermistor needle (in our case 6 mm) and ΔT is the temperature difference (K) before heating and 60 s after the heat pulse. Subscripts u and d stand for location upstream and downstream of the heater.

We corrected these values to account for wounding of the xylem tissue caused by the drilling according to the numerical model solutions for the heat pulse velocity method suggested by Burgess et al. (2001):

$$V_c = bV_{\text{sap}} + cV_{\text{sap}}^2 + dV_{\text{sap}}^3, \tag{2}$$

where V_c is the corrected sap velocity (m s^{-1}) and b, c and d are correction coefficients; for our 2 mm wounds we use $b = 1.8558$, $c = -0.0018\,\text{s m}^{-1}$ and $d = 0.0003\,\text{s}^2\,\text{m}^{-2}$ (Burgess et al., 2001). We used daytime sap velocity, averaged over a 12 h window from 08:00 to 20:00 LT. Our sensors provide measurements at three depths (5, 18 and 30 mm) within the sapwood, but because the sensors were not always ideally installed in the sapwood, we use the maximum value of these three depths' velocities for the sap velocity part of our analyses. The heat ratio method has been reported to underestimate high sap velocities (Vandegehuchte and Steppe, 2013; Fuchs et al., 2017), and the highly conductive earlywood vessels in the ring-porous oaks might exhibit locally high velocities. However, the trees in our study rarely reach the reported critical values and the oaks seem to plateau at even lower values, which is unlikely to be the result of sensor limitations.

Calculation of sap flow from sap velocity requires estimates of sapwood area and bark thickness. Sapwood area was calculated using the power law function for sapwood area based on DBH, which was originally developed by Vertessy (1995). Coefficients for beech were taken from Gebauer et al. (2008), and for oak (Q. petraea (Matt.) Liebl.) from Schmidt (2007), yielding the following equations:

$$A_{\text{S\_B}} = 0.778 \cdot \text{DBH}_B^{1.917}, \tag{3}$$

$$A_{\text{S\_O}} = 0.065 \cdot \text{DBH}_O^{2.264}, \tag{4}$$

with A_S for the sapwood area (cm^2), DBH is the diameter at breast height (cm), and subscripts are O for oak and B for beech.

The next step was to calculate the sapwood depth. From the whole-tree diameter we first subtracted an estimate for the bark to consider only the sapwood and heartwood part of the stem in the subsequent calculations. Bark thickness was estimated according to empirical relations developed by Rössler (2008):

$$d_{\text{b\_B}} = 2.61029 + 0.28522 \cdot \text{DBH}_B, \tag{5}$$

$$d_{\text{b\_O}} = 9.88855 + 0.56734 \cdot \text{DBH}_O, \tag{6}$$

where d_b is the double bark thickness (mm), DBH and subscripts are analogous to Eqs. (3) and (4).

Then we calculated the depth of the sapwood–heartwood boundary. As our sensors measure at the three depths 5, 18 and 30 mm, we assigned the corresponding velocities to the stem sections 0–15 and 15–25 mm and assumed a linear decline from 25 mm depth the up to the sapwood–heartwood boundary, using the 30 mm velocity as the maximum value of the linear decline (as used by Renner et al., 2016). The linear decline mainly applied to beech trees, as most of the smaller oaks' sap velocities at 30 mm depth were already zero. Daily sap flow for each tree was then derived by multiplying each depth's sap velocity (averaged over the 12 h from 08:00 to 20:00 LT as we did before) with the respective sapwood area sections.

We included sap flow in our analyses because, compared to sap velocity, it provides a better estimate of tree transpiration and is usually more of interest for hydrologists. It needs to be noted that the calculation relies on the published species-specific empirical relationships for sapwood area based on tree diameter (e.g. Gebauer et al., 2008; Meinzer et al., 2005; Vertessy et al., 1995). Other potential controls on transpiration such as topography or geology as proxies for rooting depth or water availability are not considered in these equations. As site characteristics can induce ecophysiological adaptations, for example in tree functional traits such as stomata density, xylem vessel diameters or hydraulic conductivity (Hajek et al., 2016; Stojnić et al., 2015), sapwood area properties might be similarly adapted to site conditions. However, to our knowledge there are no studies on the landscape scale yet which examine these adaptations.

Sap velocity is independent of these considerations; therefore, we mainly focus our analyses on this variable and use sap flow as a tentative comparison. For a more reliable way of estimating sap flow influences in a diverse landscape, the sapwood area would need to be measured directly for each tree.

2.3 Auxiliary variables: estimating potential evaporation and water availability

The main environmental limitations to sap flow are the atmospheric conditions (the solar heating of the leaves, water vapour pressure deficit, etc.) as the driving gradient for transpiration and the water supply to the trees. We assess these influences by using a thermodynamically derived measure for potential evaporation E_{pot} which has been recently developed by Kleidon and Renner (2013) as well as the soil moisture observations at each sites as a measure of water availability.

Soil moisture was measured in three profiles per site at 10, 30 and 50 cm depth. For our analyses we took the average across all depths and profiles to estimate the average soil moisture in the top 60 cm for each site.

E_{pot} was calculated as follows (Kleidon and Renner, 2013):

$$E_{pot} = \frac{1}{\lambda} \frac{s}{s + \gamma} \frac{R_{sn}}{2}, \qquad (7)$$

where $\lambda = 2.5 \times 10^6\,\mathrm{J\,kg^{-1}}$ is the latent heat of vaporisation, s is an empirical approximation of the slope of the saturation vapour pressure curve, calculated as $s = s(T) = 6.11 \cdot 5417 \cdot T^{-2} \cdot e^{19.83 - 5417/T}$, with the temperature T (K). The psychrometric constant was approximated as $\gamma \approx 65\,\mathrm{PaK^{-1}}$, and R_{sn} (W m^{-2}) is absorbed solar radiation. The air temperature was

taken from the measurements within the stands at the forest sites. This gives room for some error, as the below-canopy temperature will differ from the above-canopy temperature. However, the temperature does not have very strong leverage in Eq. (7) and we would expect even larger errors if we were to use air temperature from nearby grassland sites because of the differences in microclimate and energy balance for the different land covers. Solar radiation features more prominently in the equation and therefore needs careful estimation. We use an approach deriving the above-canopy radiation from the digital elevation model of the catchment, using the GRASS GIS package "r.sun". This method corrects for latitude, day of year and topography and corresponds well with the measured radiation at the pasture sites for cloudless days. Dividing the r.sun-estimated radiation values with the measured radiation values at each pasture site yields correction factors for actual cloud conditions for each day. We apply this cloud correction to the r.sun values for the forests, using the pasture site that is closest to the respective forest sites. These latitude-, topography- and cloud-corrected radiation estimates are then used for calculating E_{pot}. Studying transpiration along a hillslope transect, Renner et al. (2016) found E_{pot} comparable (if slightly underestimating) to a traditional Penman–Monteith approach and also tested for effects of vapour pressure deficit and wind speed. The results did not show distinct effects, and hence we used E_{pot} as a robust measure, which is appropriate to the available atmospheric measurements in our study. We also used radiation measurements within and outside the stands to determine the period when the canopy is fully developed and only use this period for our analyses. For the year 2014 this period lasted from 11 May to 20 September, amounting to 132 days.

2.4 Data analysis

We selected a dataset of continuous sap velocity measurements from 61 trees located at 24 of the altogether 29 forest sites in the CAOS dataset. Each of the monitored trees was associated with tree-, stand- and site-specific properties. Tree-specific properties were the species, diameter at breast height (DBH) and tree height, whereas the stands were characterised by the measurements undertaken in the forest inventory, namely BA and median DBH of the stand as well as the number of stems recorded on the inventory plot. Additionally, there were several landscape attributes which could be associated with the monitored trees such as their position within one of the three geological units, their location on a hillslope and the aspect of that slope. These attributes could be considered as proxies for associated soil properties and energy availability, influencing water availability and potential evaporation. The site characteristics and species entered the linear models as categorical variables. An overview of the dataset is shown in Table 1, in which the class "no-slope" for slope position refers to slopes of less than 5°, which are located in the marl area. The class "no-aspect" for aspect includes the same sites, but also flat downslope parts of four slopes in the schist and sandstone areas. Both classes probably describe landscape positions with shallower depth to groundwater than the other sites.

Potential evaporation and water availability are usually considered to be the main external dynamic controls of sap flow, so we examined their importance for the temporal variability in sap velocity by correlating the time series of sap velocity with E_{pot} and soil moisture, using the Spearman rank correlation. However, we were interested primarily in the spatial variability of sap velocity as a way to determine influences on transpiration patterns on the landscape scale. We assessed this by examining the spatially distributed dataset of daily-averaged sap velocity of the 61 trees, for each of the 132 days of our study period.

In a first step, we examined the individual influence of the different tree-, stand- and site-specific controls listed in Table 1 and of the external controls soil moisture and E_{pot} on sap velocity at the respective forest sites, averaged across the study period, separately for each tree. This first analysis ignored multivariate interactions to get a simplified overview of the data; hence, effects seen in these comparisons should not be over-interpreted. For the categorical variables (species, geology, slope position and aspect) we also looked at possible temporal changes of differences in sap velocity between the categories by testing daily datasets with the Mann–Whitney U test or the Kruskal–Wallis test, for variables consisting of three or two categories, respectively, to a significance level of $\alpha = 0.05$.

The multidimensional effect of all tree-, stand- and site-specific influences was then analysed with multiple linear regression models separately for each day. This modelling approach is meant to explore the main controls of sap velocity or sap flow patterns, but at this stage we do not aim at predicting these spatial patterns. The response variable for each of the 132 daily models was the log-transformed daily sap velocity of each tree, because the logarithmic values corresponded better to a normal distribution. The linear regression model can be expressed as

$$\ln\left(V_{sap}\right) = \beta_0 + \sum_{i=1}^{n} \beta_i x_i, \tag{8}$$

with n predictors $(x_0, \ldots x_i)$ and the regression coefficients $(\beta_0, \ldots \beta_i)$ estimated to obtain an optimum fit.

Before applying the regression models, we checked the predictors for collinearity by determining the correlation matrix. There was only one combination of predictors with a Spearman rank correlation coefficient above the widely employed critical value of collinearity, $|\rho| > 0.7$ (Dormann et al., 2008; Tannenberger et al., 2010), the number of stems and median diameter of the stand, at $\rho = -0.73$. The effects of this correlation on the linear models was tested by running the models for both the original set of predictors and again, leaving out number of stems. As the results did not

differ with respect to the variance contributions of the different predictors, we kept all predictors in the final analysis. We also did not include interaction terms in the final models because after testing with various interactions, these did not contribute much to the explained variance.

Although a step-wise simplification of the models using the Akaike information criterion led to a higher percentage of explained variance by the models, we refrained from using this simplification in order to keep the model structures similar for each day to allow comparability of the temporal, day-to-day changes in predictor importance. For prediction, the potentially best model would be more appropriate; however, in our exploratory analysis we focused on comparability. The relative importance of the predictors for explaining the observed variance of sap velocity or sap flow was assessed using the approach of Grömping (2007), made available in the R package "relaimpo". Of the different built-in methods to determine relative importance, we used "lmg". This method uses sequential sums of squares from the linear model, applies all possible orderings of regressors and obtains an overall assessment by averaging over all orders, which is deemed appropriate for causal interpretation and unknown weights of the different predictors (Grömping, 2007). The initial order of the predictors in the linear models is not relevant for the relative importance as orderings are shuffled.

Over-fitting can be a problem in linear models with many predictors. We checked for this by performing a comparison between the residual standard error (RSE) of the original models and the root mean square error (RMSE) of a 10-fold cross-validation (Fig. 2). In case of over-fitting, the RMSE of the cross-validation should be much higher than the RSE. In our case, both error measures differed only marginally and were largest when sap velocities were small. These were the days when the linear model generally failed to explain the variance in the datasets. For days with high sap velocities, the small errors as well as the small difference between RSE and RMSE indicated that the models are not over-fitted. Additionally, Fig. 2 showed that limiting the analysis to the period of fully developed canopy excludes periods of larger errors at the beginning and end of the season.

The linear models for sap velocity give an indication about possible controls of transpiration; however, a more intuitive measure for transpiration is sap flow. Therefore, as an indication of how the actual transpiration patterns are influenced by site- and stand-specific characteristics, we repeated the multiple linear regression analysis with the derived sap flow dataset. The calculation from sap velocity to sap flow is based on the species-specific relations between sapwood and DBH, so we did not use species, DBH and height as predictors in the linear models for sap flow as they are not independent anymore. We only used the remaining stand- and site-specific predictors as well as E_{pot} and soil moisture.

We analysed the temporal dynamics of the variance contributions of the individual predictors and of the proportion of variance explained by all the tree-specific predictors (only

Table 1. Overview of the characteristics associated with the trees in the sap velocity dataset. These are used as predictors in the multiple linear regression. Abbreviations are DBH for diameter at breast height, BA for basal area of the stand.

Property	Group	Class (and no. of trees in each class)	Value (25/50/75 percentile)
Species	Tree-specific	beech (39), oak (22)	
DBH (cm)	Tree-specific		34/46/63
Height (m)	Tree-specific		24/29/34
BA ($m^2\,ha^{-1}$)	Stand-specific		27/40/54
Median DBH (cm)	Stand-specific		5/14/28
No. of stems	Stand-specific		20/24/43
Geology	Site-specific	marl (13), sandstone (22), schist (26)	
Slope position	Site-specific	upslope (41), downslope (9), no-slope (11)	
Aspect	Site-specific	north (17), south (29), no-aspect (15)	

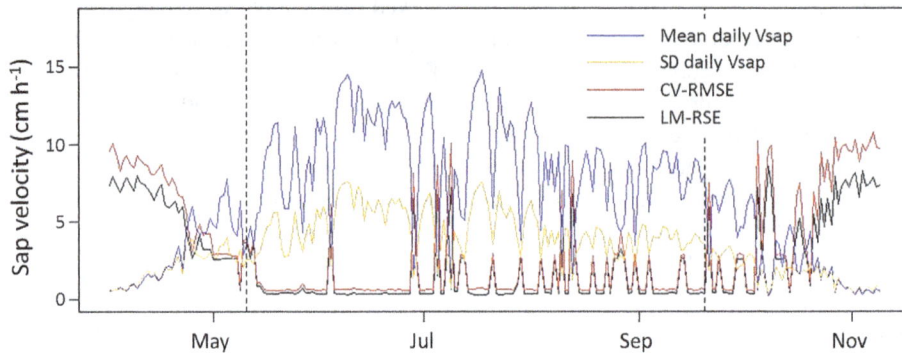

Figure 2. Comparison of residual standard error of the original linear models (LM-RSE) and the root mean square error of a 10-fold cross-validation (CV-RMSE), in relation to mean and standard deviation of daily sap velocities (V_{sap}). The dashed lines indicate the beginning and end of the focus period with a fully developed canopy.

for sap velocity) taken together, all the stand-specific predictors taken together and all the site-specific predictors taken together. We also correlated these time series to catchment-averaged time series of soil moisture and E_{pot} as indicators of general atmospheric demand and water availability, using the Spearman rank correlation. All statistical analyses were carried out in the language and environment R (R Development Core Team, 2014).

3   Results

3.1   Controls of temporal dynamics of sap velocity

Correlations of time series of sap velocity for each tree with E_{pot} and soil moisture yielded high positive and significant ($\alpha = 0.05$) Spearman rank correlations for E_{pot}, but correlations to soil moisture were slightly negative and very weak (Fig. 3).

3.2   Controls of spatial patterns in daily mean sap velocity and sap flow

A first simplified overview of the influence of the various factors on sap velocity patterns was derived from plotting sap

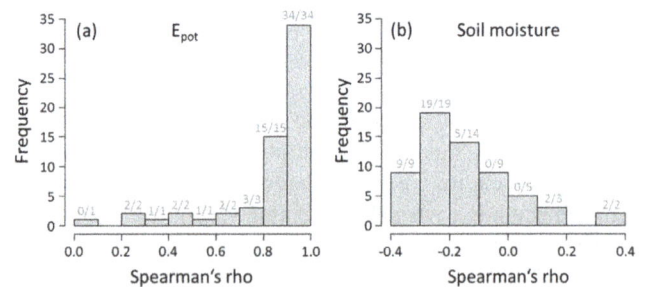

Figure 3. Histograms of temporal correlations between **(a)** E_{pot} and **(b)** soil moisture at each site with sap velocity for the 61 trees in the dataset. The small numbers in grey on top of the bars indicate how many of the correlations in the specific group are significant.

velocity, averaged over the entire study period for each tree, against the factors or their respective categories (Fig. 4). Obviously, this graph neglects the combined influence of the interplay of all these factors, but yields a first overview of the data and possible relations. For example, the difference between higher sap velocities in beech trees compared to oaks can be seen as well as a possible positive relation between sap velocity and DBH or tree height. In the category for aspect,

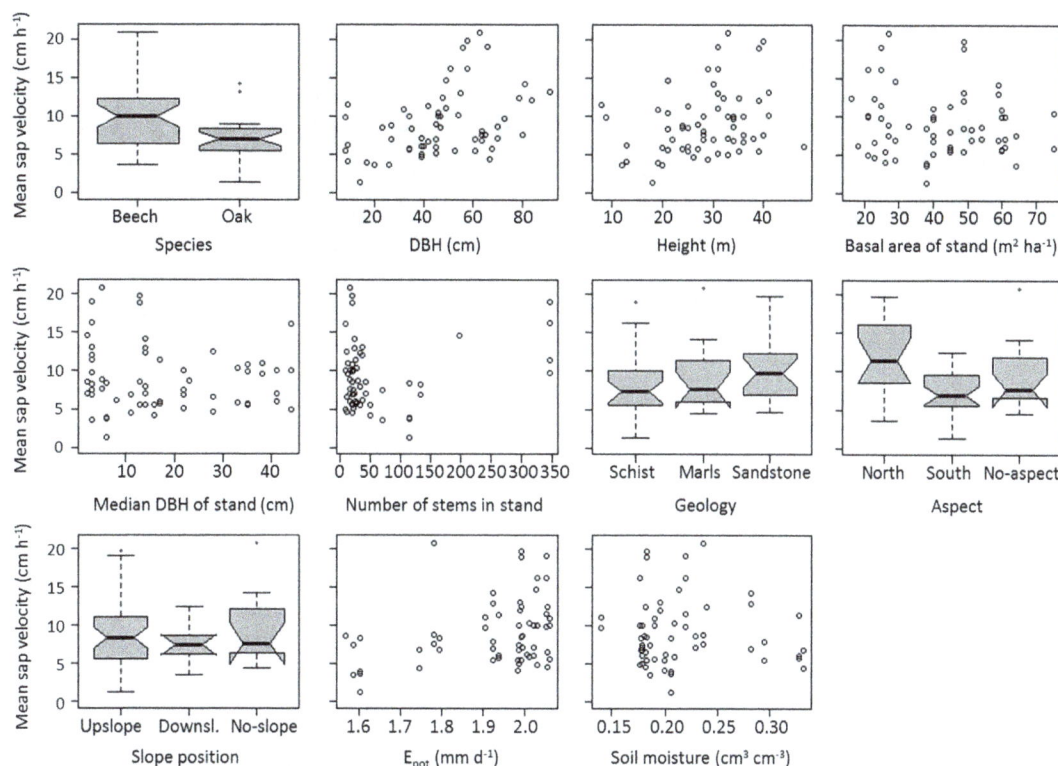

Figure 4. Univariate influence of each predictor on sap velocity means for each tree over the entire study period. Box plot parameters are as follows: the horizontal line within the box visualises the median, boxes comprise data between the 1st and 3rd quartile of the data, whiskers reach to $1.5 \times$ the interquartile range outside the box (or to the maximum/minimum value if smaller/larger), circles stand for outliers or data points outside the whiskers, notches show approximately 95 % confidence intervals around the medians.

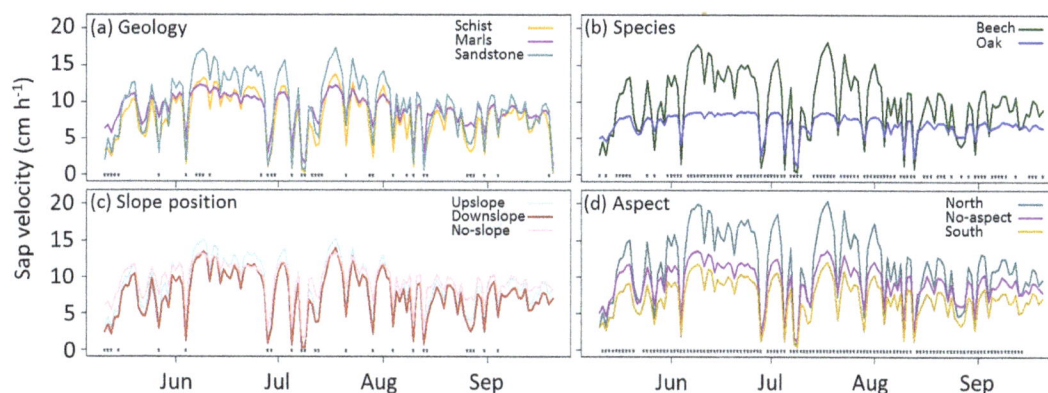

Figure 5. Differences between sap velocities depending on **(a)** geology, **(b)** species, **(c)** slope position and **(d)** aspect. Lines show the average dynamics of each class. Asterisks at the bottom of the panels indicate significant differences for that day according to Mann–Whitney U or Kruskal–Wallis tests at $\alpha = 0.05$, for differences between the two or three categories, respectively.

the box plots show a difference between higher sap velocities on north-facing slopes compared to south-facing ones, with trees located in plains having somewhat intermediate velocities.

The categorical factors were assessed in more detail looking at temporal changes in sap velocity differences. Statistical tests (Mann–Whitney U test for two and Kruskal–Wallis

for three categories, $\alpha = 0.05$) applied to the sap velocity datasets for each day for the example of the categorical factors geology, species, slope position and aspect showed significant differences (Fig. 5). There is in particular a significant difference between south- and north-facing slopes and between beech and oak trees for most days of the dataset (Fig. 5), providing a first indication of the importance of both

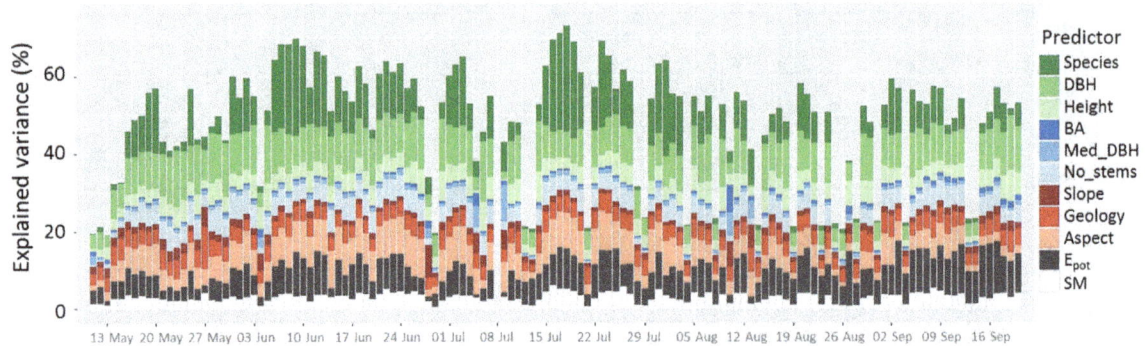

Figure 6. Proportion of variance explained by the different predictors in the daily linear models of spatial sap velocity patterns: 132 daily models from 61 trees at 24 sites.

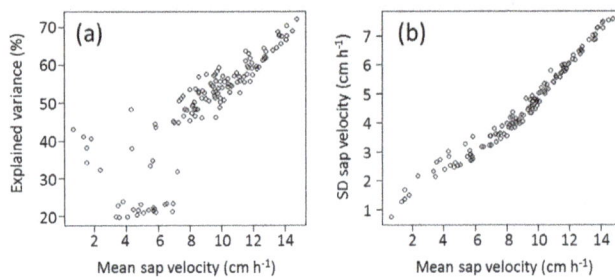

Figure 7. (a) Explained variance of the linear models in relation to mean sap velocities for all 132 days of the study period and (b) standard deviation of sap velocity depending on mean sap velocities for those 132 days.

tree- and site-specific influences. In contrast, there were only 36 of 132 days showing significant differences for geology and 25 days for slope position, occurring when sap velocities were generally low.

In a more comprehensive approach, we assessed the combined effect of the various tree-, stand- and site-specific influences on sap velocity with the help of multiple linear regression generating 132 daily models describing the spatial patterns. The total explained variance for the sap velocity models ranged from 20 %, on days when the models fail to explain the spatial variability in the dataset, to 72 %, which constitutes fairly good explanatory power (Fig. 6). The total explained variance correlated strongly with catchment averages of sap velocity (Pearson's $r = 0.84$, $p < 0.001$), especially at sap velocities $> 7\,\mathrm{cm\,h^{-1}}$ (Fig. 7). Spatial variability of sap velocity in the catchment, expressed as standard deviation of the daily values for the 61 trees, also increased with increasing mean sap velocity (Pearson's $r = 0.98$, $p < 0.001$; Fig. 7). The consistent model structure showed that the change in the proportion of explained variance over time was different for the various predictors (Fig. 6). Averaged across the 132 daily models, 9 % of the variance was explained by species, 9 % by DBH and 4 % by tree height. Characteristics of the stand yielded 1 % for BA, 1 % for me-

dian DBH and 4 % for number of stems, and the site-specific predictors amounted to 2 % for slope position, 4 % for geology and 6 % for aspect. The external dynamic controls of sap flow, E_{pot} and soil moisture, explained 7 and 3 % of the variance in daily sap velocity patterns, respectively.

The contribution of the different predictors to the overall explained variance of the linear models varied strongly from day to day. On days when average and spatial variability of sap velocity was low the models performed badly. There were some predictors which showed larger fluctuations, for example species, compared to more constant contributions from predictors like the number of stems or DBH (Fig. 6).

The multiple linear regression models for sap flow patterns explained between 18 and 56 % of the variance in the daily datasets (Fig. 8), on average 49 %. Averaged across the 132 daily models, the stand-specific predictors explained 8 % of the variance (4 % by BA, 1 % by median DBH and 3 % by the number of stems). The largest contribution came from geology with 21 %, then aspect with 10 %, while slope position only explained 3 %. E_{pot} and soil moisture explained 7 and 1 % of the variance, respectively, which is comparable to their contributions in the linear models for sap velocity patterns.

The variance contributions stayed fairly constant in time, except for days when the models failed to explain the spatial variability in the data altogether (Fig. 8). Compared to the linear models for sap velocity, the models for sap flow had less explanatory power. The contributions of the stand-specific predictors were not very important, similar to the results for sap velocity. For the site-specific controls, the largest contribution came from geology, and less from aspect; in contrast, in the sap velocity models, aspect explained a larger proportion of the variance than geology (Fig. 8).

3.3 Temporal dynamics of predictor importance

Comparing the dynamics of the proportion of variance explained by all the tree-specific predictors taken together, all the stand-specific predictors taken together and all the site-

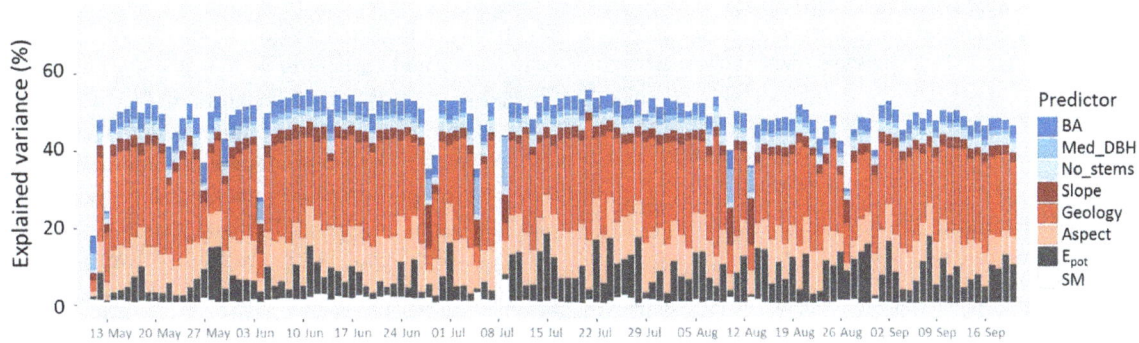

Figure 8. Proportion of variance explained by the different predictors in the daily linear models of spatial sap flow patterns: 132 daily models from 61 trees at 24 sites.

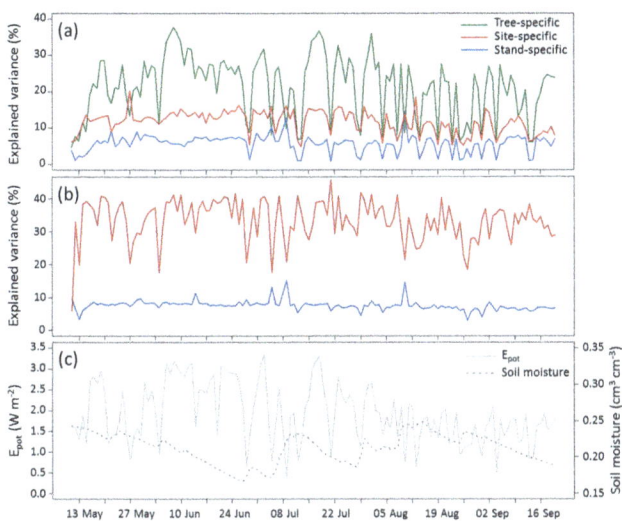

Figure 9. Explained variance of the daily linear models, separated according to the predictor groups used in the regression, **(a)** for sap velocity and **(b)** for sap flow. **(c)** Catchment average of soil moisture and potential evaporation E_{pot}.

specific predictors taken together to the catchment-average dynamics of E_{pot} and soil moisture (Fig. 9) showed that the stand- and site-specific predictors' contributions stayed relatively constant apart from the days when the model failed. This was the case for both the sap velocity and the sap flow models. In contrast, the tree-specific predictors in the sap velocity models varied to a greater extent. Visual comparison indicated a link between fluctuations of tree-specific influences and potential evaporation (E_{pot}), but not soil moisture (Fig. 9).

The Spearman rank correlations of the predictors' explained variance with E_{pot} and soil moisture, listed in Table 2, also confirmed that changes in species influence in the sap velocity were strongly linked to changes in E_{pot}, with a significant correlation of $r = 0.81$. Further weaker but significant correlations were detected between E_{pot} and number

of stems at $r = 0.50$, aspect at $r = 0.67$ and soil moisture at $r = 0.55$, respectively. Summarised into categories, the influence of tree-specific predictors strongly correlated with E_{pot} ($r = 0.86$); similarly, there was a strong correlation of the overall explained variance with E_{pot} ($r = 0.84$). Some of the correlations with soil moisture were significant, but they were mostly weak, with $|r| <= 0.37$.

In the sap flow models, the only significant correlations worth mentioning were between E_{pot} and aspect and between E_{pot} and the overall explained variance, at $r = 0.57$ and $r = 0.72$, respectively.

4 Discussion

4.1 Controls of temporal dynamics of sap velocity

The strong positive temporal correlation of sap velocity and E_{pot} (Fig. 3) confirms the well-known role of the atmospheric controls as the main external drivers for transpiration (Bovard et al., 2005; Clausnitzer et al., 2011; Granier et al., 2000; Jonard et al., 2011). Soil moisture, however, did not affect the temporal dynamics of sap velocity in a similar way (Fig. 3). One reason for this surprisingly weak relation could be that water is not a limiting factor for transpiration in this landscape, or at least not during the observed time period. In the schist area of the catchment, anecdotal evidence given by forest wardens suggests that beech trees on south-facing slopes are indeed water-stressed during dry, hot summer months, although in our data we did not see a limitation of sap velocity for the beech trees. A different explanation for the lack of correlation is that soil moisture in the top 60 cm of the soil profile is simply not a sufficiently good proxy for water availability. In the soils of the schist there might be additional water resources stored in the weathered bedrock or the schist fractures which could be accessible to roots reaching deeper than the maximum rooting depths estimated from power drill cores in the study area (Sprenger et al., 2016). In the deep sandstone soils with maximum observed rooting depths of 98 cm in the drill cores (Sprenger et al., 2016), roots could

Table 2. Spearman rank correlation between the time series of the different predictors' explained variance and the time series of potential evaporation (E_{pot}) and soil moisture. Values in bold are significant correlations (at $\alpha = 0.05$).

Predictor	Sap velocity E_{pot}	Sap velocity Soil moisture	Sap flow E_{pot}	Sap flow Soil moisture
Species	**0.81**	**−0.30**		
DBH	**0.32**	−0.14		
Height	0.11	−0.05		
BA	−0.15	−0.05	**0.44**	**−0.09**
Median DBH	**0.44**	−0.16	**0.34**	**0.00**
No. of stems	**0.50**	**−0.35**	**0.44**	−0.03
Slope	**−0.32**	**0.21**	0.22	−0.01
Geology	0.04	**0.18**	**0.35**	−0.20
Aspect	**0.67**	**−0.37**	**0.57**	−0.38
E_{pot}	**0.34**	−0.18	0.10	**0.04**
Soil moisture	**0.55**	**−0.30**	0.25	−0.21
Tree-specific	**0.86**	**−0.34**		
Stand-specific	**0.28**	**−0.33**	**0.42**	**−0.07**
Site-specific	**0.35**	−0.14	**0.46**	−0.26
Total exp. var.	**0.84**	**−0.38**	**0.72**	−0.37

also reach deeper, exploiting larger soil volumes or possibly tapping groundwater. The mostly flat marl areas exhibit shallow groundwater tables, so water limitation is unlikely for longer periods during the year. Thus, although water availability is an important boundary condition for transpiration, soil moisture measurements for the top 60 cm might not be an appropriate proxy, and including available information on groundwater levels or soil moisture in deeper layers could be useful in that regard.

4.2 Controls on spatial patterns in daily mean sap velocity and sap flow

Even from the simplified univariate assessment, the influence of characteristics such as species, DBH and aspect on spatial sap velocity patterns is visible (Fig. 4). In the more comprehensive approach applying multiple linear regression models to the daily sap velocity datasets, the combined effect of tree-, stand- and site-specific predictors surpasses by far the explanatory power of the boundary conditions, as E_{pot} and soil moisture (Fig. 6) together explained only around 10 % of the variance in sap flow patterns (Fig. 6). From the larger spatial variability of soil moisture (average of spatial standard deviation of 5 % vol, compared to an average of temporal standard deviation of 2 % vol) some influence on spatial sap velocity patterns might have been expected. But similar to the lack of temporal correlation with soil moisture, the lack of importance for spatial variability could result from the fact that measurements in the top 50 cm of the soil column were not meaningful to assess water availability at the sites or that a soil moisture limitation was not occurring in the observation period. E_{pot} held larger explanatory power, but com-

pared to the importance for temporal variability in sap velocity, the spatial effect was very small, possibly because the range of spatial variability in E_{pot} is much smaller. (The average of spatial standard deviation of E_{pot} was 0.18 W m^{-2}, whereas the average of its temporal standard deviation was 0.73 W m^{-2}). The same argument holds true for the similarly low proportion of explained variance by E_{pot} in the linear models for sap flow. This suggests that spatial patterns of (evapo)transpiration for distributed hydrological models based on meteorologically derived estimates only reflect a small part of the spatial variability of measured transpiration.

The explained variance of the tree-specific characteristics amounted to 22 % averaged over the 132 sap velocity models (Fig. 6). Mechanisms underlying the differences in sap flow related to species, tree diameter and height have been studied in great detail in the field of tree physiology. The species contrast in our case consists of higher sap velocities for beech, as beech shows physiological advantages in transpiration efficiency and outperforms oaks in sufficiently moist conditions (Hölscher et al., 2005). Sap flow contrasts are even more pronounced because the active sapwood of oaks is limited to the outermost few annuli whereas for larger beech trees it can easily reach a depth of 7 cm or more (Gebauer et al., 2008). This limitation for oaks is visible in the species-specific allometric equations for sapwood area (Eqs. 3 and 4), but additionally, sap velocity in the innermost of our sensors (at 30 mm into the tree) was frequently zero for oaks. As expected, forest species composition is a major determinant of transpiration patterns (e.g. Hernandez-Santana et al., 2015; Loranty et al., 2008).

Tree height and DBH contrasts probably reflect the differences in social status, with larger, dominant trees reaching higher transpiration values than understorey trees. That larger trees, both taller and with larger DBH, exhibit higher sap velocities is likely due to their associated larger canopy and root volume, ensuring on the one hand the exposure of the leaves to the atmospheric gradient and on the other hand having access to a larger soil volume and potential water supply (Bolte et al., 2004; Nadezhdina and Čermák, 2003). For sap flow, the contrasts are again even larger than for sap velocity, because not only is larger DBH associated with larger velocities, but according to Eqs. (3) and (4), larger DBH also entails larger sapwood area, multiplying the effect of the sap velocity differences. Implementing spatial patterns of tree sizes into hydrological models could be attempted using mapped information from forest inventories, management plans or even lidar images (Ibanez et al., 2016; Rabadán et al., 2016; Vauhkonen and Mehtätalo, 2015). The stand density, expressed as the number of stems, explained on average 4 % of the variance in the daily models for sap velocity and 3 % for sap flow. Decreasing sap velocities with increasing stand density hints at the competition for light and resources among individual trees (Cienciala et al., 2002; Dalsgaard et al., 2011; Gebauer et al., 2012; Oren and Pataki, 2001; Vincke et al., 2005). However, due to this small contribution in the linear models, the stand-specific influence should not be over-interpreted. Basal area contributed on average 4 % to sap flow models and hints at the same mechanisms as stand density.

The site-specific predictors together explained on average 12 % of the spatial variance in the sap velocity models and 34 % in the sap flow models (Figs. 6 and 8). Landscape characteristics such as topography and geology will control sap velocity patterns of otherwise homogeneous forests because they influence spatial patterns of either water or energy availability. Topography primarily controls radiation input, and to some extent water availability through depth to groundwater and soil characteristics, while geology mainly controls root distribution and water availability because it determines the depth to bedrock and depth to groundwater as well as soil type and soil depth. The effect of soil depth on transpiration, for example, has been shown by Tromp-van Meerveld and McDonnell (2006) for soils of the same type on the Panola hillslope, and the contrasts between different geological units in the Attert basin are likely to be even more pronounced. The soils in the schist area are very shallow, restricting rooting depth to an average of 68 cm (Sprenger et al., 2016). Together with moderate values for plant-available water (Jackisch, 2015), which are probably even lowered by the high rock content, this could lead to the smaller sap velocities in the schist compared to the sandstone or marls. These differences in soil depth and water retention characteristics manifest in differences in hydrological characteristics such as water storage dynamics, leading to contrasting runoff generation mechanisms for schist, marl and sandstone areas in the Attert catchment (van den Bos et al., 2006; Wrede et al., 2015). In turn, these geology-induced contrasts in depth to groundwater and water storage control tree access to these water reservoirs and favour species with adapted rooting systems (Dalsgaard et al., 2011), thus introducing a landscape-scale effect on sap flow. We see this effect in the sap velocity models and even more in those for sap flow, suggesting that including geological maps into distributed hydrological models could be helpful not only for soil and bedrock characteristics but also for transpiration patterns.

The influence of aspect (Figs. 6 and 8) was mainly due to the south-facing slopes having smaller sap velocities and sap flow values compared to north-facing slopes (Fig. 5d). This is an apparent contradiction to the expectation that the larger energy input on the south-facing slopes should induce larger transpiration values. One explanation could be that energy input is not a limiting factor for transpiration in this landscape, but on the contrary, larger energy input on south-facing slopes might make them more prone to water limitation (Holst et al., 2010), especially when combined with other limiting factors. For example, the schist area holds a large proportion of south-facing slopes in our dataset and also has shallow soils due to the geological substrate, possibly exacerbating water limitation. However, as we did not see acute signs of water limitation in our data, the contrast in sap veloc-

ities due to aspect would have to be long-term physiological adaptations (Hajek et al., 2016; Stojnić et al., 2015) to the drier conditions on south-facing slopes (as they are reported by forest wardens). The dominance of geology compared to aspect in the sap flow models could then result from a physiological effect of aspect influencing sapwood area or wood properties, which would then already be considered in the sap flow values, leaving geology as an independent predictor more important for spatial patterns in sap flow. A second explanation for the higher sap velocity and sap flow values on north-facing slopes (irrespective of species; see Appendix Fig. A1) can be seen in the exact locations of the trees. We would expect the larger E_{pot} and sap velocity values to be on the south-facing slopes due to the higher radiation input. However, the values for E_{pot} are calculated at the respective tree sites and these are not necessarily at the same relative positions on the slopes. So the E_{pot} values are probably not directly related to aspect, but to the location within the valley, shading effects, etc. Grouping the E_{pot} values according to aspect (Appendix Fig. A2), the main contrast occurs between north-facing slopes having higher values compared to no-aspect, whereas north- and south-facing slopes do not show considerable differences. Furthermore, as already mentioned before, spatial variability in the E_{pot} dataset are generally not very pronounced. Thus, a situation where the study trees are situated at more extreme locations with respect to aspect will probably induce a larger effect from E_{pot}. We suspect that the latter explanation is more relevant for the aspect differences in our data. A more targeted study including wood properties and stronger aspect contrasts would be needed to clarify this issue (and is in progress).

Slope position did not play a major role in explaining spatial sap velocity patterns (Figs. 6 and 8) although due to its possible effect on soil depth, water availability and species composition it is also the best-studied of the three site properties we included in our models (Adelman et al., 2008; Bond et al., 2002; Kumagai et al., 2007; Loranty et al., 2008; Mitchell et al., 2012). A reason for this lack of explanatory power could be that the information within this variable is partly also included in aspect because both the aspect category "no-aspect" and the slope position categories "no-slope" and "downslope" suggest sites which are close to groundwater resources. In a way, all three site-specific influences, geology, aspect and slope position, can be regarded as proxies for underlying characteristics of water availability.

4.3 Temporal dynamics of predictor importance

Understanding the feedback of spatial transpiration patterns with hydrological processes requires assessment of the temporal dynamics of the controls of these patterns, for example on a seasonal or daily basis. Our analyses indicate that spatial sap velocity patterns are governed by mainly tree- and site-specific characteristics. For the sap flow models, species and DBH were excluded as predictors because they were part of the calculation of sap flow. The resulting patterns were mainly controlled by site-specific characteristics; stand characteristics played a negligible role. The temporal shifts in these controls depend on hydro-meteorological conditions, especially potential evaporation (Fig. 9). And whereas the direct influence of hydro-meteorological variability on sap flow has been highlighted in many studies (Bovard et al., 2005; Clausnitzer et al., 2011; Granier et al., 2000; Jonard et al., 2011), the link between these conditions and spatial patterns of sap flow and their controls is still not well understood. Additionally, most studies which include site-specific controls focus on a seasonal basis or undertake plot comparisons. The temporal dynamics of the different predictor categories showed contrasting dependency on potential evaporation (Fig. 9). While the stand- and site-specific predictors as well as DBH and tree height remained fairly constant in their total explained variance, for both sap velocity and sap flow, the species-dependent temporal variance in the sap velocity models was strongly correlated to the dynamics of E_{pot} (Table 2). The species effect in this context is the contrast between oaks, which can respond to increasing E_{pot} only up to a certain threshold (Fig. 5b), and beech trees, which can reach higher sap velocities when they are not water-limited. A second predictor with considerable positive correlation with E_{pot} is aspect (Table 2), despite the fact that north- and south-facing slope locations show similar values of E_{pot}. If the trees were physiologically adapted to water limitation on south-facing slopes, under high-E_{pot} conditions the contrast between these transpiration-limited trees and those on north-facing slopes, using their full transpiration potential, could be even stronger, leading to a very temporally dynamic influence of aspect. To separate this influence from the effect that the exact tree locations might be representing shading and landscape position in general and not necessarily strong contrasts in aspect, further studies at more pronounced north- and south-facing aspects and possibly also in a very dry year could help.

Lastly, the overall explained variance of our linear models for both sap velocity and sap flow also correlates with E_{pot}, as it does with sap velocity (Fig. 7). The models can explain considerable proportions of the spatial variability in sap velocity and sap flow when those values themselves can become large, driven by high E_{pot} and thus leading to larger transpiration contrasts in the landscape, due to species or aspect, for example. At lower values of potential evaporation, the spatial variability of sap velocity is less pronounced and not primarily determined by the predictors included in our models.

For hydrological modellers this means that at low values of potential evaporation, which likely coincide with cloudy, rainy or cold days, transpiration flux is low and contains little spatial structure. On the one hand this entails smaller potential errors in the transpiration estimates, but on the other hand, it could be considered in attempts to apply dynamic model structures. During low-E_{pot} days, transpira-

tion could be implemented in a more general and aggregated way, whereas during high-E_{pot} days including the spatial patterns of tree and site (as well as stand) characteristics could markedly improve model performance and spatial representation of transpiration.

5   Conclusions

Sap flow measurements are a suitable tool to investigate the different influences that shape spatial patterns of tree transpiration in a landscape. However, there are some uncertainties involved, for example the widely applied calculation of sap flow from sap velocities includes the assumption that the tree species and size are mainly determining this relationship. As ecophysiological adaptations to site conditions have been shown in other contexts, an independent determination of sap flow for each tree in the study would need direct measurements of sapwood area and other relevant xylem characteristics. This would enable a better quantification of the different influences on spatial transpiration patterns, which would complement the more exploratory character of our study.

We examined both the influences on spatial patterns of transpiration in a landscape and their temporal dynamics, by means of sap velocity and sap flow. The spatial patterns were mainly controlled by tree- and site-specific characteristics. Temporal dynamics of the overall explained variance of the linear models and the relative importance of species was closely linked to the dynamics of potential evaporation, whereas the site-specific influences remained constant over time. This means that the abiotic characteristics of the landscape control transpiration pattern to a certain extent, and this control remains static in time. However, the importance of biotic characteristics, i.e. the landscape-scale patterns of tree species distribution, varies in time and becomes most important during days of high atmospheric demand. Our results suggest that spatial representation of landscape-scale transpiration in distributed hydrological models could be improved by including spatial patterns of tree-, stand- and site-specific characteristics. For spatial sap flow patterns, these influences were considerably larger than the obvious and widely used influences of the potential evaporation and water availability in the soil. Consequently, similar to resolving agricultural areas according to crops on a field scale, one could represent the spatial structure in forest transpiration resulting from species and size distributions, but also from patterns due to site characteristics such as geology or topography. This information can be used for model parameterisation or as a part of multi-response evaluation for soil–vegetation–atmosphere transfer and hydrological models.

Additionally, identifying phases of varying importance of the different influences, and their dependence on E_{pot}, can help modellers decide when to best include site-specific characteristics to describe spatial patterns of transpiration in models, when a classification according to species and stands might be more appropriate, or when it is not necessary at all to implement a spatially explicit transpiration estimate. Thus, the spatial representation of transpiration in hydrological models could be attempted in a temporally dynamic way, and, when spatial structure is needed, be based on information from geological maps, digital elevation models, forest inventories or remote sensing images.

Appendix A

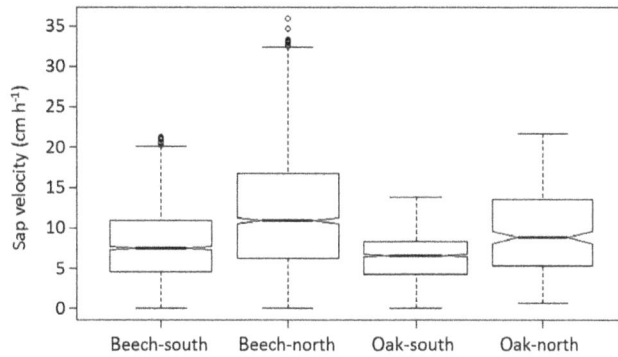

Figure A1. Sap velocities when grouped according to species and aspect.

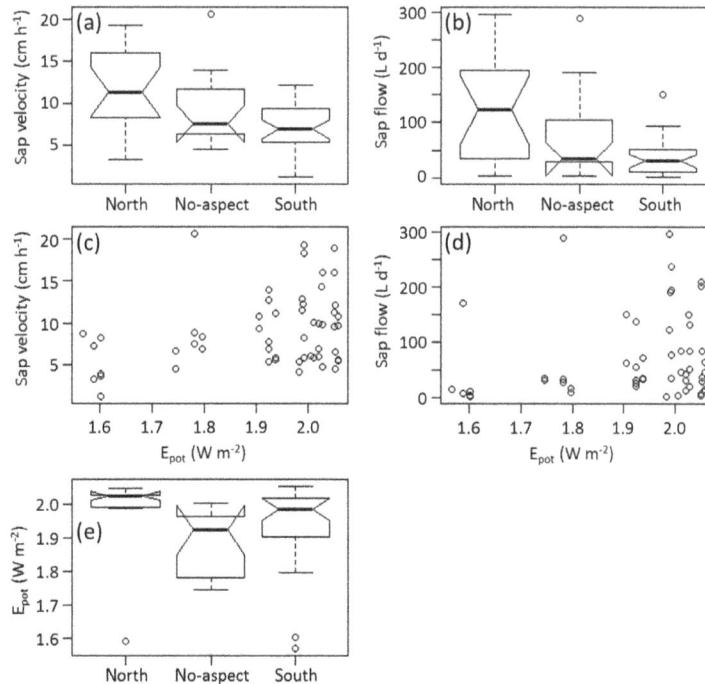

Figure A2. (a) Sap velocity and **(b)** sap flow values grouped according to aspect categories. **(c)** Sap velocities and **(d)** sap flow values of each tree with respect to E_{pot} at the tree location. **(e)** E_{pot} values at the tree locations values grouped according to aspect categories. Shown are temporal averages, for each tree (for sap velocity and sap flow) or each tree site (for E_{pot}). The uncharacteristically high sap flow values are probably due to overestimation during the upscaling to tree-level transpiration and should be treated with caution; however, absolute values are not of interest here, rather relations and spatial patterns which should be robust.

Competing interests. The authors declare that they have no conflict of interest.

Acknowledgements. We thank the German Research Foundation (DFG) for funding of the CAOS research unit FOR 1598 in which this study was undertaken. We especially acknowledge Britta Kattenstroth and Jean-François Iffly for their invaluable help in setting up and running the monitoring network, as well as countless helpers during field work. Thomas Gräff and Uwe Ehret commented on an earlier version of this paper. We also acknowledge support for open-access publishing by the Deutsche Forschungsgemeinschaft (DFG) and the Open-Access Publishing Fund of Karlsruhe Institute of Technology.

Edited by: Nunzio Romano

References

Adelman, J. D., Ewers, B. E., and MacKay, D. S.: Use of temporal patterns in vapor pressure deficit to explain spatial autocorrelation dynamics in tree transpiration, Tree Physiol., 28, 647–658, 2008.

Angstmann, J. L., Ewers, B. E., Barber, J., and Kwon, H.: Testing transpiration controls by quantifying spatial variability along a boreal black spruce forest drainage gradient, Ecohydrology, 6, 783–793, 2013.

Barnard, H. R., Graham, C. B., van Verseveld, W. J., Brooks, J. R., Bond, B. J., and McDonnell, J. J.: Mechanistic assessment of hillslope transpiration controls of diel subsurface flow: A steady-state irrigation approach, Ecohydrology, 3, 133–142, 2010.

Boer-Euser, T., McMillan, H. K., Hrachowitz, M., Winsemius, H. C., and Savenije, H. H. G.: Influence of soil and climate on root zone storage capacity, Water Resour. Res., 52, 2009–2024, 2016.

Bolte, A., Rahmann, T., Kuhr, M., Pogoda, P., Murach, D., and von Gadow, K.: Relationships between tree dimension and coarse root biomass in mixed stands of European beech (Fagus sylvatica L.) and Norway spruce (Picea abies [L.] Karst.), Plant Soil, 264, 1–11, 2004.

Bond, B. J., Jones, J. A., Moore, G., Phillips, N., Post, D., and McDonnell, J. J.: The zone of vegetation influence on baseflow revealed by diel patterns of streamflow and vegetation water use in a headwater basin, Hydrol. Process., 16, 1671–1677, 2002.

Bosch, D. D., Marshall, L. K., and Teskey, R.: Forest transpiration from sap flux density measurements in a Southeastern Coastal Plain riparian buffer system, Agr. Forest Meteorol., 187, 72–82, 2014.

Bovard, B. D., Curtis, P. S., Vogel, C. S., Su, H. B., and Schmid, H. P.: Environmental controls on sap flow in a northern hardwood forest, Tree Physiol., 25, 31–38, 2005.

Burgess, S. S. O., Adams, M. A., Turner, N. C., Beverly, C. R., Ong, C. K., Khan, A. A. H., and Bleby, T. M.: An improved heat pulse method to measure low and reverse rates of sap flow in woody plants, Tree Physiol., 21, 589–598, 2001.

Campbell, G. S., Calissendorff, C., and Williams, J. H.: Probe for measuring soil specific heat using a heat-pulse method, Soil Sci. Soc. Am. J., 55, 291–293, 1991.

Čermák, J. and Prax, A.: Water balance of a Southern Moravian floodplain forest under natural and modified soil water regimes and its ecological consequences, Ann. For. Sci., 58, 15–29, 2001.

Cienciala, E., Mellander, P.-E., Kučera, J., Opluštilová, M., Ottosson-Löfvenius, M., and Bishop, K.: The effect of a north-facing forest edge on tree water use in a boreal Scots pine stand, Can. J. Forest Res., 32, 693–702, 2002.

Clausnitzer, F., Köstner, B., Schwärzel, K., and Bernhofer, C.: Relationships between canopy transpiration, atmospheric conditions and soil water availability-Analyses of long-term sap-flow measurements in an old Norway spruce forest at the Ore Mountains/Germany, Agr. Forest Meteorol., 151, 1023–1034, 2011.

Dalsgaard, L., Mikkelsen, T. N., and Bastrup-Birk, A.: Sap flow for beech (Fagus sylvatica L.) in a natural and a managed forest – Effect of spatial heterogeneity, J. Plant Ecol., 4, 23–35, 2011.

Dormann, C. F., Purschke, O., Carča Marquéz, J. R., Lautenbach, S. and Schröder, B.: Components of uncertainty in species distribution analysis: A case study of the Great Grey Shrike, Ecology, 89, 3371–3386, 2008.

Elsner, G.: Morphological variability of oak stands (Quercus petraea and Quercus robur) in northern Germany, Ann. For. Sci., 50, 228s–232s, 1993.

Fahle, M. and Dietrich, O.: Estimation of evapotranspiration using diurnal groundwater level fluctuations: Comparison of different approaches with groundwater lysimeter data, Water Resour. Res., 50, 273–286, 2014.

Fenicia, F., Kavetski, D., Savenije, H. H. G., and Pfister, L.: From spatially variable streamflow to distributed hydrological models: Analysis of key modeling decisions, Water Resour. Res., 52, 954–989, 2016.

Fuchs, S., Leuschner, C., Link, R., Coners, H., and Schuldt, B.: Calibration and comparison of thermal dissipation, heat ratio and heat field deformation sap flow probes for diffuse-porous trees, Agr. Forest Meteorol., 244–245, 151–161, 2017.

Gebauer, T., Horna, V., and Leuschner, C.: Variability in radial sap flux density patterns and sapwood area among seven co-occurring temperate broad-leaved tree species, Tree Physiol., 28, 1821–1830, 2008.

Gebauer, T., Horna, V., and Leuschner, C.: Canopy transpiration of pure and mixed forest stands with variable abundance of European beech, J. Hydrol., 442–443, 2–14, 2012.

Ghimire, C. P., Lubczynski, M. W., Bruijnzeel, L. A., and Chavarro-Rincón, D.: Transpiration and canopy conductance of two contrasting forest types in the Lesser Himalaya of Central Nepal, Agr. Forest Meteorol., 197, 76–90, 2014.

Granier, A., Biron, P., and Lemoine, D.: Water balance, transpiration and canopy conductance in two beech stands, Agr. Forest Meteorol., 100, 291–308, 2000.

Grömping, U.: Estimators of relative importance in linear regression based on variance decomposition, Am. Stat., 61, 139–147, 2007.

Hajek, P., Kurjak, D., von Wühlisch, G., Delzon, S., and Schuldt, B.: Intraspecific variation in wood anatomical, hydraulic, and foliar traits in ten European beech provenances differing in growth yield, Front. Plant Sci., 7, 791, https://doi.org/10.3389/fpls.2016.00791, 2016.

Hernandez-Santana, V., Hernandez-Hernandez, A., Vadeboncoeur, M. A., and Asbjornsen, H.: Scaling from single-point sap velocity measurements to stand transpiration in a multispecies deciduous forest: Uncertainty sources, stand structure effect, and future scenarios, Can. J. Forest Res., 45, 1489–1497, 2015

Hölscher, D., Koch, O., Korn, S., and Leuschner, C.: Sap flux of five co-occurring tree species in a temperate broad-leaved forest during seasonal soil drought, Trees-Struct. Funct., 19, 628–637, 2005.

Holst, J., Grote, R., Offermann, C., Ferrio, J. P., Gessler, A., Mayer, H., and Rennenberg, H.: Water fluxes within beech stands in complex terrain, Int. J. Biometeorol., 54, 23–36, 2010.

Ibanez, C. A. G., Carcellar III, B. G., Paringit, E. C., Argamosa, R. J. L., Faelga, R. A. G., Posilero, M. A. V., Zaragosa, G. P., and Dimayacyac, N. A.: Estimating DBH of trees employing multiple linear regression of the best LiDAR-derived parameter combination automated in python in a natural broadleaf forest in the Philippines, Int. Arch. Photogramm., 41, 657–662, 2016.

IUSS Working Group WRB: World reference base for soil resources 2006, A framework for international classification, correlation and communication, World Soil Resources Reports, FAO, Rome, 2006.

Jackisch, C.: Linking structure and functioning of hydrological systesm – How to achieve necessary experimental and model complexity with adequate effort, Thesis, KIT Karlsruhe, https://doi.org/10.5445/IR/1000051494, 2015.

Jasechko, S., Sharp, Z. D., Gibson, J. J., Birks, S. J., Yi, Y., and Fawcett, P. J.: Terrestrial water fluxes dominated by transpiration, Nature, 496, 347–350, 2013.

Jonard, F., André, F., Ponette, Q., Vincke, C., and Jonard, M.: Sap flux density and stomatal conductance of European beech and common oak trees in pure and mixed stands during the summer drought of 2003, J. Hydrol., 409, 371–381, 2011.

Jung, E.-Y., Otieno, D., Kwon, H., Berger, S., Hauer, M., and Tenhunen, J.: Influence of elevation on canopy transpiration of temperate deciduous forests in a complex mountainous terrain of South Korea, Plant Soil, 378, 153–172, 2014.

Kleidon, A. and Renner, M.: Thermodynamic limits of hydrologic cycling within the Earth system: concepts, estimates and implications, Hydrol. Earth Syst. Sci., 17, 2873–2892, https://doi.org/10.5194/hess-17-2873-2013, 2013.

Köstner, B., Granier, A. and Cermák, J.: Sapflow measurements in forest stands: Methods and uncertainties, Ann. Sci. Forest., 55, 13–27, 1998.

Kumagai, T., Aoki, S., Shimizu, T., and Otsuki, K.: Sap flow estimates of stand transpiration at two slope positions in a Japanese cedar forest watershed, Tree Physiol., 27, 161–168, 2007.

Loranty, M. M., MacKay, D. S., Ewers, B. E., Adelman, J. D., and Kruger, E. L.: Environmental drivers of spatial variation in whole-tree transpiration in an aspen-dominated upland-to-wetland forest gradient, Water Resour. Res., 44, W02441, https://doi.org/10.1029/2007WR006272, 2008.

Loritz, R., Hassler, S. K., Jackisch, C., Allroggen, N., van Schaik, L., Wienhöfer, J., and Zehe, E.: Picturing and modeling catchments by representative hillslopes, Hydrol. Earth Syst. Sci., 21, 1225–1249, https://doi.org/10.5194/hess-21-1225-2017, 2017.

Lüttschwager, D. and Remus, R.: Radial distribution of sap flux density in trunks of a mature beech stand, Ann. Forest Sci., 64, 431–438, 2007.

Martínez-Carreras, N., Krein, A., Gallart, F., Iffly, J.-F., Hissler, C., Pfister, L., Hoffmann, L., and Owens, P. N.: The influence of sediment sources and hydrologic events on the nutrient and metal content of fine-grained sediments (attert river basin, Luxembourg), Water Air Soil Poll., 223, 5685–5705, 2012.

Meinzer, F. C., Bond, B. J., Warren, J. M., and Woodruff, D. R.: Does water transport scale universally with tree size?, Funct. Ecol., 19, 558–565, 2005.

Moore, G. W., Jones, J. A., and Bond, B. J.: How soil moisture mediates the influence of transpiration on streamflow at hourly to interannual scales in a forested catchment, Hydrol. Process., 25, 3701–3710, 2011.

Mitchell, P. J., Benyon, R. G., and Lane, P. N. J.: Responses of evapotranspiration at different topographic positions and catchment water balance following a pronounced drought in a mixed species eucalypt forest, Australia, J. Hydrol., 440–441, 62–74, 2012.

Nadezhdina, N. and Čermák, J.: Instrumental methods for studies of structure and function of root systems of large trees, J. Exp. Bot., 54, 1511–1521, 2003.

Oren, R. and Pataki, D. E.: Transpiration in response to variation in microclimate and soil moisture in southeastern deciduous forests, Oecologia, 127, 549–559, 2001.

Oren, R., Zimmermann, R., and Terbough, J.: Transpiration in Upper Amazonia Floodplain and Upland Forests in Response to Drought-Breaking Rains, Ecology, 77, 968–973, 1996.

Otieno, D., Li, Y., Ou, Y., Cheng, J., Shizhong, L., Tang, X., Zhang, Q., Jung, E.-Y., Zhang, D., and Tenhunen, J.: Stand characteristics and water use at two elevations in a sub-tropical evergreen forest in southern China, Agr. Forest Meteorol., 194, 155–166, 2014.

Pfister, L., Humbert, J., and Hoffmann, L.: Recent trends in rainfall-runoff characteristics in the Alzette River basin, Luxembourg, Climatic Change, 45, 323–337, 2000.

Pielke Sr., R. A.: Land use and climate change, Science, 310, 1625–1626, 2005.

Rabadán, M.-Á. V., Peña, J. S., and Adán, F. S.: Estimation of diameter and height of individual trees for Pinus sylvestris L. based on the individualising of crowns using airborne LiDAR and the National forest inventory data, Forest Systems, 25, e046, https://doi.org/10.5424/fs/2016251-05790, 2016.

R Development Core Team: R: A language and environment for statistical computing, R Foundation for Statistical Computing, Vienna, Austria, 2014.

Renner, M., Hassler, S. K., Blume, T., Weiler, M., Hildebrandt, A., Guderle, M., Schymanski, S. J., and Kleidon, A.: Dominant controls of transpiration along a hillslope transect inferred from eco-hydrological measurements and thermodynamic limits, Hydrol. Earth Syst. Sci., 20, 2063–2083, https://doi.org/10.5194/hess-20-2063-2016, 2016.

Rössler, G.: Rindenabzug richtig bemessen, Forstzeitung, 4, p. 21, 2008.

Schmidt, M.: Canopy transpiration of beech forests in Northern Bavaria – Structure and function in pure and mixed stands with oak at colline and montane sites, Thesis, University of Bayreuth, available at: https://epub.uni-bayreuth.de/646/ (last access: 22 December 2017), 2007.

Schume, H., Jost, G., and Hager, H.: Soil water depletion and recharge patterns in mixed and pure forest stands of European beech and Norway spruce, J. Hydrol., 289, 258–274, 2004.

Scudeler, C., Pangle, L., Pasetto, D., Niu, G.-Y., Volkmann, T., Paniconi, C., Putti, M., and Troch, P.: Multiresponse modeling of variably saturated flow and isotope tracer transport for a hillslope experiment at the Landscape Evolution Observatory, Hydrol.

Earth Syst. Sci., 20, 4061–4078, https://doi.org/10.5194/hess-20-4061-2016, 2016.

Seibert, S. P., Jackisch, C., Ehret, U., Pfister, L., and Zehe, E.: Unravelling abiotic and biotic controls on the seasonal water balance using data-driven dimensionless diagnostics, Hydrol. Earth Syst. Sci., 21, 2817–2841, https://doi.org/10.5194/hess-21-2817-2017, 2017.

Sprenger, M., Seeger, S., Blume, T., and Weiler, M.: Travel times in the vadose zone: variability in space and time, Water Resour. Res., 52, 5727–5754, https://doi.org/10.1002/2015WR018077, 2016.

Stojnić, S., Orlović, S., Miljković, D., Galić, Z., Kebert, M., and Wuehlisch, G.: Provenance plasticity of European beech leaf traits under differing environmental conditions at two Serbian common garden sites. Eur. J. For. Res., 134, 1109–1125, 2015.

Tannenberger, F., Flade, M., Preiska, Z., and Schröder, B.: Habitat selection of the globally threatened Aquatic Warbler at the western margin of the breeding range: Implications for management, Ibis, 152, 347–358, 2010.

Traver, E., Ewers, B. E., Mackay, D. S., and Loranty, M. M.: Tree transpiration varies spatially in response to atmospheric but not edaphic conditions, Funct. Ecol., 24, 273–282, 2010.

Tromp-van Meerveld, H. J. and McDonnell, J. J.: On the interrelations between topography, soil depth, soil moisture, transpiration rates and species distribution at the hillslope scale, Adv. Water Res., 29, 293–310, 2006.

van den Bos, R., Hoffmann, L., Juilleret, J., Matgen, P., and Pfister, L.: Regional runoff prediction through aggregation of first-order hydrological process knowledge: A case study, Hydrolog. Sci. J., 51, 1021–1038, 2006.

Vandegehuchte, M. W. and Steppe, K.: Sap-flux density measurement methods: working principles and applicability, Funct. Plant Biol., 40, 213–223, 2013.

Vauhkonen, J. and Mehtätalo, L.: Matching remotely sensed and field-measured tree size distributions, Can. J. Forest Res., 45, 353–363, 2015.

Vertessy, R. A., Benyon, R. G., O'Sullivan, S. K., and Gribben, P. R.: Relationships between stem diameter, sapwood area, leaf area and transpiration in a young mountain ash forest, Tree Physiol., 15, 559–567, 1995.

Vilhar, U., Starr, M., Urbančič, M., Smolej, I., and Simončič, P.: Gap evapotranspiration and drainage fluxes in a managed and a virgin dinaric silver fir-beech forest in Slovenia: A modelling study, Eur. J. Forest Res., 124, 165–175, 2005.

Vincke, C., Breda, N., Granier, A., and Devillez, F.: Evapotranspiration of a declining Quercus robur (L.) stand from 1999 to 2001. I. Trees and forest floor daily transpiration, Ann. Forest Sci., 62, 503–512, 2005.

Wrede, S., Fenicia, F., Martínez-Carreras, N., Juilleret, J., Hissler, C., Krein, A., Savenije, H. H. G., Uhlenbrook, S., Kavetski, D., and Pfister, L.: Towards more systematic perceptual model development: A case study using 3 Luxembourgish catchments, Hydrol. Process., 29, 2731–2750, 2015.

Zanetto, A., Roussel, G., and Kremer, A.: Geographic variation of inter-specific differentiation between Quercus robur L. and Quercus petraea (Matt.) Liebl., Forest Genetics, 1, 111–123, 1994.

Zehe, E., Ehret, U., Pfister, L., Blume, T., Schröder, B., Westhoff, M., Jackisch, C., Schymanski, S. J., Weiler, M., Schulz, K., Allroggen, N., Tronicke, J., van Schaik, L., Dietrich, P., Scherer, U., Eccard, J., Wulfmeyer, V., and Kleidon, A.: HESS Opinions: From response units to functional units: a thermodynamic reinterpretation of the HRU concept to link spatial organization and functioning of intermediate scale catchments, Hydrol. Earth Syst. Sci., 18, 4635–4655, https://doi.org/10.5194/hess-18-4635-2014, 2014.

Precipitation pattern in the Western Himalayas revealed by four datasets

Hong Li[1,2], Jan Erik Haugen[3], and Chong-Yu Xu[2]

[1]Norwegian Water Resources and Energy Directorate, Oslo, Norway

[2]University of Oslo, Norway

[3]Norwegian Meteorological Institute, Oslo, Norway

Correspondence: Hong Li (lihong2291@gmail.com)

Abstract. Data scarcity is the biggest problem for scientific research related to hydrology and climate studies in the Great Himalayas region. High-quality precipitation data are difficult to obtain due to a sparse network, cold climate and high heterogeneity in topography. In this paper, we examine four datasets in northern India of the Western Himalayas: interpolated gridded data based on gauge observations (IMD, $1° \times 1°$, and APHRODITE, $0.25° \times 0.25°$), reanalysis data (ERA-Interim, $0.75° \times 0.75°$) and high-resolution simulation by a regional climate model (WRF, $0.15° \times 0.15°$). The four datasets show a similar spatial pattern and temporal variation during the period 1981–2007, though the absolute values vary significantly (497–$819\,\mathrm{mm\,year^{-1}}$). The differences are particularly large in July and August at the windward slopes and high-elevation areas. Overall, the datasets show that the summer is getting wetter and the winter is getting drier, though most of the trends in monthly precipitation are not significant. Trend analysis of summer and winter precipitation at every grids confirms the changes. Wetter summers will result in more and bigger floods in the downstream areas. Warmer and drier winters will result in less glacier accumulation. All the datasets show consistency in the period 1981–2007 and can give a spatial overview of the precipitation in the region. Comparing with the Bhuntar gauge data, the WRF dataset gives the best estimates of extreme precipitation. To conclude, we recommend the APHRODITE dataset and the WRF dataset for hydrological studies for their improved spatial variation which match the scale of hydrological processes as well as accuracy in extreme precipitation for flood simulation.

1 Introduction

The Great Himalayas region is the largest cryosphere outside the polar areas and the source of many rivers which supply water to more than 800 million people (Hegdahl et al., 2016; H. Li et al., 2016). The local population depends mainly on rivers for drinking water, hygiene, industry, fishing, but also for hydro-power generation and agriculture, which is one of main sectors of local economy (Ménégoz et al., 2013). Therefore, precipitation is very important to the local society and welfare of the local people. Climate change has significant impacts on water security, where mitigation and adaption to climate change are more challenging in this area due to poverty.

Precipitation is one of the most important elements in meteorology and hydrology. Precipitation measurements at gauges are usually used as benchmark data to compare with other datasets. They are often believed to be the most reliable and accurate data. However, there are fewer gauges available in this area compared to other areas in the world. Therefore, it is tricky to look at spatial variability based on gauge data. Besides, quality of measurements is rarely high due to harsh climate and complex environment. Additionally, manual errors are very common in developing countries. These errors include, for example, error in gauge location, missing the unit of data as well as wrong position of the decimal point. Last but not least, gauge data are usually hard to obtain due to data policy and political conflict in some countries.

In recent years, with development of space-borne measurements and computing technologies, gridded precipitation

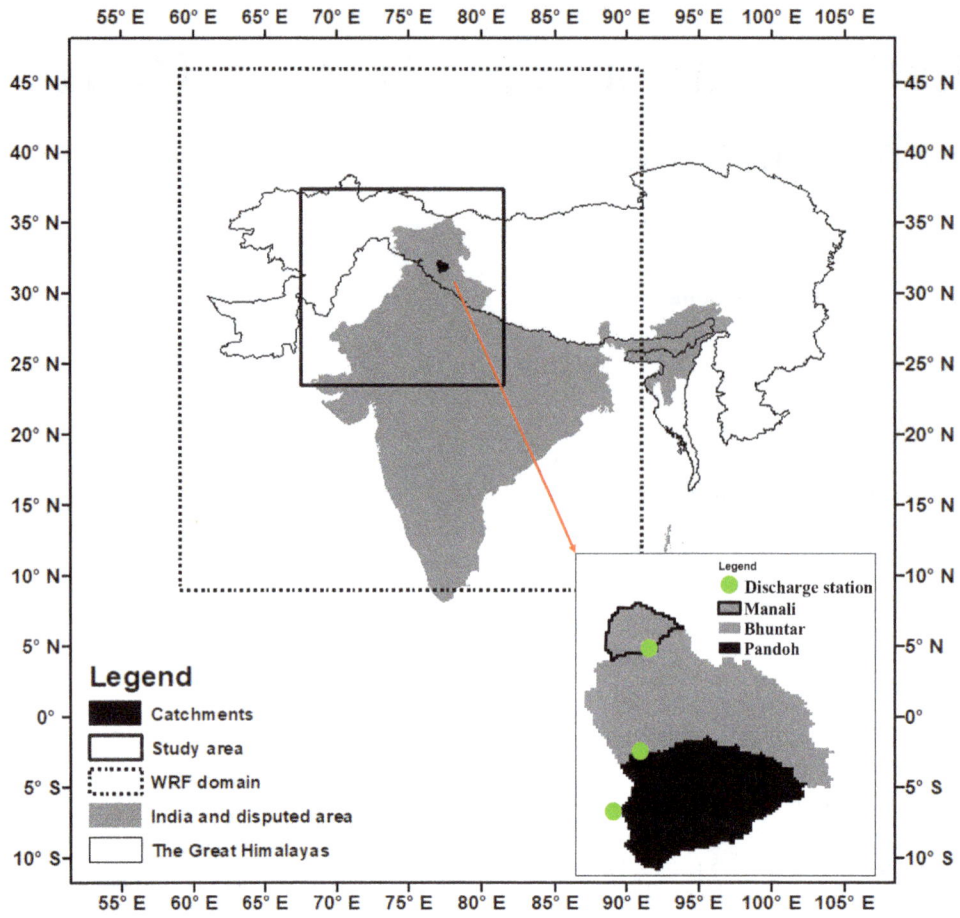

Figure 1. Location map of the study area, the Bhuntar rain gauge and three discharge stations.

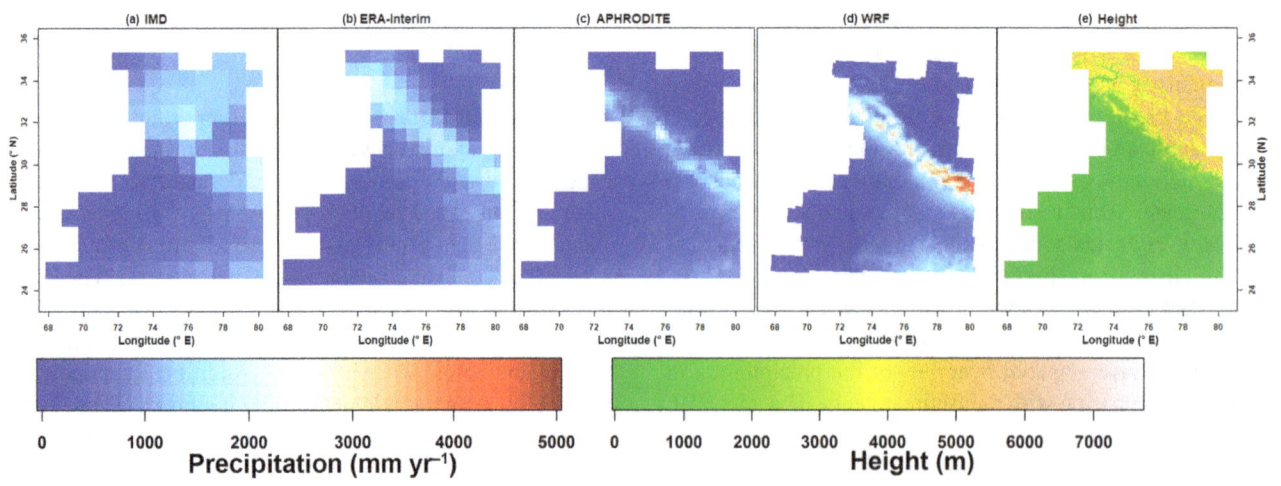

Figure 2. Mean annual precipitation (1981–2007) of the four datasets (from **a** IMD gridded observations, ERA-Interim reanalysis, APHRODITE gridded observations and WRF regional climate model simulation) and terrain height of the study area (**e**).

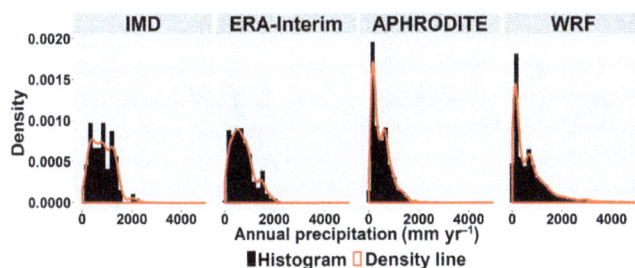

Figure 3. Density curves of mean annual precipitation values in all grid points.

datasets have been widely generated and attract much interest. Compared to measurements at traditional gauge, gridded data can cover a large area, sometimes even the globe, and disclose spatial variability at a continuous surface. Additionally, gridded data are usually produced by researchers for scientific purposes and they are free accessible to scientific research. Therefore, gridded data have been extensively used, particularly where high-quality *in situ* measurements are not available.

There have been quite a few studies on precipitation over the Great Himalayas region (Yatagai et al., 2012; Ménégoz et al., 2013; Palazzi et al., 2013). The available gridded data fall into four types: satellite data, interpolation of gauge observations, reanalysis and model simulation. However, all estimates are generally very uncertain due to the complex climate dynamics and local topography, and precipitation rates differ widely among the four types, even among different products of the same type. The satellite images show discrepancies due to platforms and characteristics of sensors. Reflectance from land surface, particularly snow and ice, can cause distinctive biases (Yin et al., 2008). The interpolated observations are usually believed the most reliable. However, great cautions have to be paid when using such data due to inadequacy of interpolating methods and unavoidable inferiors inherited from gauge measurements. For example, underestimation of precipitation could be 58 % of annual total precipitation in the cold Alaska region due to wind, wetting loss and trace precipitation (Yang et al., 1998). High-resolution climate models provide an alternative perspective and the models are competitive in the aspects of high spatio-temporal resolution, identification of precipitation forms (Ménégoz et al., 2013), and internal consistency between climate parameters. On the other hand, the simulated data may misrepresent the reality and suffer from inadequacy of boundary and forcing conditions. Reanalysis data are a combination of observations from many sources and dynamic models, but users should be cautious because of continuous changes in observing systems and systematic model errors (Dee et al., 2011). Additionally, uncertainties in reanalysis data are difficult to understand and quantify (Dee et al., 2011). The weaknesses and strength of each type are summarized in Table 1.

In this study, we select four datasets from various sources, i.e., interpolation of gauge observations, reanalysis and model simulations in northern India of the Western Himalayas as well as measurements at one rain gauge. Due to differences in availability, a common analysis is based on daily data in a long period of 27 years (1981–2007). To our knowledge, this is the first of its kind in this region in terms of number of datasets and data length. The purpose is to compare the datasets and to find their similarity and difference, as well as implications for further use in hydrological studies.

2 Study area

The study area lies in the western part of the Indian Himalayan region (Fig. 1). The highest point is 7677 m above sea level (m a.s.l.), located in the northeastern region. The low-elevation part lies in the southwestern region, which adjoins Pakistan. The climate is affected by monsoon and western disturbance. In summer, warm moisture from the Indian Ocean moves northwards and turns westward when it hits the high mountains. This interaction brings plenty of precipitation and daily precipitation can be more than 200 mm (Purohit and Kau, 2016). Precipitation in high mountains usually falls as snow in winter. Along the course of the moist wind, precipitation decreases from east to west. In winter, the climate is controlled by western turbulence. The midlatitude low-pressure systems bring some snowfall (Ménégoz et al., 2013), but winter is generally quite dry, especially in the coldest region. In this study, seasons are referred based on northern meteorological seasons (spring: March to May; summer: June to August; autumn: September to November; winter: December to February).

This area is the headwater of the Indus River and the Ganges River, which are transboundary among China, India, Pakistan and Bangladesh. Additionally, these two rivers have very high hydropower potential. How to explore hydropower is continuously negotiated among the involved countries, which makes the study area very political sensitive.

3 Data

3.1 IMD dataset

The IMD dataset is produced by the India Meteorological Department for the whole India. The time period is 1951–2007 and the spatial resolution is $1° \times 1°$. The data are interpolated from gauge measurements by using the Shepard method (Shepard, 1968). Rajeevan et al. (2006) compare the IMD dataset with the Variability Analysis of Surface Climate Observations (VASClimo) dataset and conclude that the IMD dataset is more accurate in terms of spatial variation. The IMD dataset has been extensively used in climate related research and applications, such as validation of climate models (Bollasina et al., 2011; Wiltshire, 2014) and monsoon variability and predictions (Goswami et al., 2006).

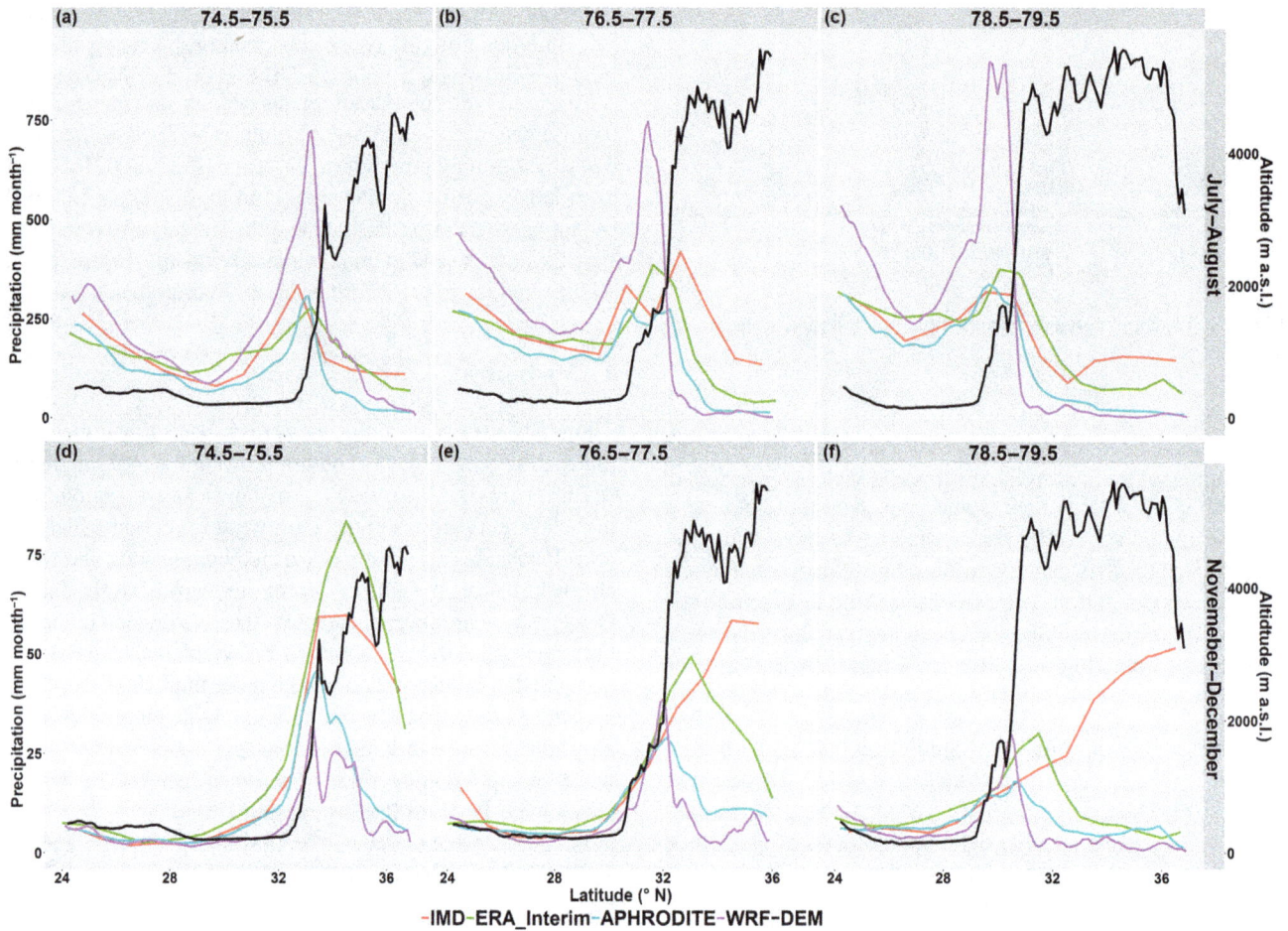

Figure 4. Precipitation (mm month^{-1}) for July–August (**a, b, c**) and November–December (**d, e, f**) and in three selected longitude bands from west to east (left to right) plotted against latitude. The longitude value of each band is indicated above the figures. The corresponding terrain height (m) in black is displays at the right axis.

Table 1. Summary of weaknesses and strength of four types of gridded precipitation data.

Data type	Strength	Weakness
gauge	original ground measurements long application	coarse distribution undercatch of snow and rain due to wind manual errors high expense or unavailability for political reasons
satellite	spatial observations quality not affected by wind or other weather conditions	dependence on platforms and sensors bias caused by snow and ice
interpolations	consistent with traditional ground observations	Inadequacy of interpolating methods unavoidable inferiors inherited from gauge measurements
output from climate models	consistent with other meteorological parameters possibility to measure uncertainties	inadequacy in algorithms, boundary and forcing
reanalysis	combination of modeling technique and many types of observations	changes in observation system model error

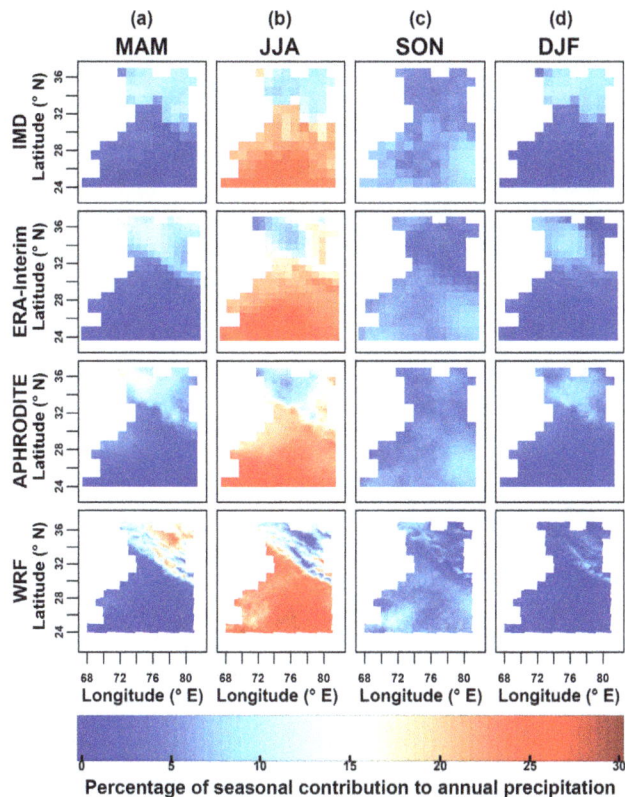

Figure 5. Seasonal contributions (in %) to annual precipitation. **(a)** spring (MAM), **(b)** summer (JJA), **(c)** autumn (SON) and **(d)** winter (DJF). From bottom to top: WRF regional climate model, APHRODITE gridded observations, ERA-Interim reanalysis and IMD gridded observations.

The number of used gauges varies during the period as well as spatially across the region. The average number of gauges per grid point is 2.99 ranging from 0.2 to 4.4 (Rajeevan et al., 2006). Spatially, more gauges are used in the central south; less gauges near the borders of India and in the northern part. No gauge measurements are available near the latitude of 35.5° N and northward.

3.2 APHRODITE dataset

The APHRODITE (Asian Precipitation – Highly Resolved Observational Data Integration Towards Evaluation of Water Resources) dataset is interpolated by the Sphere map method based on data collected at 5000–12 000 gauges (Yatagai et al., 2012). The interpolated parameter is the precipitation anomaly or ratio, instead of the precipitation amount (Yatagai et al., 2012). Elevation corrections are considered by a weighting function, which is based on the angular distance when considering topography (Yatagai et al., 2012). The dataset covers Asia over the period of 1951–2007. Different versions of the APHRODITE dataset have been used to determine Asian monsoon precipitation change, hydrolog-

ical modeling (Pechlivanidis and Arheimer, 2015; Xu et al., 2016), verification of high-resolution model simulations and satellite precipitation estimates (Kamiguchi et al., 2010). In this research, we use the latest version (V1101) for monsoon Asia at a spatial resolution of $0.25° \times 0.25°$ (Dimri et al., 2013). The APHRODITE dataset uses the largest number of gauge observations among interpolated products, and is believed to be one of the most realistic precipitation datasets for Asia (Ménégoz et al., 2013).

3.3 ERA-Interim dataset

The ERA-Interim dataset is the precipitation product of ERA-Interim (Dee et al., 2011), which is a spatially and temporally complete dataset of multiple climate variables at high spatial and temporal resolution. The data we use here are on a Gaussian grid (with a resolution of $0.7 \times 0.7°$ at the Equator) with a 3 h time resolution, and aggregated to daily time step. ERA-Interim is a global atmospheric reanalysis dataset produced by the ECMWF (European Center for Medium-Range Weather Forecasts)[1]. The dataset dates back to 1979 and is updated with approximately 1-month delay from real time. The data assimilation system is based on a 2006 release of the IFS (Cy31r2) (Dee et al., 2011). This dataset has been widely used as boundary and forcing conditions for regional climate models (Dimri et al., 2013; Katragkou et al., 2015).

3.4 WRF dataset

The WRF dataset is generated by using a regional climate model, the Weather Research & Forecasting Model (v3.7.1). The climate model is a limited-area, non-hydrostatic, primitive-equation model with multiple options for various physical parameterization schemes. The model has been used in climate simulation in Asia and other areas (Maussion et al., 2011; Li et al., 2016). Here we use the Thompson scheme for microphysics, CAM for short- and long-wave radiation, the Noah Land-Surface scheme, Mellor–Yamada–Janjic TKE for the planetary boundary layer and Kain–Fritsch (new Eta) for convection. The model is forced by 6-hourly ERA-Interim reanalysis data. To avoid error at boundary edges and to facilitate further hydrological modeling work, we set up the model at a very large domain (59–91° E, 9–46° N). The spatial resolution is around 16 km, where topography and land use are aggregated from data with an accuracy of 10 m. They are preprocessed by using the WRF Preprocessing System (WPS). We divide the atmosphere into 30 vertical layers with model top pressure 50 hPa. The height of the lowest model level varies between 15 and 27 m depending on the surface pressure. The whole simulation period is from 1979 to 2007,

---

[1]The next generation reanalysis, ERA5, featuring a higher horizontal resolution ($\sim$ 31 km) and a 10-member ensemble approach for uncertainty estimates, is released by the end of 2017. See, e.g., http://www.ecmwf.int/en/newsletter/147/news/era5-reanalysis-production (last access: 1 May 2015).

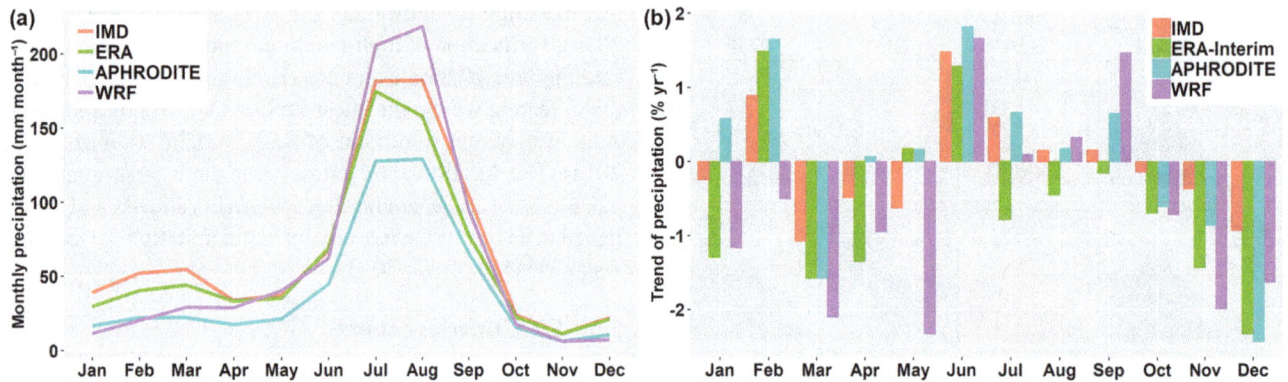

Figure 6. Monthly precipitation **(a)** and the trend during 1981–2007 **(b)**.

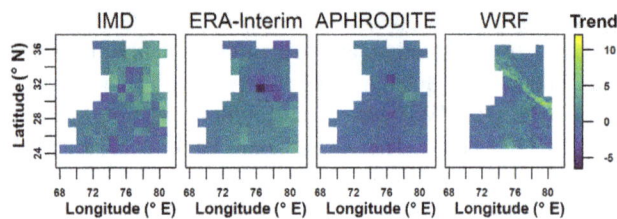

Figure 7. Trend (mm month^{-1} year^{-1}) of summer precipitation.

and the period 1979–1980 is used as model spinup. Due to the long model running time, we restart the model around every 5 years. Model setup is summarized in Table 2 and the whole setting is a file in the Supplement.

3.5 Gauge data

Rain gauge Bhuntar lies in a valley at a small town in the state of Himachal Pradesh, India (Fig. 1). The Bhuntar gauge is only 400 m down the confluence of the Parvati River with the Beas River. The altitude of the gauge is 1080 m a.s.l., and both precipitation and discharge data are used in this study. Annual precipitation is 921 mm year^{-1} based on data from 1981 to 2007, with most rainfall in July and August. Temperature is rarely below 0 °C, and only minimum temperature is occasionally below 0 °C in winters. The precipitation data have been used in hydrological modeling research for the Beas Basin (Li et al., 2015).

3.6 Discharge

Three discharge series are selected to cross validate water balance. They are respectively Pandoh (downstream), Bhuntar (middle stream) and Manali (upstream). These stations are operated by Central Water Commission regional office in India. The catchments are located in the Beas Basin, which is a main tributary of the Indus River in northern India (Fig. 1). The catchments are nested from upstream to downstream. The purpose is to reflect precipita-

Table 2. The main settings of the WRF regional climate model. The complete setting are shown in the Supplement.

Time and domain	
Period	1979–2007
Region	59–91° E, 9–46° N
Horizontal grid spacing	16 km
Dimension	(193, 241, 30)
Model top pressure	50 hPa
Height of the lowest level	15–27 m
Physics	
Microphysics	Thompson scheme
Short-wave radiation	CAM
Long-wave radiation	CAM
Surface layer	Monin–Obukhov (Janjic) scheme
Land surface	Noah Land-Surface scheme
Planetary boundary layer	Mellor–Yamada–Janjic TKE scheme
Cumulus	Kain–Fritsch (new Eta) scheme
Lateral boundaries	
Forcing	ERA-Interim 0.75° × 0.75°, 6-hourly

tion data at various elevations within a hydrological scale. Runoff is considerably influenced from glacier melting (Li et al., 2015). According to the 0.5 km MODIS-based Global Land Cover Climatology by the USGS Land Cover Institute (https://landcover.usgs.gov/global_climatology.php, last access: 1 May 2017), coverage of snow and ice is 16 % in the Pandoh catchment, 24 % in the Bhuntar catchment and 21 % in the Manali catchment. The discharge data have been manually quality controlled and missing data are filled by discharge anomaly. Discharge measurements are more qualified than precipitation in the snow and ice dominated area (Henn et al., 2015; Kretzschmar et al., 2016). Therefore, the quality of runoff simulation can infer by the forcing precipitation data. Li et al. (2016) use the WRF-Hydro (v3.5.1) modeling system in the Beas Basin, and they find that the distribution of simulated daily discharge values agrees well with observations, which reversely confirms the precipitation simulations.

Table 3. P value of the tailed Kolmogorov–Smirnov test on differences of on annual precipitation (mm year^{-1}) among the datasets. The p value indicates strong evidence against the null hypothesis. It is typically to reject the null hypothesis, which is two datasets are the same here, when the p value is not greater than 0.05.

Data	IMD	ERA-Interim	APHRODITE	WRF
IMD	–	5.4×10^{-2}	3.6×10^{-11}	3.7×10^{-10}
ERA-Interim	5.4×10^{-2}	–	5.0×10^{-11}	2.5×10^{-11}
APHRODITE	3.6×10^{-11}	5.0×10^{-11}	–	$<2.2 \times 10^{-16}$
WRF	3.7×10^{-10}	2.5×10^{-11}	$<2.2 \times 10^{-16}$	–

3.7 Evaporation

The MODIS Global Evapotranspiration Project (MOD16) (http://www.ntsg.umt.edu/project/modis/mod16.php, last access: 1 May 2017) is selected to reveal actual evaporation. The MODIS project is started in 2000, and has a short overlap period with the study period. Additionally, part of the catchments is covered by permanent snow and ice and the sensors cannot work well on this type surface. Therefore, we use annual mean amounts of 2000 to 2013 to reduce uncertainties. The missing ratios of annual mean actual evaporation are 22 % for the Pandoh catchment, 32 % for the Bhuntar catchment and 31 % for the Manali catchment.

4 Results

4.1 Spatial variations

The four datasets show similar spatial pattern of mean annual precipitation (Fig. 2). The highest precipitation is located at the foothill of the mountains and stretched from southeast to northwest. Visually, the high precipitation belt (the foothills of the mountains and the southeastern corner) is most clearly shown by the WRF dataset. The spatial variability increases from the IMD dataset to the WRF dataset. Their coefficients of variation are respectively 0.5 for the IMD data, 0.6 for the ERA-Interim data, 0.7 for the APHRODITE data and 1.1 for the WRF data. The density curves of mean annual precipitation values in all grid points (Fig. 3) and the statistics of the Kolmogorov–Smirnov test (Table 3) show the variabilities and the differences among the datasets more clearly.

Both the IMD and APHRODITE datasets are interpolated from observations at gauges. However, the APHRODITE dataset shows a rain belt at the mountains' foothills much better. Additionally, the APHRODITE dataset shows much lower estimates (less than 300 mm year^{-1}) in the northeastern corner. This area is quite high, with mean elevation at 4650 m a.s.l. and elevation ranges from 906 to 7677 m a.s.l. The temperature is $-2.35\,°C$ of annual mean and as low as $-16.81\,°C$ in January (AphroTemp, Yatagai et al., 2012). The reason for this low-precipitation area is that the APHRODITE dataset uses more gauges, particularly also observations from Nepal, Bhutan and China (Yata-

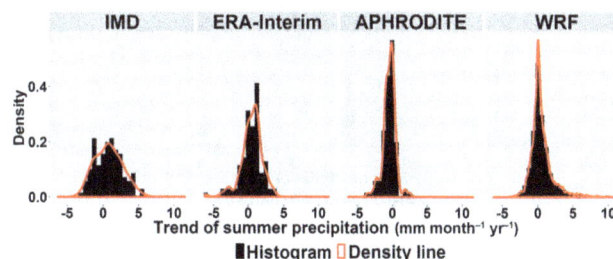

Figure 8. Density curves of trends (mm month^{-1} year^{-1}) of summer precipitation in all grid points.

gai et al., 2012). These gauges have undercatch problems, which means rain gauges could only catch part of snowfall due to wind and disturbance. In contrast, the IMD dataset uses only the gauges in the low-valley area of India and extends north by interpolation (Rajeevan et al., 2006). Eventually, the APHRODITE dataset has the lowest annual amount, only 61 % of the IMD dataset.

The ERA-Interim and WRF datasets are products with different dynamical models. The ERA-Interim data and the WRF data are similar in terms of annual total amount (ERA-Interim: 718 mm year^{-1}, WRF: 688 mm year^{-1}) and spatial pattern, partially due to the fact that in this area the observations that are assimilated into the data assimilation system are sparse and unevenly distributed. The WRF data are more realistic than the ERA-Interim data due to finer spatial resolution, especially in complex topography areas (Ménégoz et al., 2013; Dimri et al., 2013).

The effects of location and topography are shown in Fig. 4. The summer precipitation changes dramatically. Over the high flat plateau, precipitation decreases with latitude since the strength of the monsoon decreases with distance from its source. As the monsoon gets closer to the mountains, precipitation starts to increase. As the air parcel is lifted to high elevation, climate gets dry and cold. The winter precipitation occurs mainly along the upslope. The magnitude is also small and decreased along the path of the winter monsoon. The highest precipitation occurs in the windward of the upslope region, but it is 0.5 or 1.5° (around 55–110 km) far away from the mountains in summer. Bookhagen and Burbank (2006) analyze a decade of TRMM data and also find

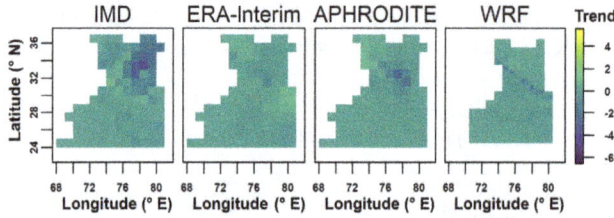

Figure 9. Trend ($mm\,month^{-1}\,year^{-1}$) of winter precipitation.

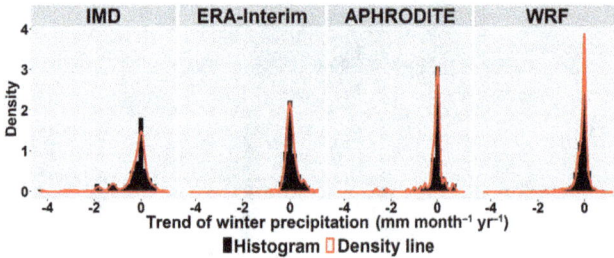

Figure 10. Density curves of trends ($mm\,month^{-1}\,year^{-1}$) of winter precipitation in all grid points.

the highest annual precipitation is offset by a few 10 s of km south of either high topography or relief. This offset has been found only over tall and broad mountain regions rather than narrow mountain peaks (Dimri and Niyogi, 2013).

The differences among the datasets are more obvious in summer at the mountain foot. The WRF dataset gives much more precipitation ($700\,mm\,month^{-1}$) in July and August at the mountain foot, almost 2 times that of other datasets ($300\,mm\,month^{-1}$). This is reported as a moisture bias in summer (Srinivas et al., 2013; Li et al., 2016). It is often cited as orographic bias which describes as strong overprediction of precipitation rates along windward slopes while predicted snowfall lies under measured values along leeward slopes (Maussion et al., 2011).

4.2 Temporal variations and changes

The inter-annual patterns are very similar as indicated by high correlations between pairs of datasets, shown in Table 4. The correlation between the IMD and APHRODITE datasets is the highest, reaching 0.91. The WRF dataset has low correlation with all other datasets. Spatially, the four datasets show a similar seasonal distribution, and the WRF dataset has the highest variability (Fig. 5). The intra-annual cycle is also similar as shown in Fig. 6. The WRF and APHRODITE datasets have respectively the highest and lowest precipitation in summer.

To look at changes over time, we select the Theil–Sen median method to calculate trends due to its robustness and the non-parametric Mann–Kendall test for the significance test. The trend analysis and significance test are done for the areal mean of each month (Fig. 6), and every individual grid for

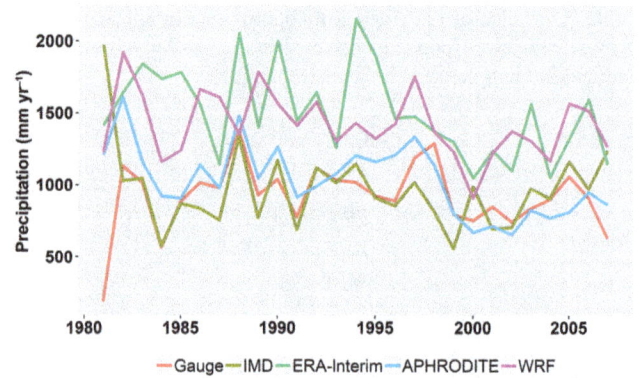

Figure 11. Annual precipitation at the Bhuntar gauge. Data at the nearest point to the Bhuntar gauge are extracted from the gridded datasets.

Table 4. Pearson's correlation of annual precipitation series. The italics indicate the minimum values by row and by column.

Data	IMD	ERA-Interim	APHRODITE	WRF
IMD	–	0.86	0.91	*0.64*
ERA-Interim	0.86	–	0.86	*0.64*
APHRODITE	0.91	0.86	–	*0.59*
WRF	*0.64*	*0.54*	*0.59*	–

summer (Figs. 7 and 8) and winter (Figs. 9 and 10). The figures show an increase in summer precipitation and a decrease in winter precipitation, although both increase and decrease exist in each dataset. Three of the areal mean trends (May by the WRF dataset; June by the IMD and ERA-Interim datasets) are statistically significant at the 95 % confidence level. The spatial distribution of trends in summer precipitation varies a lot. Most decreasing trends of winter precipitation occur in the northern part. Approximately 10 % of grids are significant at the 10 % confidence level.

It is difficult to conclude why northern India of the Western Himalayas shows an increase in summer precipitation. However, Bollasina et al. (2011) find the same increasing monsoon precipitation in northern India but decreasing monsoon precipitation in central Asia. They use a series of climate model experiments, and conclude that such pattern is a robust outcome of a slowdown of the tropical meridional overturning circulation, which could be attributed mainly to human-influenced aerosol emissions. The trends will continue and become more significant with time if greenhouse gas emission continues as usual. Such trends would lead to strong negative mass balance conditions of glaciers, which is discussed in the next section.

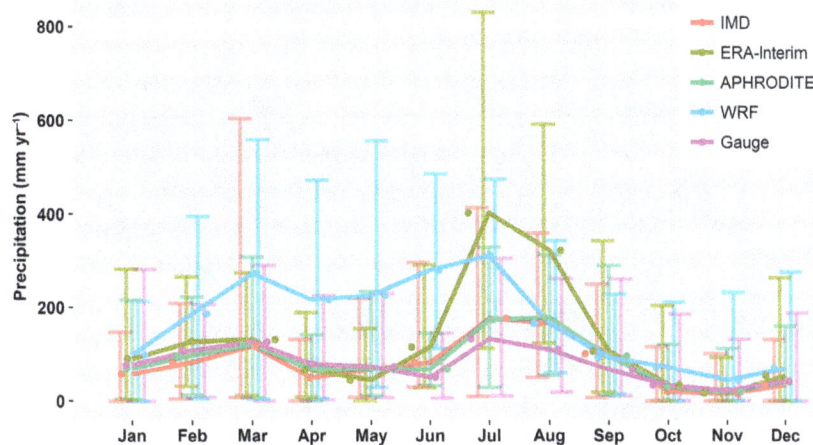

Figure 12. Monthly anomaly at the Bhuntar gauge. Data at the nearest point to the Bhuntar gauge are extracted from the gridded datasets.

Table 5. Statistics of annual maximum daily precipitation (mm day^{-1}) at the Bhuntar gauge. Data of the nearest point are extracted from the gridded datasets. Bold indicates the value closest to the data of the Bhuntar gauge.

quantiles Data	minimum	0.05	0.25	median	0.75	0.95	maximum
Gauge	38.0	41.1	57.8	69.6	82.2	104.3	106.0
IMD	26.4	30.2	36.1	52.1	67.3	116.8	147.3
ERA	59.2	67.7	80.2	94.0	121.7	149.5	154.4
APHRO	28.2	29.8	38.6	52.9	58.0	70.9	**103.1**
WRF	**43.5**	**51.0**	**57.3**	**65.7**	**78.3**	**93.5**	99.8

5 Discussions

5.1 Comparison of gridded precipitation datasets with gauge data

To compare the gridded datasets with measurements at the Bhuntar gauge, we extract the time series at the nearest point to the Bhuntar gauge. We look at annual precipitation (Fig. 11), monthly anomaly (Fig. 12) as well as extreme precipitation, i.e., annual maximum daily precipitation (Table 5). As shown in Fig. 11, all gridded datasets are comparable with the Bhuntar gauge data. The interpolated datasets, IMD and APHRODITE are quantitatively closest to the Bhuntar gauge data. The ERA-Interim and WRF datasets generally give 2 or 3 times higher precipitation than the Bhuntar gauge data. Figure 12 shows the differences are mainly from March to July in the WRF dataset, and from July and August in the ERA-Interim dataset. In addition, the WRF dataset shows large variations from February to June, and the ERA-Interim dataset shows large variations in July and August. Table 5 shows the statistics of annual maximum daily precipitation. Notably, the WRF dataset gives the closest estimate to the Bhuntar data in five quantiles, and the APHRODITE dataset gives the best estimate of the maximum precipitation over the whole period.

5.2 Comparison of gridded precipitation datasets with runoff data

The annual actual evaporation from MODIS data is $614\,\text{mm year}^{-1}$ at the Pandoh catchment, $639\,\text{mm year}^{-1}$ at the Bhuntar catchment and $649\,\text{mm year}^{-1}$ at the Manali catchment. The values are too high compared with $64\,\text{mm year}^{-1}$ at the Pandoh catchment for the period from 1990 to 2004 calculated by Kumar et al. (2007) using potential evaporation, mean and maximum temperature. The Pandoh catchment covers the lower and middle parts, and should have the highest evaporation due to warm climate among three catchments. The MODIS data are not qualified at the catchments and at small catchment scales for the study period.

The precipitation and runoff relationship is shown in Fig. 13 as accumulation of monthly precipitation and runoff. Though the lines have different slopes, but they share very similar linear relationships. They are consistent in terms of temporal changes. Errors are systematic within each dataset. Runoff is generally less than precipitation due to evaporation loss. However, runoff could possibly exceed precipitation at glacierized catchments due to glacier melting. In the Manali catchment, runoff is much more than precipitation. In the Bhuntar catchment, only the ERA-Interim data show less

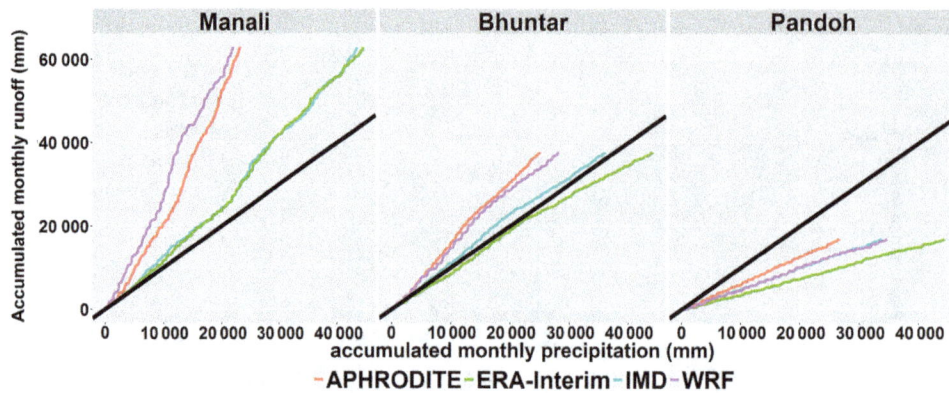

Figure 13. Accumulated precipitation and discharge.

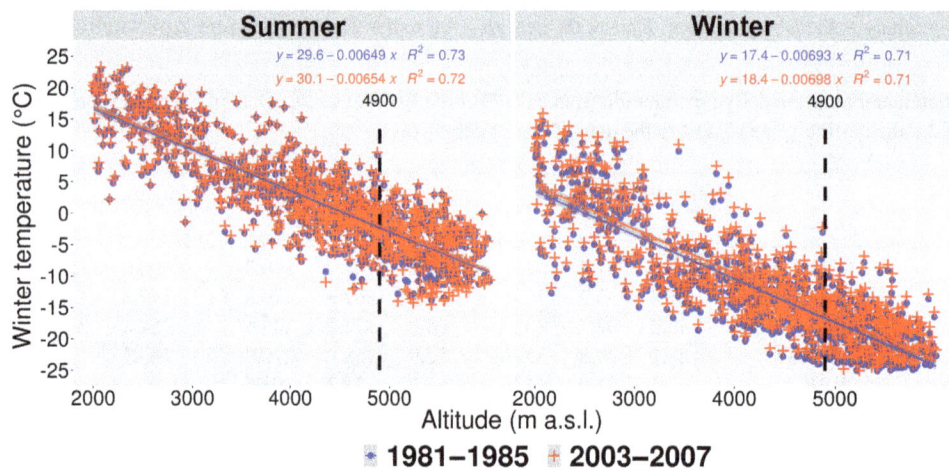

Figure 14. Mean temperature and its regression lines for the periods of 1981–1985 and 2003–2007 by the WRF simulation.

runoff than precipitation. All datasets show less runoff than precipitation in the Pandoh catchment. Precipitation is definitely underestimated in higher-elevation areas, especially in the Manali catchment. Azam et al. (2014) reconstruct annual mass balance of Chhota Shigri glacier since 1969. The Chhota Shigri glacier lies in the Western Himalayas, India and it is representative in terms of mass balance for the Western Himalayas glaciers (Azam et al., 2014). The mass loss rates are 0.36 ± 0.36 for 1969 to 1985 and 0.57 ± 0.36 m water equivalent per year (m w.e. a^{-1}) for 2001 to 2015. The runoff contribution from glacier melting is only 3306 mm within 29 years with assumptions of 20 % glacier coverage and -0.57 m w.e. a^{-1}.

5.3 Implications for glaciers

In the Great Himalayas region, there are many glaciers, and they are key indicators of regional climate change and water resources. Temperature in combination with precipitation controls survival of glaciers. Therefore, we also look at changes in temperature by comparing the temperature results by the same simulation of the WRF precipitation dataset

for the first and last 5 years, namely 1981–1985 and 2003–2007. We skip the trend analysis and significance test, because it is already well known that temperature has been increasing quickly in the Great Himalayas region since the 1980s (Ren et al., 2017). Temperature is well measured and simulated. Therefore, there is no need to go through many datasets. We are particularly interested in temperature at the equilibrium line altitude (ELA). As the slope of the regression lines shown in Fig. 14, the WRF model is able to reproduce the lapse rates. Between the two 5-year periods, temperature increases by 0.91 °C in winter and by 0.26 °C in summer. Such changes lead to an increase in the elevation of the freezing point (0 °C) of 125 m in winter and 32 m in summer. As shown in Sect. 4.2, precipitation overall decreases in winter. In combination with increasing temperature, this is an unfavourable condition for the glaciers with less accumulation and faster melting. Moreover, the area between 4900 m a.s.l., which is the equilibrium line altitude (ELA) of the Chhota Shigri glacier (see Azam et al., 2012, Fig. 2), and 5200 m a.s.l. is large. Therefore, as the climate gets warmer, the ELA will further move up. Such a nonlinear characteristic of elevation distribution results in a potential large reduction in the accumulation area and small storage buffer of permanent snow and ice.

6 Conclusions

Data scarcity is a major problem for hydrological research in the Great Himalayas region. High-quality precipitation data are difficult to obtain due to the sparse network, cold climate and high heterogeneity in topography. This paper investigates the spatial and temporal pattern of precipitation in this region based on four datasets: interpolated gridded data based on gauge observations (IMD, $1° \times 1°$ and APHRODITE, $0.25° \times 0.25°$), reanalysis data (ERA-Interim, $0.75° \times 0.75°$) and high-resolution simulation by a regional climate model (WRF, $0.15° \times 0.15°$) in northern India of the Western Himalayas during the period 1981–2007.

The four datasets are similar in terms of spatial pattern and temporal variation and changes, though the absolute values vary a lot (497–819 mm year^{-1}) due to the data source and the methods of data generation. The differences are particularly large in July and August and at the windward slopes and the high-elevation areas. The datasets reveal that summer gets wetter and winter gets drier, though most of the trends are not statistically significant. Wetter summer results in more and bigger floods at the downstream areas. Warmer and drier winter results in less glaciers accumulation. The four datasets are able to give a good overview of spatial pattern and temporal changes. Comparison with measurements at the Bhuntar gauge shows that the WRF and APHRODITE datasets give the best estimate of extreme precipitation amounts. To conclude, the APHRODITE and WRF datasets are recommended for hydrological studies due to their improved spatial variations which match the scale of hydrological processes as well as accuracy in extreme precipitation for flood simulation. However, careful local correction is definitely required.

Appendix A

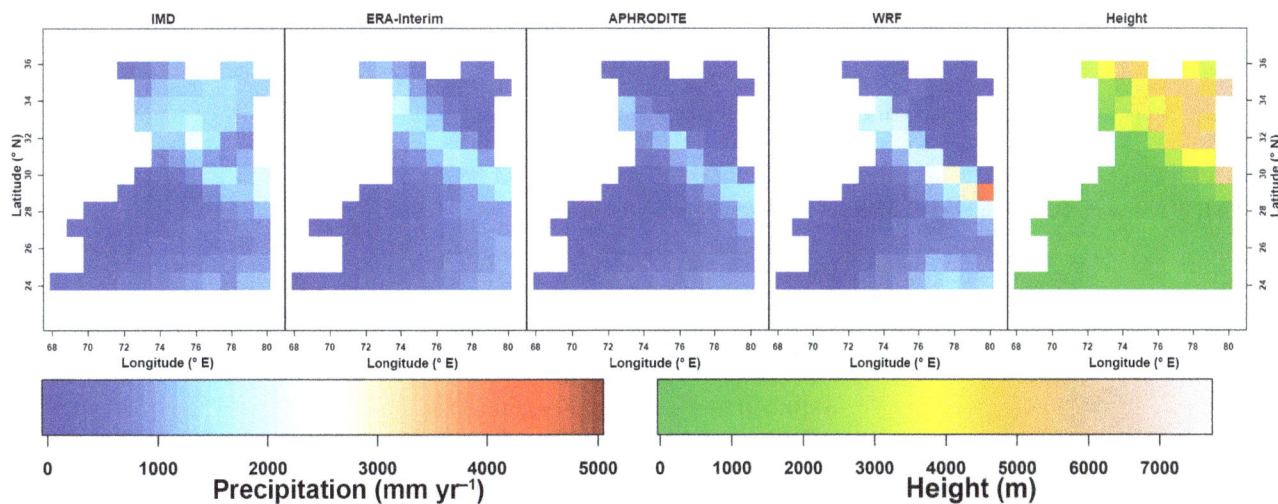

Figure A1. Precipitation of the gridded datasets and terrain height of the study area resampled to the IMD grid by bilinear interpolation.

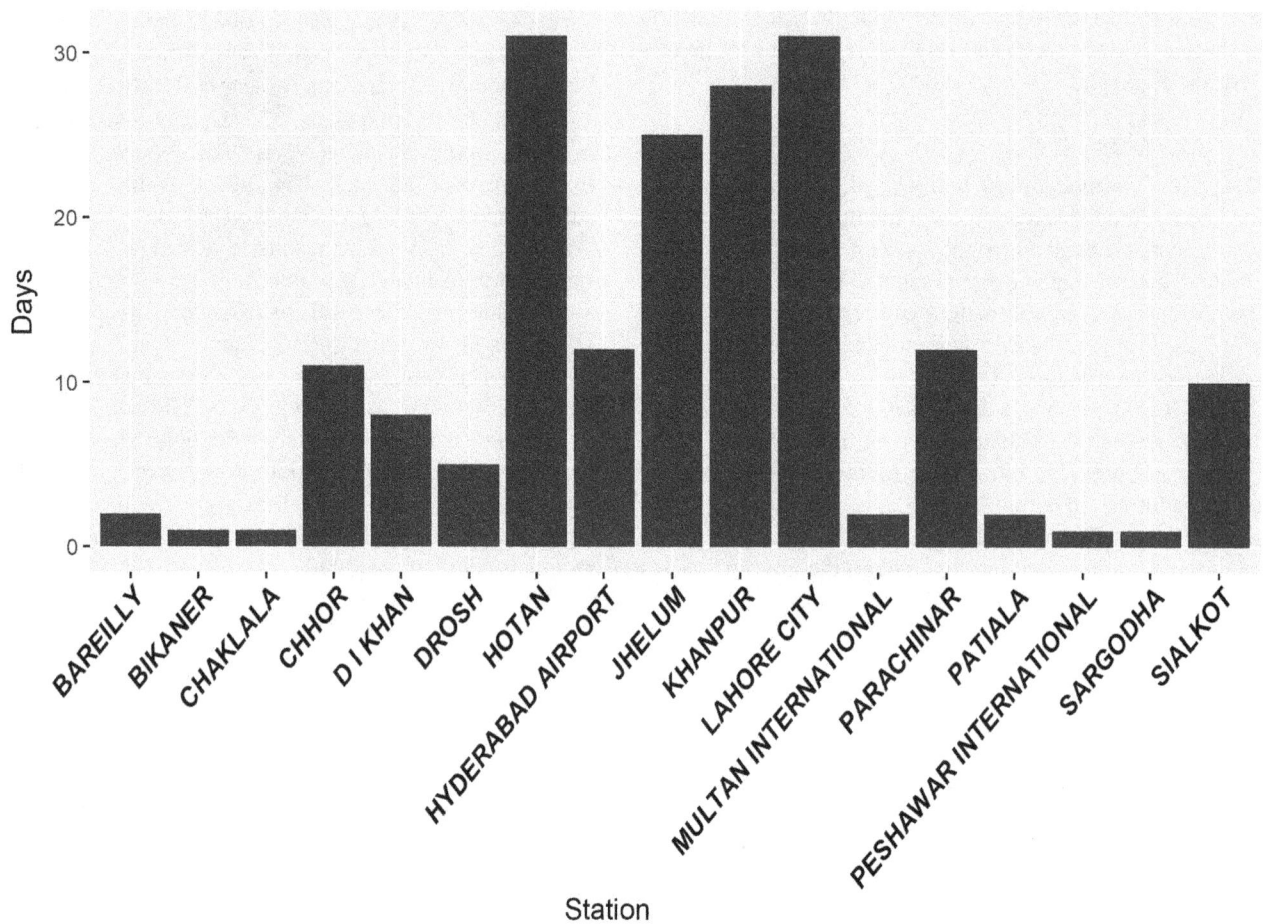

Figure A2. Data missing rate in the National Climatic Data Center in the study areas.

Table A1. Statistics of the non-parametric Mann–Kendall test of precipitation.

Data	Trend (mm year^{-1})	p value
IMD	−2.01	0.509
ERA-Interim	0.86	0.994
APHRODITE	−3.59	0.098
WRF	−0.70	0.819

Author contributions. HL: model simulation, data analysis and writing. JEH: model simulation, analysis of results, review and revision. CYX: review and revision.

Competing interests. The authors declare that they have no conflict of interest.

Special issue statement. This article is part of the special issue "The changing water cycle of the Indo-Gangetic Plain". It is not associated with a conference.

Acknowledgements. This study is funded by the Research Council of Norway through research program NORKLIMA under grant project 216546. We thank the India Meteorological Department and Sonia Grover at the Water Resources Division at TERI (India), the European Centre for Medium-Range Weather Forecasts and APHRODITE (http://www.chikyu.ac.jp/precip/english/products.html, last access: 1 May 2017), and research program JOINTINDNOR under grant project 203867 for provision of data. We thank Oskar Landgren at the Norwegian Meteorological Institute for help in modeling and data analysis as well as review of the manuscript. The model simulation was done when the first author worked at the Norwegian Meteorological Institute.

Edited by: Ian Holman

References

Azam, M. F., Wagon, P., Ramanathan, A., Vincent, C., Sharma, P., Arnaud, Y., Linda, A., Pottakkal, J. G., Chevallier, P., Singh, V. B., and Berthier, E.: From balance to imbalance: a shift in the dynamic behaviour of Chhota Shigri glacier, western Himalaya, India, J. Glaciol., 58, 315–324, https://doi.org/10.3189/2012JoG11J123, 2012.

Azam, M. F., Wagnon, P., Vincent, C., Ramanathan, A., Linda, A., and Singh, V. B.: Reconstruction of the annual mass balance of Chhota Shigri glacier, Western Himalaya, India, since 1969, Ann. Glaciol., 55, 69–80, https://doi.org/10.3189/2014AoG66A104, 2014.

Bollasina, M. A., Ming, Y., and Ramaswamy, V.: Anthropogenic Aerosols and the Weakening of the South Asian Summer Monsoon, Science, 334, 502–505, 2011.

Bookhagen, B. and Burbank, D. W.: Topography, relief, and TRMM-derived rainfall variations along the Himalaya, Geophys. Res. Lett., 33, L08405, https://doi.org/10.1029/2006GL026037, 2006.

Dee, D. P., Uppala, S. M., Simmons, A. J., Berrisford, P., Poli, P., Kobayashi, S., Andrae, U., Balmaseda, M. A., Balsamo, G., Bauer, P., Bechtold, P., Beljaars, A. C. M., van de Berg, L., Bidlot, J., Bormann, N., Delsol, C., Dragani, R., Fuentes, M., Geer, A. J., Haimberger, L., Healy, S. B., Hersbach, H., Hólm, E. V., Isaksen, L., Kållberg, P., Köhler, M., Matricardi, M., McNally, A. P., Monge-Sanz, B. M., Morcrette, J.-J., Park, B.-K., Peubey, C., de Rosnay, P., Tavolato, C., Thépaut, J.-N., and Vitart, F.: The ERA-Interim reanalysis: configuration and performance of the data assimilation system, Q. J. Roy. Meteor. Soc., 137, 553–597, https://doi.org/10.1002/qj.828, 2011.

Dimri, A. P. and Niyogi, D.: Regional climate model application at subgrid scale on Indian winter monsoon over the western Himalayas, Int. J. Climatol., 33, 2185–2205, https://doi.org/10.1002/joc.3584, 2013.

Dimri, A. P., Yasunari, T., Wiltshire, A., Kumar, P., Mathison, C., Ridley, J., and Jacob, D.: Application of regional climate models to the Indian winter monsoon over the western Himalayas, Sci. Total Environ., 468–469, 36–47, https://doi.org/10.1016/j.scitotenv.2013.01.040, 2013.

Goswami, B. N., Venugopal, V., Sengupta, D., Madhusoodanan, M. S., and Xavier, P. K.: Increasing Trend of Extreme Rain Events Over India in a Warming Environment, Science, 314, 1442–1445, 2006.

Hegdahl, T. J., Tallaksen, L. M., Engeland, K., Burkhart, J. F., and Xu, C.-Y.: Discharge sensitivity to snowmelt parameterization: a case study for Upper Beas basin in Himachal Pradesh, India, Hydrol. Res., https://doi.org/10.2166/nh.2016.047, 2016.

Henn, B., Clark, M. P., Kavetski, D., and Lundquist, J. D.: Estimating mountain basin-mean precipitation from streamflow using Bayesian inference, Water Resour. Res., 51, 8012–8033, https://doi.org/10.1002/2014WR016736, 2015.

Kamiguchi, K., Arakawa, O., Kitoh, A., Yatagai, A., Hamada, A., and Yasutomi, N.: Development of APHRO_JP, the first Japanese high-resolution daily precipitation product for more than 100 years, Hydrological Research Letters, 4, 60–64, https://doi.org/10.3178/hrl.4.60, 2010.

Katragkou, E., García-Díez, M., Vautard, R., Sobolowski, S., Zanis, P., Alexandri, G., Cardoso, R. M., Colette, A., Fernandez, J., Gobiet, A., Goergen, K., Karacostas, T., Knist, S., Mayer, S., Soares, P. M. M., Pytharoulis, I., Tegoulias, I., Tsikerdekis, A., and Jacob, D.: Regional climate hindcast simulations within EURO-CORDEX: evaluation of a WRF multi-physics ensemble, Geosci. Model Dev., 8, 603–618, https://doi.org/10.5194/gmd-8-603-2015, 2015.

Kretzschmar, A., Tych, W., Chappell, N. A., and Beven, K. J.: Reversing hydrology: quantifying the temporal aggregation effect of catchment rainfall estimation using sub-hourly data, Hydrol. Res., 47, 630–645, 2016.

Kumar, V., Singh, P., and Singh, V.: Snow and glacier melt contribution in the Beas River at Pandoh Dam, Himachal Pradesh, India, Hydrolog. Sci. J., 52, 376–388, https://doi.org/10.1623/hysj.52.2.376, 2007.

Li, H., Beldring, S., Xu, C.-Y., Huss, M., Melvold, K., and Jain, S. K.: Integrating a glacier retreat model into a hydrological model – Case studies of three glacierised catchments in Norway and Himalayan region, J. Hydrol., 527, 656–667, https://doi.org/10.1016/j.jhydrol.2015.05.017, 2015.

Li, H., Xu, C.-Y., Beldring, S., Tallaksen, L. M., and Jain, S. K.: Water Resources under Climate Change in Himalayan Basins, Water Resour. Manag., 30, 843–859, 2016.

Li, L., Gochis, D. J., Sobolowski, S., and Mesquita, M. d. S.: Evaluating the present annual water budget of a Himalayan headwater river basin using a high-resolution atmosphere-hydrology model, in: EGU General Assembly Conference Abstracts, vol. 18, p. 2480, 2016.

Maussion, F., Scherer, D., Finkelnburg, R., Richters, J., Yang, W., and Yao, T.: WRF simulation of a precipitation event over the Tibetan Plateau, China – an assessment using remote sensing and ground observations, Hydrol. Earth Syst. Sci., 15, 1795–1817, https://doi.org/10.5194/hess-15-1795-2011, 2011.

Ménégoz, M., Gallée, H., and Jacobi, H. W.: Precipitation and snow cover in the Himalaya: from reanalysis to regional climate simulations, Hydrol. Earth Syst. Sci., 17, 3921–3936, https://doi.org/10.5194/hess-17-3921-2013, 2013.

Palazzi, E., von Hardenberg, J., and Provenzale, A.: Precipitation in the Hindu-Kush Karakoram Himalaya: Observations and future scenarios, J. Geophys. Res.-Atmos., 118, 85–100, https://doi.org/10.1029/2012JD018697, 2013.

Pechlivanidis, I. G. and Arheimer, B.: Large-scale hydrological modelling by using modified PUB recommendations: the India-HYPE case, Hydrol. Earth Syst. Sci., 19, 4559–4579, https://doi.org/10.5194/hess-19-4559-2015, 2015.

Purohit, M. K. and Kau, S.: Rainfall Statistics of India – 2016, Tech. rep., India Meteorological Department, available at: http://hydro.imd.gov.in/hydrometweb/(S(0ymurl55bikbhgzupnyvnny0))/PRODUCTS/Publications/RainfallStatisticsofIndia-2016/RainfallStatisticsofIndia-2016.pdf (last access: 26 September 2018), 2016.

Rajeevan, M., Bhate, J., Kale, J. D., and Lal, B.: High resolution daily gridded rainfall data for the Indian region: Analysis of break and active monsoon spells, Curr. Sci., 91, 296–306, 2006.

Ren, Y.-Y., Ren, G.-Y., Sun, X.-B., Shrestha, A. B., You, Q.-L., Zhan, Y.-J., Rajbhandari, R., Zhang, P.-F., and Wen, K.-M.: Observed changes in surface air temperature and precipitation in the Hindu Kush Himalayan region over the last 100-plus years, Advances in Climate Change Research, 8, 148–156, https://doi.org/10.1016/j.accre.2017.08.001, 2017.

Shepard, D.: A two-dimensional interpolation function for irregularly-spaced data, in: Proceedings of the 1968 23rd ACM national conference, 517–524, ACM, New York, USA, 1968.

Srinivas, C. V., Hariprasad, D., Bhaskar Rao, D. V., Anjaneyulu, Y., Baskaran, R., and Venkatraman, B.: Simulation of the Indian summer monsoon regional climate using advanced research WRF model, Int. J. Climatol., 33, 1195–1210, https://doi.org/10.1002/joc.3505, 2013.

Wiltshire, A. J.: Climate change implications for the glaciers of the Hindu Kush, Karakoram and Himalayan region, The Cryosphere, 8, 941–958, https://doi.org/10.5194/tc-8-941-2014, 2014.

Xu, H., Xu, C.-Y., Chen, S., and Chen, H.: Similarity and difference of global reanalysis datasets (WFD and APHRODITE) in driving lumped and distributed hydrological models in a humid region of China, J. Hydrol., 542, 343–356, https://doi.org/10.1016/j.jhydrol.2016.09.011, 2016.

Yang, D., Goodison, B. E., Ishida, S., and Benson, C. S.: Adjustment of daily precipitation data at 10 climate stations in Alaska: Application of World Meteorological Organization intercomparison results, Water Resour. Res., 34, 241–256, https://doi.org/10.1029/97WR02681, 1998.

Yatagai, A., Kamiguchi, K., Arakawa, O., Hamada, A., Yasutomi, N., and Kitoh, A.: APHRODITE: Constructing a Long-Term Daily Gridded Precipitation Dataset for Asia Based on a Dense Network of Rain Gauges, B. Am. Meteorol. Soc., 93, 1401–1415, https://doi.org/10.1175/BAMS-D-11-00122.1, 2012.

Yin, Z.-Y., Zhang, X., Liu, X., Colella, M., and Chen, X.: An Assessment of the Biases of Satellite Rainfall Estimates over the Tibetan Plateau and Correction Methods Based on Topographic Analysis, J. Hydrometeorol., 9, 301–326, https://doi.org/10.1175/2007JHM903.1, 2008.

10

Spatial and temporal variability of rainfall and their effects on hydrological response in urban areas

Elena Cristiano, Marie-claire ten Veldhuis, and Nick van de Giesen

Department of Water Management, Delft University of Technology, P.O. Box 5048, 2600 GA, Delft, the Netherlands

Correspondence to: Elena Cristiano (e.cristiano@tudelft.nl)

Abstract. In urban areas, hydrological processes are characterized by high variability in space and time, making them sensitive to small-scale temporal and spatial rainfall variability. In the last decades new instruments, techniques, and methods have been developed to capture rainfall and hydrological processes at high resolution. Weather radars have been introduced to estimate high spatial and temporal rainfall variability. At the same time, new models have been proposed to reproduce hydrological response, based on small-scale representation of urban catchment spatial variability. Despite these efforts, interactions between rainfall variability, catchment heterogeneity, and hydrological response remain poorly understood. This paper presents a review of our current understanding of hydrological processes in urban environments as reported in the literature, focusing on their spatial and temporal variability aspects. We review recent findings on the effects of rainfall variability on hydrological response and identify gaps where knowledge needs to be further developed to improve our understanding of and capability to predict urban hydrological response.

1 Introduction

The lack of sufficient information about spatial distribution of short-term rainfall has always been one of the most important sources of errors in urban runoff estimation (Niemczynowicz, 1988). In the last decades considerable advances in quantitative estimation of distributed rainfall have been made, thanks to new technologies, in particular weather radars (Leijnse et al., 2007; van de Beek et al., 2010; Otto and Russchenberg, 2011). These developments have been applied in urban hydrology researches; see Einfalt et al. (2004) and Thorndahl et al. (2017) for a review. The hydrological response is sensitive to small-scale rainfall variability in both space and time (Faures et al., 1995; Emmanuel et al., 2012; Smith et al., 2012; Ochoa-Rodriguez et al., 2015b), due to a typically high degree of imperviousness and to a high spatial variability of urban land use.

Progress in rainfall estimation is accompanied by increasing availability of high-resolution topographical data, especially digital terrain models and land use distribution maps (Mayer, 1999; Fonstad et al., 2013; Tokarczyk et al., 2015). High-resolution topographical datasets have promoted development of more detailed and more complex numerical models for predicting flows (Gironás et al., 2010; Smith et al., 2013). However, model complexity and resolution need to be balanced with the availability and quality of rainfall input data and datasets for catchment representation (Morin et al., 2001; Rafieeinasab et al., 2015; Rico-Ramirez et al., 2015; Rafieeinasab et al., 2015; Pina et al., 2016). This is particularly critical in small catchments, where flows are sensitive to variations at small space and timescales as a result of the fast hydrological response and the high catchment variability (Fabry et al., 1994; Singh, 1997). Alterations of natural flows introduced by human interventions, especially artificial drainage networks, sewer pipe networks, detention and control facilities, such as reservoirs, pumps, and weirs, are additional elements to take into account for flow predictions. Recently, various authors investigated the sensitivity of spatial and temporal rainfall variability on the hydrological response for urban areas (Bruni et al., 2015; Ochoa-Rodriguez et al., 2015b; Rafieeinasab et al., 2015). Despite these efforts, many aspects of hydrological processes in urban areas remain poorly understood, especially in the interaction between rainfall and runoff.

It is timely to review recent progress in understanding of interactions between rainfall spatial and temporal resolution, variability of catchment properties and their representation in hydrological models. Section 2 of this paper is dedicated to definitions of spatial and temporal scales and catchments in hydrology and methods to characterize these. Section 3 focuses on rainfall, analysing the most used rainfall measurement techniques, their capability to accurately measure small-scale spatial and temporal variability, with particular attention to applications in urban areas. Hydrological processes are described in Sect. 4, highlighting their variability and characteristics in urban areas. Thereafter, the state of the art of hydrological models, as well as their strengths and limitations to account for spatial and temporal variability, are discussed. Section 6 presents recent approaches to understand the effect of rainfall variability in space and time on hydrological response. In Sect. 7, main knowledge gaps are identified with respect to accurate prediction of urban hydrological response in relation to spatial and temporal variability of rainfall and catchment properties in urban areas.

2   Scales in urban hydrology

2.1   Spatial and temporal scale definitions

Hydrological processes occur over a wide range of scales in space and time, varying from 1 mm to 10 000 km in space and from seconds up to 100 years in time. A scale is defined here as the characteristic region in space or period in time at which processes take place or the resolution in space or time at which processes are best measured (Salvadore et al., 2015).

Several authors have classified hydrological process scales and variability, focusing in particular on the interaction between rainfall and the other hydrological processes (Blöschl and Sivapalan, 1995; Bergstrom and Graham, 1998). Blöschl and Sivapalan (1995) presented a graphical representation of spatial and temporal variability of the main hydrological processes on a logarithmic plane. The plot has been updated by other authors, each focusing on specific aspects. For example, Salvadore et al. (2015) analysed phenomena related to urban processes, focusing on small spatial scale, while Van Loon (2015), added scales of some hydrological problems, such as flood and drought. Figure 1 presents an updated version of the plot that integrates the information contributed by Berndtsson and Niemczynowicz (1986), Blöschl and Sivapalan (1995), Stahl and Hisdal (2004), and Salvadore et al. (2015). Figure 1 shows that in urban hydrology attention is mainly focused on small scales. Characteristic processes, such as storm drainage, infiltration, and evaporation, vary at a small temporal and spatial scale, from seconds to hours and from centimetres to hundreds of metres. Many processes are driven by rainfall, that varies over a wide range of scales.

Blöschl and Sivapalan (1995) highlighted the importance of making a distinction between two types of scales: the "process scale", i.e. the proper scale of the considered phenomenon, and the "observation scale", related to the measurement and depending on techniques and instruments used. Under the best scenario, process and observation scale should match, but this is not always the case, and transformations based on downscaling and upscaling techniques (Fig. 2) might be necessary to obtain the required match between scales. These techniques are discussed in Sect. 2.2.

2.2   Rainfall downscaling

The term downscaling usually refers to methods used to take information known at large scale and make predictions at small scale. There are two main downscaling approaches: dynamic or physically based and statistical methods (Xu, 1999). Dynamic downscaling approaches solve the process-based physics dynamics of the system. In statistical downscaling, a statistical relationship is defined between local variables and large-scale prediction and this relationship is applied to simulate local variables (Xu, 1999). Dynamical downscaling is widely used in climate modelling and numerical weather prediction, while statistical models are often used in hydrometeorology, for example rainfall downscaling. Dynamic downscaling models have the advantage of being physically based, but they require a lot of computational power compared to statistical downscaling models. Statistical approaches require historical data and knowledge of local conditions (Xu, 1999).

Ferraris et al. (2003) presented a review of three common stochastic downscaling models, mainly used for spatial rainfall downscaling: multifractal cascades, autoregressive processes, and point-process models based on the presence of individual cells. The first were introduced in the 1970s and are widely used to reproduce the spatial and temporal variability (see Schertzer and Lovejoy, 2011 for a review). Autoregressive methods, also nowadays often referred to as "rainfall generator models", are used to generate multidimensional random fields while preserving the rainfall spatial autocorrelation, for natural (Paschalis et al., 2013; Peleg and Morin, 2014; Niemi et al., 2016) and urban (Sørup et al., 2016) areas. Point-process models are used when the spatial structure of intense rainfall is defined by convective rainfall cells (see McRobie et al., 2013 for an example). It incorporates local information and requires a more detailed storm cell identification.

Statistical downscaling and upscaling approaches are reported in the literature for a wide variety of variables (Rummukainen, 1997; Deidda, 2000; Ferraris et al., 2003; Gires et al., 2012; Wang et al., 2015b; Muthusamy et al., 2017) and techniques such as regression methods, weather pattern-based approaches and stochastic rainfall generators (see Wilby and Wigley, 1997; Wilks and Wilby, 1999 for a review). Some recent studies about downscaling and upscaling focus mainly on urban areas (Gires et al., 2012; Wang et al.,

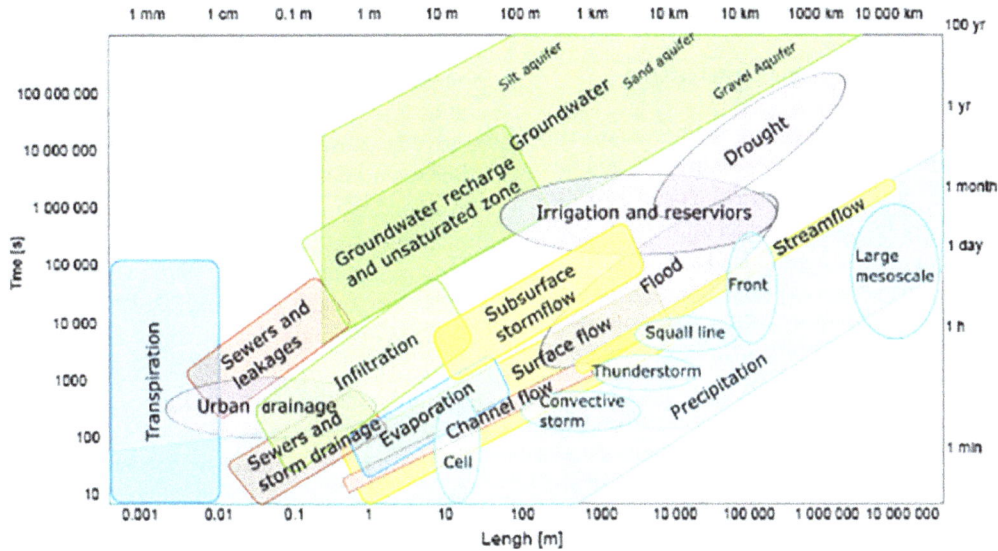

Figure 1. Spatial and temporal scale variability of hydrological processes, adapted from Berndtsson and Niemczynowicz (1986), Blöschl and Sivapalan (1995), Stahl and Hisdal (2004), and Salvadore et al. (2015). Colours represent different groups of physical processes: blue for processes related to the atmosphere, yellow for surface processes, green for underground processes, red highlights typical urban processes, and grey indicates problems hydrological processes can pose to society.

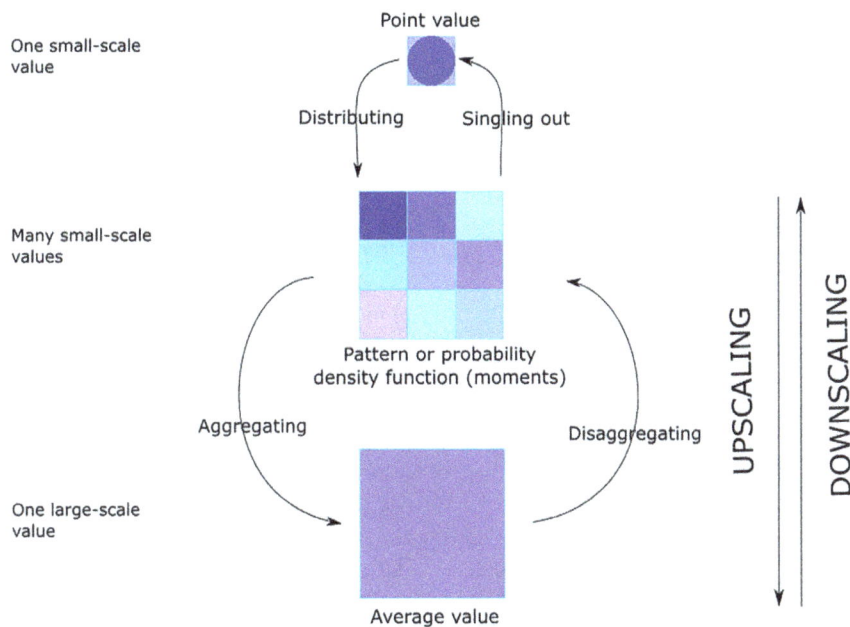

Figure 2. Downscaling and upscaling processes (modified from Blöschl and Sivapalan, 1995).

2015b, a; Muthusamy et al., 2017): Wang et al. (2015a), for example, presented a gauge-based radar-rainfall adjustment method sensitive to singularities, characteristic of small scale.

The importance of using downscaling methods was discussed by Fowler et al. (2007), in a work where they investigated what can be learned from downscaling method comparison studies, what new methods can be used together with downscaling to assess uncertainties in hydrological response and how downscaling methods can be better utilized within the hydrological community. They highlighted that the importance given to the applied research is still too little, and manager and stakeholders should be more aware of uncertainties within the modelling system.

2.3 Methods to characterize hydrological process scales

2.3.1 Spatial variability of basin characteristics

Slope, degree of imperviousness, soil properties, and many other catchment characteristics are variable in space and time and this variability affects the hydrological response (Singh, 1997). This is especially the case of urban areas, where spatial variability and temporal changes in land use are typically high.

Julien and Moglen (1990) gave a first definition of the catchment length scale L_s as part of a theoretical framework applied to a natural catchment, where they analysed 8400 dimensionless hydrographs obtained from one-dimensional finite element models under spatially varied input. Length scale was presented as a function of rainfall duration d, spatially averaged rainfall intensity i, average slope s_0, and average roughness n:

$$L_s = \frac{d^{\frac{5}{6}} s_0^{\frac{1}{2}} i^{\frac{2}{3}}}{n}. \tag{1}$$

In urban catchments, the concept of catchment length, defined as the squared root of the (sub)catchment or runoff area, has been used (Bruni et al., 2015; Ochoa-Rodriguez et al., 2015a). Additionally, Bruni et al. (2015) introduced the sewer length or inter-pipes sewer distance, as the ratio between the catchment area and the total length of the sewer, to characterize the spatial scale of sewer networks. Ogden et al. (2011) used the width function, defined as the number of channel segments at a specific distance from the outlet, to represent the spatial variability of the drainage network. This parameter describes the network geomorphology by counting all stream links located at the same distance from the outlet, but it does not give an accurate description of the spatial variability of hydrodynamic parameters.

2.3.2 Timescale characteristics

In this section, we present a brief overview of timescales reported in the literature and discuss approaches to estimate characteristic timescales that have been specifically developed for urban areas. A summary of timescale characteristics is presented in Table 1.

The first method to investigate the hydrological response is the rational method, presented more than a century ago by (Kuichling, 1889) for urban areas. This method was later adapted for rural areas. The rational method requires the estimation of the time of concentration in order to define the runoff volume.

Time of concentration t_c is one of the most common hydrological characteristic timescales and it is defined as the time that a drop that falls on the most remote part of the basin needs to reach the basin outlet (Singh, 1997; Musy and Higy, 2010). Several equations to estimate this parameter are available in the literature for natural (Gericke and Smithers, 2014) and urban (McCuen et al., 1984) catchments. The time of concentration is difficult to measure, because it assumes that initial losses are already satisfied and the rainfall event intensity is constant for a period at least as long as the time of concentration. Different theoretical definitions have been developed in order to estimate the time of concentration as function of basin length, slope and other characteristics (see for some examples Singh, 1976; Morin et al., 2001; USDA, 2010; Gericke and Smithers, 2014).

Due to difficulties related to the estimation of time of concentration, Larson (1965) introduced the time of virtual equilibrium t_{ve}, defined as the time until response is 97 % of runoff supply.

When a given rainfall rate persists on a region for enough time to reach the equilibrium, this time is called time to equilibrium t_e (Ogden et al., 1995; Ogden and Dawdy, 2003; van de Giesen et al., 2005). Time of equilibrium for a turbulent flow on a rectangular runoff plane given rainfall intensity i, with given roughness n, length L_p and slope S can be written as (Ogden et al., 1995)

$$t_e = \left[\frac{n L_p}{S^{\frac{1}{2}} i^{\frac{2}{3}}} \right]^{\frac{3}{5}}. \tag{2}$$

Another commonly used hydrological characteristic timescale or response time is the lag time t_{lag}. It represents the delay between rainfall and runoff generation. t_{lag} is defined as the distance between the hyetograph and hydrograph centre of mass of (Berne et al., 2004), or between the time of rainfall peak and time of flow peak (Marchi et al., 2010; Yao et al., 2016). t_{lag} can be considered characteristic of a basin, and is dependent on drainage area, imperviousness and slope (Morin et al., 2001; Berne et al., 2004; Yao et al., 2016). Berne et al. (2004), including the results of Schaake et al. (1967) and Morin et al. (2001), defined a relation between the dimension of the catchment area S (in ha) and the lag time t_{lag} (in millimetres): $t_{lag} = 3S^{0.3}$ for urban areas. Empirical relations between t_{lag} and t_c are presented in the literature (USDA, 2010; Gericke and Smithers, 2014).

Another characteristic timescale is the "response timescale" T_s, presented for the first time by Morin et al. (2001). It is defined as the timescale at which the pattern of the time averaged and basin averaged radar-rainfall hyetograph is most similar to the pattern of the measured hydrograph at the outlet of the basin. This definition was updated by Morin et al. (2002), which used an objective and automatic algorithm to analyse the smoothness of the hyetograph and hydrograph instead of the general behaviour, and by Shamir et al. (2005), who related the number of peaks with the total duration of the rising and declining limbs of hyetographs and hydrographs.

In urban areas, where most of the surface is directly connected to the drainage system, concentration time is given by

Table 1. Timescale parameters.

Characteristic	Reference	Description
Time of concentration t_c	Singh (1997) Gericke and Smithers (2014)	The time that a drop that falls on the most remote part of the drainage basin needs to reach the basin outlet
Time of equilibrium t_e	Ogden et al. (1995) Ogden and Dawdy (2003) van de Giesen et al. (2005)	Minimum time needed for a given stationary uniform rainfall to persist until equilibrium runoff flow is reached
Lag time t_{lag}	Berne et al. (2004) Marchi et al. (2010) Gericke and Smithers (2014)	The time difference between the gravity centre of the hyetograph of catchment mean rainfall and the gravity centre of the generated hydrograph
Response timescale T_s	Morin et al. (2001) Morin et al. (2002) Morin et al. (2003) Shamir et al. (2005)	The timescale at which the pattern of time averaged radar hyetograph is most similar to the pattern of the measured hydrograph at the outlet of the basin

the time the rainfall needs to enter the sewer system and the travel time through the sewer system.

3 Rainfall measurement and variability in urban regions

Rainfall is an important driver for many hydrological processes and represents one of the main sources of uncertainty in studying hydrological response (Niemczynowicz, 1988; Einfalt et al., 2004; Thorndahl et al., 2017; Rico-Ramirez et al., 2015).

Urban areas affect the local hydrological system, not only by increasing the imperviousness degree of the soil but also by changing rainfall generation and intensity patterns. Several studies show that increase in heat and pollution produced by human activities and changes in surface roughness influence rainfall and wind generation (Huff and Changno, 1973; Shepherd et al., 2002; Givati and Rosenfeld, 2004; Shepherd, 2006; Smith et al., 2012; Daniels et al., 2015; Salvadore et al., 2015). This phenomenon is not deeply investigated in this paper, but it is an important aspect to consider.

In this section instruments and technologies for rainfall measurement are described, pointing out their opportunities and limitations for measuring spatial and temporal variability in urban environments. Subsequently, methods to characterize rainfall events according to their space and time variability are described.

3.1 Rainfall estimation

Rain gauges were the first instrument used to measure rainfall and are still commonly used, because they are relatively low in cost and easy to install (WMO, 2008).

Afterwards, weather radars were introduced to estimate the rainfall spatial distribution. These instruments allow one to get measurements of rainfall spatially distributed over the area, instead of a point measurement as in the case of rain gauges. Rainfall data obtained from weather radars are used to study the hydrological response in natural watersheds and urban catchments (Einfalt et al., 2004; Berne et al., 2004; Sangati et al., 2009; Smith et al., 2013; Ochoa-Rodriguez et al., 2015b; Thorndahl et al., 2017) often combined with rainfall measurement from rain gauge networks (Winchell et al., 1998; Smith et al., 2005; Segond et al., 2007; Smith et al., 2012), as well as to improve short-term weather forecasting and nowcasting (Montanari and Grossi, 2008; Liguori and Rico-Ramirez, 2013; Dai et al., 2015; Foresti et al., 2016).

More recently, commercial microwave links have been used to estimate the spatial and temporal rainfall variability (Leijnse et al., 2007; Fencl et al., 2015, 2017). Rainfall estimates are obtained from the attenuation of the signal caused by rain along microwave link paths. This approach can be particularly useful in cities that are not well equipped with rain gauges or radars, but where the commercial cellular communication network is typically dense (Leijnse et al., 2007).

3.1.1 Rain gauges networks

Several types of rain gauges have been developed, such as weighing gauges, tipping bucket gauges and pluviographs (Lanza and Stagi, 2009; Lanza and Vuerich, 2009). They are able to constantly register accumulation of rainfall volume over time, thus providing a measurement of temporal variability of rainfall intensity. Rain gauge measurements are sensitive to wind exposure and the error caused by wind field above the rain gauge is 2–10 % for rainfall and up to 50 % for solid precipitations (WMO, 2008). Other errors can be due to tipping bucket losses during the rotation, to wetting losses on the internal walls of the collector, to evaporation (especially in hot climates), or water splashing into and out of the collector (WMO, 2008). The main disadvantage of rain gauges is that the obtained data are point measurements and, due to the high spatial variability of rainfall events, measurements from a single rain gauges are often not representative of a larger area. Rainfall fields, however, present a spatial organization and, by interpolating data from a rain gauge networks, it is possible to obtain distributed rainfall fields (Villarini et al., 2008; Muthusamy et al., 2017). Uncertainty induced by interpolation strongly depends on the density of the rain gauge network and on homogeneity of the rainfall field (Wang et al., 2015b).

In urban areas, rainfall measurements with rain gauges present specific challenges associated with microclimatic effects introduced by the building envelope. WMO (2008) recommended minimum distances between rain gauges and obstacles of 1 to 2 times the height of the nearest obstacle, a condition that is hard to fulfil in densely built areas. A second problem is introduced by hard surfaces, that may cause water splashing into the gauges, if it is not placed at an elevation of at least 1.2 m. Rain gauges in cities are often mounted on roofs for reasons of space availability and safety from vandalism. This means they are affected by the wind envelope of the building, unless they are elevated to a sufficient height above the building.

Rain gauge measurement error can be 30 % or more depending on the type of instrument used for the measurement and local conditions (van de Ven, 1990; WMO, 2008).

3.1.2 Weather radars

In the last decades, weather radars have been increasingly used to measure rainfall (Niemczynowicz, 1999; Krajewski and Smith, 2005; Otto and Russchenberg, 2011; Berne and Krajewski, 2013). Radars transmit pulses of microwave signals and measure the power of the signal reflected back by raindrops, snowflakes, and hailstones (backscatter). Rainfall rate R [$L\,T^{-1}$] is estimated using the reflectivity Z [$L^6\,L^{-3}$] measured from the radar through a power law:

$$R = aZ^b, \tag{3}$$

where a and b depend on type of precipitation, raindrop distribution, climate characteristics and spatial and temporal scales considered (Marshall and Palmer, 1948; van de Beek et al., 2010; Smith et al., 2013). Weather radars present different wavelengths λ, frequencies ν and sizes of the antenna l. Characteristics of commonly used weather radars are reported in Table 2. X-band radars can be beneficial for urban areas; they are low cost and they can be mounted on existing buildings and measure rainfall closer to ground at higher resolution than national weather radar networks (Einfalt et al., 2004). Polarimetric weather radars transmit signals polarized in different directions (Otto and Russchenberg, 2011), enabling it to distinguish between horizontal and vertical dimension, thus between rain drops and snowflakes

Table 2. Weather radar characteristics.

	λ cm	ν GHz	l m
S-band	8–15	2–4	6–10
C-band	4–8	4–8	3–5
X-band	2.5–4	8–12	1–2

as well as between smaller or larger oblate rain drops. A specific strength of polarimetric radars is the use of differential phase K_{dp}, which allows one to correct signal attenuation thus solving an important problem generally associated with X-band radars (Otto and Russchenberg, 2011; Ochoa-Rodriguez et al., 2015b; Thorndahl et al., 2017).

3.1.3 Opportunities and limitations of weather radars

Berne and Krajewski (2013) presented a comprehensive analysis of the advantages, limitations and challenges in rainfall estimation using weather radars. One of the main problems is that an indirect relation is used (Eq. 3) to estimate rainfall. Rainfall measurements have to be adjusted based on rain gauges and disdrometers. Various techniques have been studied to calibrate radars (Wood et al., 2000), to combine radar-rainfall measurements with rain gauge data for ground truthing (Cole and Moore, 2008; Smith et al., 2012; Wang et al., 2013; Gires et al., 2014; Nielsen et al., 2014; Wang et al., 2015b) and to define the uncertainty related to radar-rainfall estimation (Ciach and Krajewski, 1999; Quirmbach and Schultz, 2016; Villarini et al., 2008; Mandapaka et al., 2009; Peleg et al., 2013; Villarini et al., 2014). These studies show that in most of the cases, radar measurements underestimate the rainfall compared to rain gauge measurements (Smith et al., 2012; Overeem et al., 2009a; Overeem et al., 2009b; van de Beek et al., 2010).

Another downside of radars is their installation at high locations to have a clear view without obstacles, while rainfall intensities can change before reaching the ground (Smith et al., 2012). Moreover, radar measurements need to be combined with a rain drop size distribution to obtain an accurate rainfall estimation. Berne and Krajewski (2013) pointed out additional aspects that have to be taken into account, e.g., management and storage of the high quantity of data that are measured, possibility to use the weather radars to estimate snowfall and the uncertainty related to it, and problems related to rainfall measurement in mountain areas.

Rain gauge measurements in urban areas tend to be prone to errors due to microclimatic effects introduced by the building envelope. In this context, the use of weather radar could represent a big improvement to obtain a more accurate rainfall information for studying hydrological response.

A promising application of radar is their combination with nowcasting models to obtain short-term rainfall forecasts. Liguori and Rico-Ramirez (2013) presented a review of different nowcasting models, which benefit from radar data. This work focused in particular on a hybrid model, able to merge the benefits of radar nowcasting and numerical weather prediction models. Radar data can provide an accurate short-term forecast and recent studies have presented nowcasting systems able to reduce errors in rainfall estimation (e.g. Foresti et al., 2016).

3.2 Characterizing rainfall events according to their spatial and temporal scale

Rainfall events are characterized by several elements, such as duration, intensity, velocity and their spatial and temporal variability, and many possible classifications are presented in the literature. Some of the most used examples of rainfall classification considering the rainfall variability, are described in this section.

Characterizations and classifications of intense rainfall events have been proposed by various authors, combining rain gauges and radar-rainfall data. In particular, weather radars are used as main tools to analyse rainfall spatial and temporal scale in urban areas. An example of characterization of rainfall structure was given by Smith et al. (1994), who presented an empirical analysis of four extreme rainstorms in the Southern Plains (USA), using data from two networks of more than 200 rain gauges and from a weather radar. They defined *major rainfall event* as storms for which 25 mm of rain covered an area larger than $12\,500\,km^2$. Thorndahl et al. (2014) presented a storm catalogue of heavy rainfall, over a study area of $73\,500\,km^2$ in southern Wisconsin, and key elements of storm evolution that control the scale. The catalogue contains the 50 largest rainfall events recorded during a 16-year period by WSR-88D radar with spatial and temporal resolution of $1\,km \times 1\,km$ and 15 min respectively. Over the 50 events, there is 0.60 probability that rainfall exceeds 25 mm of daily accumulation in a $1\,km^2$ pixel and 0.14 probability of exceeding 100 mm. Results showed that there is a clear relation between the characteristic length and timescale of the events. The length scale increased with timescale; a length scale of $35 \pm 20\,km$ was found for a time step of 15 min, up to $160 \pm 25\,km$ for a 12 h aggregation time.

3.3 Rainfall variability at the urban scale

Rainfall events are often described and classified considering they variability in space and time. Spatial variability can be defined, following Peleg et al. (2017), as "the variability derived from having multiple spatially distributed rainfall fields for a given point in time". Peleg et al. (2017) introduced also the definition of climatological variability as the variability obtained from multiple climate trajectories that produce different storm distributions and rainfall intensities in time.

Studying rainfall variability at the urban scale, Emmanuel et al. (2012) classified 24 rain periods, recorded by the weather radar located in Treillieres (France), with a spatial

and temporal resolution of $250\,m \times 250\,m$ and 5 min respectively. They classified the events into four groups, based on variogram analysis: light rain period, shower periods, storms organized into rain bands and unorganized storms. These groups are defined considering the decorrelation distance (and decorrelation time), defined as distance (and time) from which two points show independent statistical behaviour, and it is obtained as the range of the climatological variogram (Emmanuel et al., 2012). The first group, characterized by light rainfall events, presented very high decorrelation distance and time (17 km and 15 min) compared to the second group, with a decorrelation distance and time of 5 km and a decorrelation time of 5 min. The last two groups presented a double structure, where small and intense clusters, with low decorrelation distance and time (less than 5 km and 5 min) are located, in a random or organized way, inside areas with a lower variability (decorrelation of 15 km and 15 min).

Jensen and Pedersen (2005) presented a study about variability in accumulated rainfall within a single radar pixel of $500\,m \times 500\,m$, comparing it with 9 rain gauges located in the same area. The results showed a variation of up to 100 % at a maximum distance of about 150 m, due to the rainfall spatial variability. This study suggested that a huge quantity of rain gauges is needed to have a powerful rain gauge network capable of representing small-scale variability. An alternative solution is to consider the variance reduction factor method, a numerical method to represent the uncertainty from averaging a number of rain gauges per pixel, taking into account their spatial distribution and the correlation between them. The variance reduction factor method was introduced for the first time by Rodriguez-Iturbe and Mejía (1974) and lately applied in various studies (Krajewski et al., 2000; Villarini et al., 2008; Peleg et al., 2013).

Gires et al. (2014) focused on the gap between rain gauges and radar spatial scale, considering that a rain gauge usually collects rainfall over 20 cm of surface and the spatial resolution of most used radars is of $1\,km \times 1\,km$. They evaluate the impact of small-scale rainfall variability using a universal multifractal downscaling method. The downscaling process was validated with a dense rain gauge and disdrometer network, with 16 instruments located in $1\,km \times 1\,km$. They showed two effects of small-scale rainfall variability that are often not taken into account; high rainfall variability occurred below $1\,km^2$ spatial scale and the random position of the point measurement within a pixel influenced measured rainfall events. Similar results are confirmed by Peleg et al. (2016), who studied the spatial variability of extreme rainfall at radar subpixel scale. Comparing a radar pixel of $1\,km \times 1\,km$ with high-resolution rainfall data, obtained by applying the stochastic rainfall generator STREAP (Paschalis et al., 2013) to simulate rain fields, this study highlights that subpixel variability is high and increases with increasing of return period and with shorter duration.

In Table 3 four types of rainfall events are presented with their characterization and typical spatial and temporal decor-

Table 3. Characterization of rainfall events, spatial and temporal scales, and rainfall estimation uncertainty. From van de Beek et al. (2010), Smith et al. (1994), and Emmanuel et al. (2012).

	Characterization and Intensity	Spatial Range	Temporal Range	Radar Estimation
Light rainfall	$1\,\mathrm{mm\,h^{-1}}$	17 km	15 min	Underestimation rainfall values often below the threshold ($0.17\,\mathrm{mm\,h^{-1}}$)
Convective cells	short and intense from 25 mm	5 km	5 min	Overestimation
Organized stratiform	up to $17\,\mathrm{mm\,h^{-1}}$	<5 km	< 5 min	General underestimation, good representation of the hyetograph behaviour
Unorganized stratiform	high-intensity core, combined with low-intensity areas	15 km	15 min	Underestimation of the peaks, good representation of the hyetograph behaviour

relation lengths, based on van de Beek et al. (2010), Emmanuel et al. (2012), and Smith et al. (1994). Considering that the minimal rainfall measurement resolution required for urban hydrological modelling is 0.4 the decorrelation length (Julien and Moglen, 1990; Berne et al., 2004; Ochoa-Rodriguez et al., 2015b), operational radars are not able to satisfy this requirement.

4 Hydrological processes

In this section, general characteristics and parametrizations of hydrological processes are presented, highlighting their spatial and temporal variability and characteristics specific to urban environments.

4.1 Precipitation losses

4.1.1 Infiltration, interception and storage

The term infiltration is usually used to describe the physical processes by which rain enters the soil (Horton, 1933). Different equations and models have been proposed to describe infiltration. The most commonly used is Richards equation (Richards, 1931), which represents this phenomenon using a partial differential equation with non-linear coefficients.

Another possibility to estimate the infiltration capacity is given by the empirical equation presented by Horton (1939). In Horton's equation hydraulic conductivity and diffusivity are constant and do not depend on water content or on depth.

If water cannot infiltrate, as is the case in impervious areas, it can be stored in local depressions, where it does not contribute to runoff flow. This is the case of local depressions on streets or flat roofs, where water accumulates until the storage capacity is reached. Before reaching the ground, rainfall can be intercepted by vegetation cover or buildings. Interception can constitute up to 20 % of rainfall at the start of a rainfall event (Mansell, 2003), and decreases quickly to zero, once surfaces are wetted.

During the process of transformation of rainfall in runoff, part of the water is lost due to several phenomena, such as infiltration, storage or evaporation. Ragab et al. (2003) presented an experimental study of water fluxes in a residential area, in which they estimated infiltration and evaporation in urban areas, showing that the assumption that all rainfall becomes runoff is not correct and that it leads to an overestimation of runoff. Ramier et al. (2011) studied the hydrological behaviour of urban streets over a 38-month period to estimate runoff losses and to better define rainfall runoff transformations. They estimated losses due to evaporation and infiltration inside the road structure between 30 and 40 % of the total rainfall.

Spatial scale of precipitation losses is strongly influenced by land cover variation. In urban areas, land cover variability typically occurs at a spatial scale of 100 to 1000 m. Timescale is associated with local storage accumulation volume, sorptivity, and hydraulic conductivity, which in turn depend on soil type and soil compaction.

4.1.2 Groundwater recharge and subsurface processes in urban areas

Groundwater recharge mechanisms change due to human activities and urbanization, both in terms of volume and quality of the water. The increase of imperviousness of land cover leads to a decrease in infiltration of rainfall into soil, reducing direct recharge to groundwater. The presence of leakage from drinking water and sewer networks can increase infiltration to groundwater and amount of contaminants that is spread from the sewer system into the soil (Salvadore et al., 2015).

Although it is well known that not all rainfall turns into runoff (Boogaard et al., 2013; Lucke et al., 2014), it is common to consider the losses from impervious areas so small that they can be assumed negligible compared to the total runoff volume (Ragab et al., 2003; Ramier et al., 2011). Ragab et al. (2003) tried to emphasize the importance of accounting for infiltration in the urban water balance, and found that infiltration through the road surface can constitute between 6 and 9 % of annual rainfall. Due to high spatial variability of infiltration, representative measurements are difficult to obtain and require a large amount of point-scale measurements (Boogaard et al., 2013; Lucke et al., 2014).

Several types of pervious pavements are used in urban areas. They can generally be divided into monolithic and modular structures. Monolithic structures consist of a combination of impermeable blocks of concrete and open joints or apertures that allow water to infiltrate. In modular structures, gaps between two blocks are not filled with sand, as with conventional pavements, but with 2–5 mm of bedding aggregate, that facilitate infiltration (Boogaard et al., 2013). Following European standards, minimum infiltration capacity for permeable pavements is $270 \, \mathrm{L\,s^{-1}\,ha^{-1}}$, equal to $97.2 \, \mathrm{mm\,h^{-1}}$ (Opzoekingscentrum voor de Wegenbouw, 2008).

Pervious areas in cities can effectively act as semi-impervious areas, because within the soil column there is a shallow layer that presents a low hydraulic conductivity at saturation, caused by soil compaction during the building process. Smith et al. (2015) studied the influence of this phenomenon on peak runoff flow by applying 21 storm events on a physically based, minimally calibrated model of the dead run urban area (USA) with and without the compacted soil layer. Results showed that the compacted soil layer reduced infiltration by 70–90 % and increased peak discharge by 6.8 %.

4.2 Surface runoff

When rainfall intensity exceeds infiltration capacity of the soil, water starts to accumulate on the surface and flows following the slope of the ground. This process is generally called Hortonian runoff (Horton, 1933) or infiltration capacity excess flow. It is usually contrasted with saturation excess flow, or Dunne flow (Dunne, 1978), that occurs when the soil is saturated and rainfall can no longer be stored (van de Giesen et al., 2011).

In urban areas, runoff is generated when the surface is impervious and water can not infiltrate, or when infiltration capacity is exceeded by rainfall intensity. Water flows over the surface and can reach natural drainage channels or be intercepted by the drainage network through gullies and manholes. If the drainage network capacity is exceeded, the system become pressurized, and water starts to flow out from gullies, increasing runoff on the street (Ochoa-Rodriguez et al., 2015a).

It is important to pay attention to some elements that characterize the runoff in urban environments: sharp corners or obstacles can, for example, deviate the flow and introduce additional hydraulic losses. Interactions between surface flow and subsurface sewer systems through sewer inlets and gully pots are hydraulically complex and their influence on overland and in sewer flows remains poorly understood. Runoff

flows are often characterized by very small water depths that are often alternated with dry surfaces, especially when rainfall intensities vary strongly in space and time.

4.3 Impact of land cover on overland flow in urban areas

In urban areas, the land cover, represented by an alternation of impervious surfaces, such as roads and roofs, and small pervious areas, such as gardens, vegetation and parks, shows a high variability in space.

The impact of increase of imperviousness on hydrological response was studied by Cheng (2002), who analysed the effects of urban development in Wu-Tu (Taiwan's catchment) considering 28 rainfall events (1966–1997). Results showed that response peak increased by 27 % and the time to peak decreased from 9.8 to 5.9 h, due to an increase of imperviousness from 4.78 to 11.03 %.

In a similar study, Smith et al. (2002) analysed the effects of imperviousness on flood peak in the Charlotte metropolitan region (USA), analysing a 74-year discharge record. Results showed that different land covers were associated with large differences in timing and magnitude of flood peak, while there were not significant differences in the total runoff volume. Hortonian runoff was the dominant runoff mechanism. Antecedent soil moisture played an important role in this watershed, even in the most urbanized catchment.

The influence of antecedent soil moisture is, however, not always so evident. Smith et al. (2013) showed that in nine watersheds, located in the Baltimore metropolitan area, the antecedent soil moisture, defined as 5-day antecedent rainfall, seemed not to affect the hydrological response. Introduction of storm water management infrastructure played an important role in reducing flood peaks and increasing runoff ratios. Results showed that rainfall variability may have important effects on spatial and temporal variation in flood hazard in this area.

Analysing the effects of a moderate extreme and an extreme rainstorm on the same area presented by Smith et al. (2013), Ogden et al. (2011) highlighted the importance of changes in imperviousness on flood peaks. They found that for extreme rainfall event, imperviousness had a small impact on runoff volume and runoff generation efficiency.

4.4 Evaporation

Evaporation plays an important role in the hydrological cycle: in forested catchment around 60–95 % of total annual rainfall evaporates or is absorbed by the vegetation (Fletcher et al., 2013). In an urban catchment, evaporation is drastically reduced (Oke, 2006; Fletcher et al., 2013; Salvadore et al., 2015). Evaporation is often neglected in analysis of fast and intense rainfall events (Cui and Li, 2006); the order of magnitude of evaporation is very small compared to the total amount of rainfall. Some studies have shown that evaporation is not always negligible in urban areas and can constitute up to 40 % of the annual total losses (Grimmond and Oke, 1991; Salvadore et al., 2015).

In their experimental study, Ragab et al. (2003) showed that evaporation represents 21–24 % of annual rainfall, with more evaporation taking place during summer than winter. It is particularly important to have measurements with high resolution because a coarse spatial description can hide heterogeneous land covers and consequently, heterogeneous evaporation losses (Salvadore et al., 2015).

Evaporation measurements in urban areas are one of the weak points of the water balance (van de Ven, 1990) and they present many problems and challenges (Oke, 2006). It is quite hard in fact to find a site, representative of the area, far enough from obstacles and not unduly shaded. Errors in estimation of annual evaporation in urban areas may still be higher than 20 % (van de Ven, 1990).

Different techniques and approaches have been developed to measure and estimate the impact of evaporation, from the standard lysimeter to the use of remote sensing (Nouri et al., 2013), to the combined used of remote sensing and ground measurements (Hart et al., 2009). Different models to estimate evaporation in urban areas have been proposed (Marasco et al., 2015; Litvak et al., 2017). Litvak et al. (2017) estimated evaporation in the urban area of Los Angeles, as combination of empirical models of turfgrass evaporation and tree transpiration derived from in situ measurements. Evaporation from non-vegetated areas appears to be negligible compared with the vegetation, and turfgrass was responsible for 70 % of evaporation from vegetated areas.

4.5 Flow in sewer systems

In urban areas, part of the surface runoff enters in the sewer system through gully inlets, depending on the capacity of these elements, on their maintenance (Leitão et al., 2016) and the sewer system itself.

Storm water flow in sewer systems is highly non-uniform and unsteady, it can be considered as one dimensional, assuming that depth and velocity vary only in the longitudinal direction of the channel. Flow in sewer pipes is usually free surface, but during intense rainfall events the system can become full and temporarily behave as a pressurized system, a phenomenon called surcharge. In particular conditions, as for example in flat catchments, inversion of the flow direction in pipes can occur during filling and emptying of the system. The most common form to model flow in sewer pipes is based on a one-dimensional form of the de Saint-Venant equations.

Sewer system density influences runoff generation (Ogden et al., 2011; Yang et al., 2016): a dense pipe network can, in fact, reduce the runoff generation, increasing the storage capacity of the system (Yang et al., 2016). Ogden et al. (2011) presented a study about the importance of drainage density on flood runoff in urban catchments. Defin-

ing the drainage density as channel length per total catchment area, they studied the hydrological response of the same basin modelled with drainage density that varied from 0.4 to 3.9 km km^{-2}. Results showed a significant increase in peak discharge and runoff volume for drainage density between 0.4 and 0.9 km km^{-2}, while for values higher than 0.9 km km^{-2}, effects were negligible. When the storage and transport capacity of a system is not sufficient to prevent flooding, detention basins are effective tools to reduce peak flows, and they can reduce the superficial runoff up to 11 % (Smith et al., 2015).

Similarly, green roofs can significantly decrease and slow peak discharge and reduce runoff volume. Versini et al. (2014) presented a study on the impact of green roofs at urban scale using a distributed rainfall model. They showed that green roofs can reduce runoff generation in terms of peak discharge, depending on the rainfall event and initial conditions. The reduction can be up to 80 % for small events, with an intensity lower than 6 mm.

5 Urban hydrological models

Urban hydrological models were developed since the 1970s to better understand the behaviour of the components of the water cycle in urban areas (Zoppou, 2000). Since then, many models, with different characteristics, principles, and complexity have been built. These models are used for several purposes, such as to study and predict the effects of urbanization increase on the hydrological cycle, to support flood risk management, to ensure clean and fresh drinking water for the population, and to support improvement of waste water networks and treatments (see Zoppou, 2000; Fletcher et al., 2013 for a review). A good summary of the most used urban hydrological models has been recently proposed by Salvadore et al. (2015), where a table with the most used hydrological models is presented and discussed.

Hydrological models have shown to be useful to compensate partially for the lack of measurements (Salvadore et al., 2015), but all models present errors and uncertainties of different nature and magnitude (Rafieeinasab et al., 2015). In this chapter, different classifications and characterizations of hydrological models are presented.

5.1 Urban hydrological model characterization

Hydrological models can be characterized and classified in different ways. A first distinction can be made according to the representation of spatial variability of the catchment. A lumped model does not consider spatial variability of the input, and uses spatial averaging to represent catchment behaviour. In contrast, distributed models describe spatial variability, usually using a node-link structure to describe subcatchment components (Zoppou, 2000; Fletcher et al., 2013). The choice of a suitable model depends on many factors and it is generally related to the applications and final objective. For example Berne et al. (2004) suggested a guideline for choosing between lumped and distributed modelling considering the representative surface associated to a single rain gauge S_r. This characteristic, defined in relation to the rainfall spatial resolution r as $S_r = \pi [r/2]^2$, is compared with the surface area of a catchment S. If $S_r > S$ or $S_r \sim S$ a lumped modelling approach is suggested, while for $S_r < S$, a distributed model is recommended, as well as collecting measurements at the subcatchment scale. Different sub-categories are presented to characterize model spatial variability. Distributed models can be divided into fully distributed and semi-distributed models. Fully distributed models present a detailed discretization of the surface, using a grid or a mesh of regular or irregular elements, and apply the rainfall input to each grid element, generating grid-point runoff. The flow can be estimated at any location within the basin and not only at the catchment outlet. This is, however, possible only if the rainfall is provided with an appropriate spatial resolution. Semi-distributed models are based on subcatchment units, through which rainfall is applied. Each subcatchment is modelled in a lumped way, with uniform characteristics and a unique discharge point (Pina et al., 2014). Salvadore et al. (2015) proposed a model classification based on spatial variability with five categories: lumped, semi-distributed, Hydrological response unit based (semi-distributed with a specific way to define the subcatchment area), grid-based spatially distributed, and urban hydrological element based (mainly focused on the urban fluxes).

Another distinction is between conceptual and physically based (or process based) models, depending on whether the model is based on physical laws or not. Recently, Fatichi et al. (2016) presented an overview of the advantages and limitations of physically based models in hydrology. They defined a physically based hydrological model as "a set of process descriptions that are defined depending on the objectives". The downsides of using a physically based model are related to over-complexity and over-parametrization: conceptual models are much easier to manage and they are usually less affected by numerical instability. Physically based models usually require high computational power and time and a large number of parameters, but there are situations in which it is important to keep the complexity to better understand system mechanisms. They are also necessary to deal with system variability and allow one to include a stochastic component to represent uncertainty in parameter and input values (Del Giudice et al., 2015).

5.2 Spatial and temporal variability in urban hydrological models

Depending on their characteristics, models can be very sensitive to spatial and temporal rainfall variability or not be able to correctly reproduce effects of this variability. Spatial variability of land cover and soil characteristics is an im-

portant element in hydrological models. Choosing between a lumped, semi-distributed, or fully distributed hydrological model leads to different representation of catchment characteristics and, consequently, to a different output (Meselhe et al., 2009; Salvadore et al., 2015; Pina et al., 2016).

A comparison between semi-distributed and fully distributed urban storm water models was made by Pina et al. (2016). Two small urban catchments, Cranbrook (London, UK) and the centre of Coimbra (Portugal), were modelled with a semi- and a fully distributed model. Flow and depth in the sewer system of the different models were compared with observations and, in general, semi-distributed models predicted sewer flow patterns and peak flows more accurately, while fully distributed models had a tendency to underestimate flows. This was mainly due to the presence small-scale surface depressions, building singularities or lack of knowledge about private pipe connections. Although fully distributed models are more realistic and able to better represent spatial variability of the land cover, they need a higher resolution and accuracy to define module connections. Calibration of detailed, distributed models remains a complex issue that is not yet well resolved. The authors suggested to use a semi-distributed model approach in cases of low data resolution and accuracy.

To study the hydrological response Aronica and Canarozzo (2000) presented the Urban Drainage Topological Model (UDTM), a model that represents subcatchments of a semi-distributed model with two conceptual linear elements: a reservoir and a channel. In a more recent study (Aronica et al., 2005), this model was compared to the Storm Water Management Model (EPA SWMM model; Rossman, 2010), that allows the user to choose different conceptual models to simulate runoff and sewer flow. Results showed that model structure and sensitivity to parameters influence the sensitivity to the rainfall input resolution.

6 Interaction of spatial and temporal rainfall variability with hydrological response in urban basins

Storm structure and motion play an important role in the variability of the hydrological response (Smith et al., 1994; Bacchi and Kottegoda, 1995; Ogden et al., 1995; Singh, 1997; Emmanuel et al., 2012; Nikolopoulos et al., 2014; Emmanuel et al., 2015), especially for small catchments (Faures et al., 1995; Fabry et al., 1994). The characterization and the influence of spatial and temporal rainfall variability on runoff response is still not well understood (Emmanuel et al., 2015).

Recent studies address the impact of rainfall variability, focusing on urban catchments (Berne et al., 2004; Ochoa-Rodriguez et al., 2015b; Rafieeinasab et al., 2015; Yang et al., 2016). The main results and conclusions are presented in the following sections. It is discussed how basin characteristics impact the sensitivity of hydrological response to rainfall variability and how the interaction between spatial and temporal rainfall variability influences hydrological response.

6.1 Interaction between rainfall resolution and urban hydrological processes

Many studies highlight the importance of high-resolution rainfall data (Notaro et al., 2013; Emmanuel et al., 2012; Bruni et al., 2015) and how their use could improve runoff estimation, especially in an urban scenario, where drainage areas are small and spatial variability is high (Schilling, 1991; Schellart et al., 2011; Smith et al., 2013). These studies have shown how catchments act as filters in space and time for hydrological response to rainfall, delaying peaks and smoothing the intensity. However, the influence of spatial variability of rainfall on catchment response in urban areas is complex and remains an open research subject.

A theoretical study, conducted by Schilling (1991), emphasized the necessity to use rainfall data with a higher resolution for urban catchments compared to rural areas, and suggested to choose a minimum temporal resolution of 1–5 min and a spatial resolution of 1 km. The effects of temporal and spatial rainfall variability below 5 min and 1 km scale were subsequently studied by Gires et al. (2012). They investigated the urban catchment of Cranbrook (London, UK), with the aim of quantifying uncertainty in urban runoff estimation associated with unmeasured small-scale rainfall variability. Rainfall data were obtained from the national C-band radar with a resolution of 1 km^2 and 5 min and were downscaled with a multifractal process, to obtain a resolution 9–8 times higher in space and 4–1 in time. Uncertainty in simulated peak flow associated with small-scale rainfall variability was found to be significant, reaching 25 and 40 % respectively for frontal and convective events.

To investigate the effects of spatial and climatological variability on urban hydrological response, Peleg et al. (2017) used a stochastic rainfall generator to obtain high-resolution spatially variable rainfall as input for a calibrated hydrodynamic model. They compared the contributions of climatological rainfall variability and spatial rainfall variability on peak flow variability, over a period of 30 years. They found that peak flow variability is mainly influenced by climatological rainfall, while the effects of spatial rainfall variability increase for longer return periods.

Required rainfall resolution for urban hydrological modelling strongly depends on the characteristics of the catchment. Several researchers have studied the sensitivity of urban hydrological response to different rainfall resolutions, highlighting correlations between rainfall resolution and catchment dimensions, such as drained area (Berne et al., 2004; Ochoa-Rodriguez et al., 2015b) or catchment scale length (Ogden and Julien, 1994; Chirico et al., 2001; Bruni et al., 2015).

6.2 Influence of spatial and temporal rainfall variability in relation to catchment dimensions

Drainage area dimensions influence hydrological response and their sensitivities to spatial and temporal rainfall resolution have recently been investigated.

Wright et al. (2014) presented a flood frequency analysis, based on stochastic storm transposition (Wright et al., 2013) coupled with high-resolution radar rainfall measurements, with the aim to examine the effects of rainfall time and length scale on the flood response. Rainfall data were used as input for a physics-based hydrological model representative of 4 urbanized subcatchemnts. This study showed that there is an interaction between rainfall and basin characteristics, such as drainage area and drainage system location, that strongly affects the runoff.

Berne et al. (2004) studied the hydrological response of six urban catchments located in the south-east of the French Mediterranean coast. Rainfall data and runoff measurements were collected using two X-band weather radars, one vertically pointing radar, and one radar performing vertical plane cuts of the atmosphere, with a spatial resolution of 7.5 and 250 m and a temporal resolution of 4 s and 1 min respectively. The minimum temporal resolution required Δt was defined as $\Delta t = t_{\text{c}}/4$, where t_{c} is the characteristic time of a system and the value 4 depends on catchment properties (Schilling, 1991). By considering lag time t_{lag} equal to the characteristic time t_{c}, it was possible to write the minimum required temporal resolution as a function of surface area S, based on the relationship $t_{\text{lag}} = 3S^{0.3}$: $\Delta t = 0.75\,S^{0.3}$. Spatial resolution was studied considering rainfall data collected from the X-band weather radar performing vertical plane cuts of the atmosphere, combined with measurements of rain gauges. Two spatial climatological variograms were built with a time resolution of 1 min (from radar) and 6 min (from a network of 25 rain gauges). Based on variogram analysis, it was possible to define the relation between range r and time resolution Δt as $(r = 4.5\sqrt{\Delta t})$. The minimum required spatial resolution Δs was defined by the authors as $\Delta s = r/3$, and it can also be expressed as a function of Δt:

$$\Delta s = 1.5\Delta t. \tag{4}$$

In this way, both spatial and temporal resolution requirements were defined as a function of surface dimensions of a catchment. Required resolutions for urban catchments of 100 ha are 3 min and 2 km, but common operational rain gauge networks are usually less dense, while radars seldom provide data at this temporal resolution. Results presented are valid for catchments with characteristics similar to the catchments studied, such as surface area (from 10 to 10 000 ha), slope (1 to 10 %), imperviousness degree (10 to 60 %), and exposed to climatic conditions similar to those of Mediterranean area.

Ochoa-Rodriguez et al. (2015b) analysed the impact of spatial and temporal rainfall resolution on hydrological response in seven urban catchments, located in areas with different geomorphological characteristics. Using rainfall data measured by a dual polarimetric X-band weather radar with spatial resolution of 100 m $\times$ 100 m and temporal resolution of 1 min, they investigated the effects of combinations of different resolutions, with the aim to identify critical rainfall resolutions. They investigated the impact of 16 combinations of 4 different spatial resolutions (100 m $\times$ 100 m, 500 m $\times$ 500 m, 1000 m $\times$ 1000 m, and 3000 m $\times$ 3000 m) combined with four different temporal resolutions (1, 3, 5, and 10 min). Resolution combinations were chosen considering different aspects, such as the operational resolution of radar and rain gauges networks, characteristics temporal and spatial scale. A strong relation between drainage area and critical rainfall resolution and between spatial and temporal resolutions was found. Sensitivity to different rainfall resolutions decreased when the size of the subcatchment considered increased, especially for catchment size above 1 km^2. This study highlighted the importance of high-resolution rainfall data as input. Spatial resolution of 3 km $\times$ 3 km is not adequate for urban catchments and temporal resolution should be lower than 5 min. Most operational radars present a temporal resolution of 5 min, not sufficient to correctly represent the effects of temporal rainfall variability.

The sensitivity to rainfall variability on 5 urban catchments of different sizes, located in the City of Arlington and Grand Prairie (USA), was studied with a distributed hydrological model (HLRDHM, Hydrology Laboratory Research Distributed Hydrological Model) by Rafieeinasab et al. (2015). Rainfall data were provided by the Collaborative Adaptive Sensing Atmosphere (CASA) X-band radar with spatial resolution of 250 m $\times$ 250 m and temporal resolution of 1 min and upscaled in various steps to 2 km $\times$ 2 km and 1 h. Results showed peak intensity and time to peak error to be sensitive to spatial rainfall variability. The model was able to represent observed variability for all catchments except the smallest (3.4 km^2) at a temporal resolution of 15 min or lower, combined with spatial variability of 250 km $\times$ 250 m and capture variability in streamflow.

Resolution required to measure rainfall for small basins is usually high, as in the case of urban catchments. The influence of slope, imperviousness degree or soil type were not separately investigated, but the relationships between catchment area and rainfall resolution are expected to depend on these characteristics as well.

Sensitivity of hydrological response to different spatial and temporal rainfall resolutions has been investigated with dimensionless parameters to represent the length scales of storm events, catchments and of sewer networks.

Ogden and Julien (1994) identified dimensionless parameters to analyse correlations between catchment and storm characteristics and to study sensitivity of runoff models to radar-rainfall resolution. Rainfall data of a convective storm event, measured by a polarimetric radar with a spatial resolution of 1 km $\times$ 1 km, were applied on two basins. The storm smearing was defined as the ratio between rainfall data grid size and rainfall decorrelation length. Storm smearing occurs when rainfall data length is equal to or longer than the rainfall decorrelation length. The watershed smearing was described as the ratio between rainfall data grid size and basin length scale. When infiltration is negligible, watershed smearing is an important source of hydrological modelling errors, if the watershed ratio (rainfall measurement length/basin length) is higher than 0.4.

A similar approach, with dimensionless parameters, was recently applied by Bruni et al. (2015) to urban catchments. Rainfall data from a X-band dual polarimetric weather radar were applied to an hydrodynamic model, to investigate sensitivity of urban model outputs to different rainfall resolutions. The runoff sampling number was defined as ratio between rainfall length and runoff area length. Results confirm what was found by Ogden and Julien (1994). A third dimensionless parameter, called runoff sampling number, was identified. Small-scale rainfall variability at the 100 m $\times$ 100 m affects hydrological response and the effect of spatial resolution coarsening on rainfall values strongly depends on the movement of storm cells relative to the catchment.

Using dimensionless parameters is a productive approach to study sensitivity of hydrological response to spatial and temporal rainfall variability. Effects of other catchment characteristics, such as slope or imperviousness, were so far neglected, but they need a deeper investigation.

6.3 Spatial vs. temporal resolution

As it was already discussed in previous sections, there is a dependency between spatial and temporal rainfall required resolution and they affect in a different way the hydrological response (Marsan et al., 1996; Singh, 1997; Berne et al., 2004; Gires et al., 2011; Ochoa-Rodriguez et al., 2015b).

A first interaction between spatial and temporal rainfall scale was defined based on the assumption that atmospheric properties are valid also for rainfall. Following this assumption, Kolgomorov's theory (Kolgomorov, 1962) was combined with the scaling properties of the Navier–Stokes equation, in order to define a relation between space and time variability. For large Reynolds numbers, in fact, the Navier–Stokes equation is invariant under scale transformations (Marsan et al., 1996; Deidda, 2000; Gires et al., 2011), and in this way temporal and spatial "scale changing" operator can be defined by dividing space and time (s and t) by scaling factors λ_s and λ_t relatively: $s \longmapsto s/\lambda_s$ and $t \longmapsto t/\lambda_t$. For scaling processes, there is a relation between scaling factors in time and space to take into account, that is represented the anisotropy coefficient H_t: $\lambda_t = \lambda_s^{(1-H_t)}$. H_t is a priori unknown for rainfall, but it can be assumed equal to 1/3, a value that characterize atmospheric turbulence (Marsan et al., 1996; Gires et al., 2011, 2012). Lovejoy and Schertzer (1991) estimated $H_t = 0.5 \pm 0.3$ for raindrops. An example of application of this theory in a rainfall downscaling process is given by Gires et al. (2012): here, the rainfall is measured with a certain spatial resolution s and temporal resolution t. They hypothesized to downscale the radar pixels, dividing the length by a scaling factor $\lambda_s = 3$, to obtain nine pixels out of one. In this case, to keep the relation between spatial and temporal resolution, the duration of the time step has to be divided by a scaling factor $\lambda_t = \lambda_s^{1-1/3} = 2^{2/3} \simeq 2$.

Studying the hydrological response of the south-east French Mediterranean coast, Berne et al. (2004) proposed another relationship between spatial Δs and temporal Δt resolution used to measure rainfall, as $\Delta s = 1.5\sqrt{\Delta t}$ (see Sect. 6.2 for the formula derivation).

Ochoa-Rodriguez et al. (2015b) derived the theoretically required spatial rainfall resolution for urban hydrological modelling starting from a climatological variogram, which characterized average spatial structure of rainfall fields over the peak storm period, fitted with an exponential variogram model. They defined characteristic length scale r_c of a storm event as $r_c = (\sqrt{2\pi}/3)r$, where r is the variogram range. The minimum required spatial resolution for adequate modelling of urban hydrological response was defined as half characteristic length scale of the storm: $\Delta s = r_c/2 \cong 0.418r$. The theoretically required temporal resolution Δt, was defined based on the time needed for a storm to move over distance equal to the characteristic length scale of the storm event r_c. It can be written as $\Delta t = r_c/v$, where v is the magnitude of the mean storm velocity, obtained from average of the velocity vectors (magnitude and direction) estimated at each time step. Ochoa-Rodriguez et al. (2015b) investigated also the impact of different combinations of spatial and temporal resolutions as described in Sect. 6.2. One of the criteria used to choose some of the resolution combination was the already discussed in the literature (Berne et al., 2004), and according to Kolgomorov's scaling theory (Kolgomorov, 1962). Results showed that hydrodynamic models are more sensitive to the coarsening of temporal resolution of rainfall inputs than to the coarsening of spatial resolution, especially for fast moving storms.

In this work, the authors presented also a relation between spatial and temporal critical rainfall resolutions depending on drainage area (Table 4). For small catchments, with area smaller than 1 ha, was found to be equal to 100 m $\times$ 100 m and 1 min, while for areas between 1 and 100 ha, a spatial resolution of 500 m $\times$ 500 m can be sufficient to estimate the hydrological response. The critical spatial resolution found is lower than 5 min, for catchment size from about 250 to 900 ha. Results were confirmed by Yang et al. (2016), that presented an analysis of flash flooding in two small urban subcatchments of Harry's Brook (Princeton, New Jersey, USA), focusing on the influence of rainfall variability of storm events on hydrological response.

Table 4. Critical resolutions in relation with the drainage area.

Drainage Area DA (ha)	Critical spatial resolution (m $\times$ m)	Critical temporal resolution (min)
DA < 1	100	1
1 < DA < 100	500	1
250 < DA < 900	1000	< 5

Spatial variability seems to influence timing of runoff hydrograph, while temporal variability mainly influences peak value Singh (1997).

Ochoa-Rodriguez et al. (2015b) investigated the influence of spatial and temporal scaling factor introduced at the beginning of this section, on runoff estimation from different input, introducing also a combined spatio-temporal factor Θ_{st}. This factor was defined using the anisotropy coefficient as $\Theta_{st} = (\frac{\Delta S_r}{\Delta S})(\frac{\Delta t}{\Delta t_r})^{(\frac{1}{1-H_t})}$, where ΔS and Δt_r are the required spatial and temporal resolutions, ΔS and Δt are the space and time resolutions used as input for model simulations and H_t is the scaling anisotropy factor. The stronger relation between drainage area and combined spatio-temporal factor Θ_{st} compared to the relation with singular spatial or temporal scaling factor suggests that the effects of space and time has to be considered together. However, the combined effects of spatial and temporal resolution on the sensitivity to hydrological response requires future works and deeper investigations.

These studies highlighted the relatively more important role of temporal variability compared to spatial variability, for extreme rainfall events. The impact of the spatial variability, seemed to decrease with increase of total rainfall accumulation.

7   Summary and future directions

In this article, the state of the art of spatial and temporal variability impacts of rainfall and catchment characteristics on hydrological response in urban areas has been presented. The main key points and conclusion of this study are the following.

A first aspect that has been highlighted is the high variability in space and time of hydrological processes and phenomena in urban environments. Measuring, understanding and effectively characterizing temporal and spatial variability at small-scales is therefore of utmost importance. High-resolution data are essential given the high variability of catchment characteristics and hydrological processes, such as infiltration, evaporation and surface runoff. An important role in urban areas is played by drainage infrastructures that highly affect the hydrological response, while in some cases the effects of these structures are not perfectly understood. Current methods and instruments often have insufficient capability to measure the considered process at their relevant scales.

Several definitions to classify timescale characteristics are available in the literature, such as time of concentration, lag time, time of equilibrium and response timescale. However, measurement or estimation of those parameters is often ambiguous, which implies a high level of uncertainty. Thus far, no common agreement has emerged on a unique set of parameters able to characterize small-scale variability of urban catchments in a way that enhances our understanding of urban hydrological response. Improved rainfall measurements have also allowed to investigate the relations between temporal and spatial rainfall scale. Relations have been presented, mostly adapting the Kolgomorov?s theory to rainfall, to define the interaction between spatial and temporal scale in atmosphere. A unique relationship has not yet been found. This highlights the need for methods that can better characterize spatial and temporal scale parameters of rainfall and urban catchments in an effective way.

Uncertainty associated with rainfall spatial and temporal variability is one of the main sources of error in the estimation of hydrological response in urban areas. New technologies have been developed to measure rainfall spatial and temporal variability more accurately and at higher resolution. While rain gauges remain the most common used rainfall measurement instruments, weather radars are a promising example of recently developed instruments, able to estimate rainfall variability at high resolution. However, they still need to be combined with rain gauge networks in order to improve their accuracy. Rain gauges applied in urban areas present many limitations due to strong microclimatic variability, complicating identification of suitable locations for representative rainfall measurements. Polarimetric X-band radars combine high-resolution and high-accuracy measurement capability with the advantages of local installation thus avoiding overshooting and resolution loss with distance associated with large radar network. They constitute a promising direction for future urban hydrological research and rainfall and flood forecasting applications.

Many studies are reported in the literature using hydrological models with different characteristics and different representations of the catchment spatial variability. Different types of hydrological models have been developed in order to represent the spatial variability of catchment properties, such as land cover and imperviousness degree. Models can be classified based on their ability to represent the spatial variability of the catchment into lumped, semi-distributed and fully distributed models. These models have become more and more detailed, reaching high levels of spatial resolution. However, unless they are driven by similarly high-resolution rainfall data, increasing model resolution cannot fundamentally improve understanding of hydrological processes or improve reliability of hydrological predictions. Infiltration, local storage, interception and evaporation are quite difficult to measure, especially in urban areas, because of the strong heterogeneity of urban land use.

The impact of spatial and temporal rainfall variability on the hydrological response in urban areas and the role of drainage infrastructure and man-made control structures herein still remains poorly understood. It was found that sensitivity of hydrological response to spatial and temporal rainfall variability varies with catchment size, catchment shape, storm scale and storm velocity. So far, findings are mainly based on sensitivity studies using theoretical model scenarios. A wider range of conditions and scenarios based on observational datasets for urban hydrological basins need to be analysed in order to characterize better the hydrological response and its sensitivity to different spatial and temporal rainfall resolutions.

Competing interests. The authors declare that they have no conflict of interest.

Special issue statement. This article is part of the special issue "Rainfall and urban hydrology".

Acknowledgements. This work has been supported by the EU INTERREG IVB through funding of the RainGain Project (http://www.raingain.eu).

Edited by: Peter Molnar

References

Aronica, G. and Canarozzo, M.: Studying the hydrological response of urban catchments using a semi-distributed linear non-linear model, J. Hydrol., 238, 35–43, 2000.

Aronica, G., Freni, G., and Oliveri, E.: Uncertainty analysis of the influence of rainfall time resolution in the modelling of urban drainage systems, Hydrol. Process., 19, 1055–1071, 2005.

Bacchi, B. and Kottegoda, N.: Identification and calibration of spatial correlation patterns of rainfall, J. Hydrol., 165, 311–348, 1995.

Bergstrom, S. and Graham, L. P.: On the scale problem in hydrological modelling, J. Hydrol., 211, 253–265, 1998.

Berndtsson, R. and Niemczynowicz, J.: Spatial and temporal scales in rainfall analysis – some aspects and future perspective, J. Hydrol., 100, 293–313, 1986.

Berne, A. and Krajewski, W.: Radar for hydrology: Unfulfilled promise or unrecognized potential?, Adv. Water Resour., 51, 357–366, 2013.

Berne, A., Delrieu, G., Creutin, J., and Obled, C.: Temporal and spatial resolution of rainfall measurements required for urban hydrology, J. Hydrol., 299, 166–179, 2004.

Blöschl, G. and Sivapalan, M.: Scale issues in hydrological modelling: a review, Hydrol. Process., 9, 251–290, 1995.

Boogaard, F., Lucke, T., and Beecham, S.: Effect of Age of Permeable Pavements on Their Infiltration Function, Clean Soil Air Waters, 41, 146–152, 2013.

Bruni, G., Reinoso, R., van de Giesen, N. C., Clemens, F. H. L. R., and ten Veldhuis, J. A. E.: On the sensitivity of urban hydrodynamic modelling to rainfall spatial and temporal resolution, Hydrol. Earth Syst. Sci., 19, 691–709, https://doi.org/10.5194/hess-19-691-2015, 2015.

Cheng, S.and Wang, R.: An approach for evaluating the hydrological effects of urbanization and its application, Hydrol. Process., 16, 1403–1418, 2002.

Chirico, G. B., Grayson, R. B., Western, A. W., Woods, R., and Seed, A.: Sensitivity of simulated catchment response to the spatial resolution of rainfall, Proceedings of conference: MODSIM 2001, Modelling and Simulation Society of Australia and New Zealand inc, 10–13 December 2001, Canberra, Australia, 377–388, 2001.

Ciach, G. J. and Krajewski, W. F.: On the estimation of radar rainfall error variance, Adv. Water Resour., 22, 585–595, https://doi.org/10.1016/s0309-1708(98)00043-8, 1999.

Cole, S. and Moore, R.: Hydrological modelling using rain gauge- and radar-based estimators of areal rainfall, J. Hydrol., 358, 159–181, 2008.

Cui, X. and Li, X.: Role of surface evaporation in surface rainfall processes, J. Geophys. Res., 111, D17112, https://doi.org/10.1029/2005JD006876, 2006.

Dai, Q., Rico-Ramirez, M. A., Han, D., Islam, T., and Liguori, S.: Probabilistic radar rainfall nowcast using empirical and theoretical uncertainty models, Hydrol. Process., 29, 66–79, 2015.

Daniels, E. E., Lenderink, G., Hutjes, R. W. A., and Holtslag, A. A. M.: Observed urban effects on precipitation along the Dutch West coast, Int. J. Climatol., 36, 2111–2119, https://doi.org/10.1002/joc.4458, 2015.

Deidda, R.: Rainfall downscaling in a space time multifractal framework, Water Resour. Res., 36, 1779–1794, 2000.

Del Giudice, D., Löwe, R., Madsen, H., Mikkelsen, P. S., and Rieckermann, J.: Comparison of two stochastic techniques for reliable urban runoff prediction by modeling systematic errors, Water Resour. Res., 51, 5004–5022, 2015.

Dunne, T.: Field studies of hillslope flow processes, in: Hillslope Hydrology, edited by: Kirkby, M. J., Wiley, New York, USA, 227–293, 1978.

Einfalt, T., Arnbjerg-Nielsen, K., Golz, C., Jensen, N., Quirmbach, M., Vaes, G., and Vieux, B.: Towards a Roadmap for Use of Radar Rainfall data use in Urban Drainage, J. Hydrol., 299, 186–202, 2004.

Emmanuel, I., Andrieu, H., Leblois, E., and Flahaut, B.: Temporal and spatial variability of rainfall at the urban hydrological scale, J. Hydrol., 430–431, 162–172, 2012.

Emmanuel, I., Andrieu, H., Leblois, E., Janey, N., and Payrastre, O.: Influence of rainfall spatial variability on rainfall-runoff modelling: benefit of a simulation approach?, J. Hydrol., 531, 337–348, 2015.

Fabry, F., Bellon, A., Duncan, M. R., and Austin, G. L.: High resolution rainfall measurements by radar for very small basins: the sampling problem reexamined, J. Hydrol., 161, 415–428, 1994.

Fatichi, S., Vivoni, E. R., Ogden, F. L., Ivanov, V. Y., Mirus, B., Gochis, D., Downer, C. W., Camporese, M., Davison, J. H., and Ebel, B. E. A.: An overview of current applications,challenges, and future trends in distributed process-based models in hydrology, J. Hydrol., 537, 45–60, 2016.

Faures, J., Goodrich, D. C., Woolhiser, D. A., and Sorooshian, S.: Impact of small scale spatial rainfall variability on runoff modelling, J. Hydrol., 173, 309–326, 1995.

Fencl, M., Rieckermann, J., Sýkora, P., Stránský, D., and Bareš, V.: Commercial microwave links instead of rain gauges: fiction or reality?, Water Sci. Technol., 71, 31–37, https://doi.org/10.2166/wst.2014.466, 2015.

Fencl, M., Dohnal, M., Rieckermann, J., and Bareš, V.: Gauge-adjusted rainfall estimates from commercial microwave links, Hydrol. Earth Syst. Sci., 21, 617–634, https://doi.org/10.5194/hess-21-617-2017, 2017.

Ferraris, L., Gabellani, S., Rebora, N., and Provenzale, A.: A comparison of stochastic models for spatial rainfall downscaling, Water Resour. Res., 39, 1368, https://doi.org/10.1029/2003WR002504, 2003.

Fletcher, T. D., Andrieu, H., and Hamel, P.: Understanding, management and modelling of urban hydrology and its consequences for receiving waters: a state of the art, Adv. Water Resour., 51, 261–279, 2013.

Fonstad, M. A., Dietrich, J. T., Courville, B. C., Jensen, J. L., and Carbonneau, P. E.: Topographic structure from motion: a new development in photogrammetric measurement, Earth Surf. Proc. Land., 38, 421–430, 2013.

Foresti, L., Reyniers, M., Seed, A., and Delobbe, L.: Development and verification of a real-time stochastic precipitation nowcasting system for urban hydrology in Belgium, Hydrol. Earth Syst. Sci., 20, 505–527, https://doi.org/10.5194/hess-20-505-2016, 2016.

Fowler, H. J., Blenkinsop, S., and Tebaldi, C.: Linking climate change modelling to impacts studies: recent advances in downscaling techniques for hydrological modelling, Int. J. Climatol., 27, 1547–1578, https://doi.org/10.1002/joc.1556, 2007.

Gericke, O. J. and Smithers, J. C.: Review of methods used to estimate catchment response time for the purpose of peak discharge estimation, Hydrolog. Sci. J., 59, 1935–1971, 2014.

Gires, A., Onof, C., Tchiguirinskaia, I., Schertzer, D., and Lovejoy, S.: Analyses multifractales et spatio-temporelles des précipitations du modèle Méso-NH et des données radar, Hydrolog. Sci. J., 56, 380–396, 2011.

Gires, A., Onof, C., Maksimovic, C., Schertzer, D., Tchiguirinskaia, I., and Simoes, N.: Quantifying the impact of small scale unmeasured rainfall variability on urban hydrology through multifractal downscaling: a case study, J. Hydrol., 442-443, 117–128, 2012.

Gires, A., Tchiguirinskaia, I., Schertzer, D., Schellart, A., Berne, A., and Lovejoy, S.: Influence of small scale rainfall variability on standard comparison tools between radar and rain gauge data, Atmos. Res., 138, 125–138, 2014.

Gironás, J., Niemann, J., Roesner, L., Rodriguez, F., and Andrieu, H.: Evaluation of Methods for Representing Urban Terrain in Storm-Water Modeling, J. Hydrol. Eng., 15, 1–14, 2010.

Givati, A. and Rosenfeld, D.: Quantifying precipitation suppression due to air pollution, J. Appl. Meteorol., 43, 1038–1056, 2004.

Grimmond, C. and Oke, T.: An evapotranspiration-interception model for urban areas, Water Resour. Res., 27, 1739–1755, 1991.

Hart, Q. J., Brugnach, M., Temesgen, B., Rueda, C., Ustin, S. L., and Frame, K.: Daily reference evapotranspiration for California using satellite imagery and weather station measurement interpolation, Civ. Eng. Environ. Syst., 26, 19–33, https://doi.org/10.1080/10286600802003500, 2009.

Horton, R.: The role of infiltration in the hydrologic cycle, Eos Trans. AGU, 14, 446–460, 1933.

Horton, R.: Analysis of runoff-plat experiments with varing infiltration-capacity, EOS Earth and Space Science News, 20, 693–711, 1939.

Huff, F. A. and Changno, S. A. J.: Precipitation Modification By Major Urban Areas, B. Am. Meteorol. Soc., 54, 1220–1232, 1973.

Jensen, N. E. and Pedersen, L.: Spatial variability of rainfall: Variations within a single radar pixel, Atmos. Res., 77, 269–277, 2005.

Julien, P. Y. and Moglen, G. E.: Similarity and length scalefor spatially varied overland flow, Water Resour. Res., 26, 1819–1832, 1990.

Kolgomorov, A. N.: A refinement of previous hypotheses concerning the local structure of turbulence in a viscous incompressible fluid at high Reynolds number, J. Fluid Mech., 13, 82–85, 1962.

Krajewski, W. F. and Smith, J. A.: Radar hydrology: rainfall estimation, Adv. Water Resour., 25, 1387–1394, 2005.

Krajewski, W. F., Ciach, G. J., McCollum, J. R., and Bacotiu, C.: Initial validation of the global precipitation climatology project monthly rainfall over the United States, J. Appl. Meteorol., 39, 1071–1086, https://doi.org/10.1175/1520-0450(2000)039<1071:ivotgp>2.0.co;2, 2000.

Kuichling, E.: The Relation between the Rainfall and the Discharge of Sewers in Populous Districts, Transactions of ASCE, 20, 1–60, 1889.

Lanza, L. and Stagi, L.: High resolution performance of catching type rain gauges from the laboratory phase of the WMO Field Intercomparison of Rain Intensity Gauges, Atmos. Res., 94, 555–563, 2009.

Lanza, L. and Vuerich, E.: The WMO Field Intercomparison of Rain Intensity Gauges, Atmos. Res., 94, 534–543, 2009.

Larson, C. L.: A two phase approach to the prediction of peak rates and frequencies of runoff for small ungauged watersheds, Technical Report n. 53, Department of Civil Engineering, Standford University, Stanford, USA, 1965.

Leijnse, H., Uijlenhoet, R., and Stricker, J.: Rainfall measurement using radio links from cellular communication networks, Water Resour. Res., 43, W03201, https://doi.org/10.1029/2006WR005631, 2007.

Leitão, P. J., Simões, N. E., Pina, R. D., Ochoa-Rodriguez, S., Onof, C., and Marques, A. S.: Stochastic evaluation of the impact of sewer inlets' hydraulic capacity on urban pluvial flooding, Stoch. Environ. Res. Risk Assess., 1–16, https://doi.org/10.1007/s00477-016-1283-x, 2016.

Liguori, S. and Rico-Ramirez, M. A.: A review of current approaches to radar based Quantitative Precipitation Forecasts, International Journal of River Basin Management, 12, 391–402, 2013.

Litvak, E., Manago, K. F., Hogue, T. S., and Pataki, D. E.: Evapotranspiration of urban landscapes in Los Angeles, California at the municipal scale, Water Resour. Res., 53, 4236–4252, https://doi.org/10.1002/2016WR020254, 2017.

Lovejoy, S. and Schertzer, D.: Multifractal analysis techniques and the rain and cloud fields from 10^{-3} to 10^6 m, Non linear variability in geophysics, 111–144, https://doi.org/10.1007/978-94-009-2147-4_8, 1991.

Lucke, T., Boogaard, F., and van de Ven, F.: Evaluation of a new experimental test procedure to more accurately determine the surface infiltration rate of permeable pavement systems, Urban, Planning and Transport Research: An Open Access Journal, 2, 22–35, 2014.

Mandapaka, P. V., Krajeski, W. F., Ciach, G. J., Villarini, G., and Smith, J. A.: Estimation of radar-rainfall error spatial correlation, Adv. Water Resour., 32, 1020–1030, 2009.

Mansell, M. G.: Rural and urban hydrology, Thomas Telford Ltd, London, UK, 2003.

Marasco, D. E., Culligan, P. J., and McGillis, W.: Evaluation of common evapotranspiration models based on measurements from two extensive green roofs in New York City, Ecol. Eng., 84, 451–462, https://doi.org/10.1016/j.ecoleng.2015.09.001, 2015.

Marchi, L., Borga, M., Preciso, E., and Gaume, E.: Characterisation of selected extreme flash floods in Europe and implications for flood risk management, Hydrol. Process., 23, 2714–2727, 2010.

Marsan, D., Schertzer, D., and Lovejoy, S.: Causal space-time multifractal processes: Predictability and forecasting of rain fields, J. Geophys. Res., 101, 26333–26346, 1996.

Marshall, J. S. and Palmer, W. M. K.: The distribution of raindrops with size, McGill University, https://doi.org/10.1175/1520-0469(1948)005<0165:TDORWS>2.0.CO;2, 1948.

Mayer, H.: Automatic Object Extraction from Aerial Imagery – A Survey Focusing on Buildings, Comput. Vis. Image Und., 74, 138–149, 1999.

McCuen, R. H., Wong, S. L., and Rawls, W. J.: Estimating Urban Time of Concentration, J. Hydraul. Eng., 110, 887–904, https://doi.org/10.1061/(ASCE)0733-9429(1984)110:7(887), 1984.

McRobie, F. H., Wang, L.-P., Onof, C., and Kenney, S.: A spatial-temporal rainfall generator for urban drainage design, Water Sci. Technol., 68, 240–249, https://doi.org/10.2166/wst.2013.241, 2013.

Meselhe, E., Habib, E., Oche, O., and Gautam, S.: Sensitivity of conceptual and physically based hydrologic models to temporal and spatial rainfall sampling, J. Hydrol. Eng., 14, 711–720, 2009.

Montanari, A. and Grossi, G.: Estimating the uncertainty of hydrological forecasts: A statistical approach, Water Resour. Res., 44, W00B08, https://doi.org/10.1029/2008WR006897, 2008.

Morin, E., Enzel, Y., Shamir, U., and Garti, R.: The characteristic time scale for basin hydrological response using radar data, J. Hydrol., 252, 85–99, 2001.

Morin, E., Georgakakos, K. P., Shamir, U., Garti, R., and Enzel, Y.: Objective, observations-based, automatic estimation of the catchment response timescale, Water Resour. Res., 38, 1212, https://doi.org/10.1029/2001WR000808, 2002.

Morin, E., Georgakakos, K. P., Shamir, U., Garti, R., and Enzel, Y.: Investigating the effect of catchment characteristics on the response time scale using a distributed model and weather radar information, Weather Radar Information and Distributed Hydrological Modelling, Proceedings of symposium I-IS03 held during IUOG2003 at Sapporo, Japan, 30 June–11 July 2003, IAHS Publ., 282, 177–185, 2003.

Musy, A. and Higy, C.: Hydrology A Science of Nature, Science Publishers, CRC Press, Boca Raton, Florida, USA, 2010.

Muthusamy, M., Schellart, A., Tait, S., and Heuvelink, G. B. M.: Geostatistical upscaling of rain gauge data to support uncertainty analysis of lumped urban hydrological models, Hydrol. Earth Syst. Sci., 21, 1077–1091, https://doi.org/10.5194/hess-21-1077-2017, 2017.

Nielsen, J. E., Thorndahl, S., and Rasmussen, M. R.: Improving weather radar precipitation estimates by combining two types of radars, Atmos. Res., 139, 36–45, https://doi.org/10.1016/j.atmosres.2013.12.013, 2014.

Niemczynowicz, J.: The rainfall movement – A valuable complement to short-term rainfall data, J. Hydrol., 104, 311–326, 1988.

Niemczynowicz, J.: Urban hydrology and water management – present and future challenges, Urban Water, 1, 1–14, 1999.

Niemi, T. J., Guillaume, J. H. A., Kokkonen, T., Hoang, T. M. T., and Seed, A. W.: Role of spatial anisotropy in design storm generation: Experiment and interpretation, Water Resour. Res., 52, 69–89, https://doi.org/10.1002/2015WR017521, 2016.

Nikolopoulos, E., Borga, M., Zoccatelli, D., and Anagnostou, E. N.: Catchment scale storm velocity: quantification, scale dependence and effect on flood response, Hydrolog. Sci. J., 59, 1363–1376, 2014.

Notaro, V., Fontanazza, C. M., Freni, G., and Puleo, V.: Impact of rainfall data resolution in time and space on the urban flooding evaluation, Water Sci. Technol., 68, 1984–1993, 2013.

Nouri, H., Beecham, S., Kazemi, F., and Hassanli, A.: A review of ET measurement techniques for estimating the water requirements of urban landscape vegetation, Urban Water, 10, 247–259, 2013.

Ochoa-Rodriguez, S., Onof, C., Maksimovic, C., Wang, L., Willems, P., Assel, J., Gires, A., Ichiba, A., Bruni, G., and ten Veldhuis, A. E. J.: Urban pluvial flood modelling: current theory and practice. Review document related to Work Package 3, RainGain Project, WP3 review document, available at: http://www.raingain.eu (last access: 31 May 2017), 2015a.

Ochoa-Rodriguez, S., Wang, L., Gires, A., Pina, R., Reinoso-Rondinel, R., Bruni, G., Ichiba, A., Gaitan, S., Cristiano, E., Assel, J., Kroll, S., Murlà-Tuyls, D., Tisserand, B., Schertzer, D., Tchiguirinskaia, I., Onof, C., Willems, P., and ten Veldhuis, A. E. J.: Impact of Spatial and Temporal Resolution of Rainfall Inputs on Urban Hydrodynamic Modelling Outputs: A Multi-Catchment Investigation, J. Hydrol., 531, 389–407, 2015b.

Ogden, F. L. and Dawdy, D. R.: Peak discharge scaling in small hortonian watershed, J. Hydrol. Eng., 8, 64–73, 2003.

Ogden, F. L. and Julien, P. Y.: Runoff model sensitivity to radar rainfall resolution, J. Hydrol., 158, 1–18, 1994.

Ogden, F. L., Richardson, J. R., and Julien, P. Y.: Similarity in catchment response, Water Resour. Res., 31, 1543–1547, 1995.

Ogden, F. L., Pradhan, N. R., Downer, C. W., and Zahner, J. A.: Relative Importance of Impervious Area, Drainage Density, width Function, and Subsurface Storm Drainage on Flood Runoff from an Urbanized Catchment, Water Resour. Res., 47, 1–12, 2011.

Oke, T. R.: Initial Guidance to Obtain Representative Meteorological Observations at Urban Sites, OM Report No.81, WMO/TD No. 1250, World Meteorological Organization, Geneva, Switzerland, 2006.

Opzoekingscentrum voor de Wegenbouw: Waterdoorlatende Verhardingen met Betonstraatstenen, Report on pervious pavements, Brussel, Belgium, 2008.

Otto, T. and Russchenberg, H. W.: Estimation of Specific Differential Phase Backscatter Phase From Polarimetric Weather Radar Measurement of Rain, IEEE Geosci. Remote S., 5, 988–922, 2011.

Overeem, A., Holleman, I., and Bruihand, A.: Derivation of a 10 year radar based climatology of rainfall, J. Appl. Meteorol. Clim., 48, 1448–1463, 2009a.

Overeem, A., Buishand, A., and Holleman, I.: Extreme rainfall analysis and estimation of depth duration frequency curves using weather radar, Water Resour. Res., 45, W10424, https://doi.org/10.1029/2009WR007869, 2009b.

Paschalis, A., Molnar, P., Fatichi, S., and Burlando, P.: A stochastic model for high resolution space-time precipitation simulation, Water Resour. Res., 49, 8400–8417, https://doi.org/10.1002/2013WR014437, 2013.

Peleg, N. and Morin, E.: Stochastic convective rain-field simulation using a high-resolution synoptically conditioned weather generator (HiReS-WG), Water Resour. Res., 50, 2124–2139, https://doi.org/10.1002/2013WR014836, 2014.

Peleg, N., Ben-Asher, M., and Morin, E.: Radar subpixel-scale rainfall variability and uncertainty: lessons learned from observations of a dense rain-gauge network, Hydrol. Earth Syst. Sci., 17, 2195–2208, https://doi.org/10.5194/hess-17-2195-2013, 2013.

Peleg, N., Marra, F., Fatichi, S., Paschalis, A., Molnar, P., and Burlando, P.: Spatial variability of extreme rainfall at radar subpixel scale, J. Hydrol., in press, https://doi.org/10.1016/j.jhydrol.2016.05.033, 2016.

Peleg, N., Blumensaat, F., Molnar, P., Fatichi, S., and Burlando, P.: Partitioning the impacts of spatial and climatological rainfall variability in urban drainage modeling, Hydrol. Earth Syst. Sci., 21, 1559–1572, https://doi.org/10.5194/hess-21-1559-2017, 2017.

Pina, R., Ochoa-Rodriguez, S., Simones, N.and Mijic, A., Sa Marques, A., and Maksimovik, C.: Semi-distributed or fully distributed rainfall-runoff models for urban pluvial flood modelling?, 13th International Conference on Urban Drainage, 7–12 September 2014, Sarawak, Malaysia, 2014.

Pina, R., Ochoa-Rodriguez, S., Simones, N., Mijic, A., Sa Marques, A., and Maksimovik, C.: Semi- vs fully- distributed urban stormwater models: model set up and comparison with two real case studies, Water, 8, 58, https://doi.org/10.3390/w8020058, 2016.

Quirmbach, M. and Schultz, G. A.: Comparison of rain gauge and radar data as input to an urban rainfall-runoff model, Water Sci. Technol., 45, 27–33, 2016.

Rafieeinasab, A., Norouzi, A., Kim, S., Habibi, H., Nazari, B., Seo, D., Lee, H., Cosgrove, B., and Cui, Z.: Toward high-resolution flash flood prediction in large urban areas – Analysis of sensitivity to spatiotemporal resolution of rainfall input and hydrologic modeling, J. Hydrol., 531, 370–388, 2015.

Ragab, R., Rosier, P., Dixon, A., Bromley, J., and Cooper, J. D.: Experimental study of water fluxes in a residential area: 2. Road infiltration, runoff and evaporation, Hydrol. Process., 17, 2423–2437, 2003.

Ramier, D., Berthier, E., and Andrieu, H.: The hydrological behaviour of urban streets: long-term observations and modelling of runoff losses and rainfall runoff transformation, Hydrol. Process., 25, 2161–2178, 2011.

Richards, L. A.: Capillary Conduction of Liquids Through Porous Mediums, Physics, 1, 318, https://doi.org/10.1063/1.1745010, 1931.

Rico-Ramirez, M. A., Liguori, S., and Schellart, A.: Quantifying radar rainfall uncertainties in urban drainage flow modelling, J. Hydrol., 528, 17–28, 2015.

Rodriguez-Iturbe, I. and Mejıa, J. M.: The design of rainfall networks in time and space, Water Resour. Res., 10, 713–728, https://doi.org/10.1029/WR010i004p00713, 1974.

Rossman, L. A.: Storm water management model user's manual, version 5.0, National Risk Management Research Laboratory, Office of Research and Development, US Environmental Protection Agency, Cincinnati, OH, USA, 2010.

Rummukainen, M.: Methods for statistical downscaling of GCM simulation, SWECLIM report, Rossby Centre, SMHI, Norrköping, Sweden, 1997.

Salvadore, E., Bronders, J., and Batelaan, O.: Hydrological modelling of urbanized catchments: A review and future directions, J. Hydrol., 529, 61–81, 2015.

Sangati, M., Borga, M., Rabuffeti, D., and Bechini, R.: Influence of rainfall and soil properties spatial aggregation on extreme flash flood response modelling: an evaluation based on the Sesia river basin, North Western Italy, Adv. Water Resour., 32, 1090–1106, 2009.

Schaake, J., Geyer, J., and Knapp, J.: Experimental examination of the rational method, Journal of Hydrological Division, 93, 353–370, 1967.

Schellart, A., Shepherd, W., and Saul, A.: Influence of rainfall estimation error and spatial variability on sewer flow prediction at a small urban scale, Adv. Water Resour., 45, 65–75, https://doi.org/10.1016/j.advwatres.2011.10.012, 2011.

Schertzer, D. and Lovejoy, S.: Multifractals, generalized scale invariance and complexity in geophysics, Int. J. Bifurcation Chaos, 21, 3417–3456, 2011.

Schilling, W.: Rainfall data for urban hydrology: What do we need?, Atmos. Res., 27, 5–21, 1991.

Segond, M.-L., Wheater, H. S., and Onof, C.: The significance of spatial rainfall representation for flood runoff estimation: A numerical evaluation based on the Lee catchment, UK, J. Hydrol., 347, 116–131, 2007.

Shamir, E., Imam, B., Morin, E., Gupta, H. V., and Sorooshian, S.: The role of hydrograph indices in parameter estimation of rainfall-runoff models, Hydrol. Process., 19, 2187–2207, 2005.

Shepherd, J. M.: Evidence of urban-induced precipitation variability in arid climate regimes, J. Arid Environ., 67, 607–628, 2006.

Shepherd, J. M., Pierce, H., and Negr, i. A. J.: Rainfall Modification by Major Urban Areas: Observations from Spaceborne Rain Radar on the TRMM Satellite, J. Appl. Meteor., 41, 689–701, 2002.

Singh, V. P.: Derivation of time of concentration, J. Hydrol., 30, 147–165, 1976.

Singh, V. P.: Effect of spatial and temporal variability in rainfall and watershed characteristics on stream flow hydrographs, Hydrol. Process., 11, 1649–1669, 1997.

Smith, A. J., Baeck, M. L., Morrison, J. E., Sturevant-Rees, P., Turner-Gillespie, D. F., and Bates, P. D.: The regional hydrology of extreme floods in an urbanizing drainage basin, American Meterological Society, 3, 267–282, 2002.

Smith, A. J., Baeck, M. L., Villarini, G., Welty, C., Miller, A. J., and Krajewski, W. F.: Analyses of a long term, high resolution radar rainfall data set for the Baltimore metropolitan region, Water Resour. Res., 48, W04504, https://doi.org/10.1029/2011WR010641, 2012.

Smith, B. K., Smith, A. J., Baeck, M. L., Villarini, G., and Wright, D. B.: Spectrum of storm event hydrologic response in urban watersheds, Water Resour. Res., 49, 2649–2663, 2013.

Smith, B. K., Smith, A. J., Baeck, M. L., and Miller, A. J.: Exploring storage and runoff generation processes for urban flooding through a physically based watershed model, Water Resour. Res., 51, 1552–1569, 2015.

Smith, J. A., Bradley, A. A., and Baeck, M. L.: The space-time structure of extreme storm rainfall on the Southern Plains, J. Appl. Meteorol., 33, 1402–1417, 1994.

Smith, J. A., Baeck, M. L., Meierdiercks, K. L., Nelson, P. A., Miller, A. J., and Holland, E. J.: Field studies of the storm event hydrologic response in an urbanizing watershed, Water Resour. Res., 41, W10413, https://doi.org/10.1029/2004WR003712, 2005.

Sørup, H. J. D., Christensen, O. B., Arnbjerg-Nielsen, K., and Mikkelsen, P. S.: Downscaling future precipitation extremes to urban hydrology scales using a spatio-temporal Neyman–Scott weather generator, Hydrol. Earth Syst. Sci., 20, 1387–1403, https://doi.org/10.5194/hess-20-1387-2016, 2016.

Stahl, K. and Hisdal, H.: Hydroclimatology in: Hydrological Drought – Processes and Estimation Methods, Elvisier, Amsterdam, the Netherlands, 19–51, 2004.

Thorndahl, S., Smith, J. A., Baeck, M. L., and Krajewski, W. F.: Analyses of the temporal and spatial structures of heavy rainfall from a catalog of high-resolution radar rainfall fields, Atmos. Res., 144, 111–125, 2014.

Thorndahl, S., Einfalt, T., Willems, P., Nielsen, J. E., ten Veldhuis, M.-C., Arnbjerg-Nielsen, K., Rasmussen, M. R., and Molnar, P.: Weather radar rainfall data in urban hydrology, Hydrol. Earth Syst. Sci., 21, 1359–1380, https://doi.org/10.5194/hess-21-1359-2017, 2017.

Tokarczyk, P., Leitao, J. P., Rieckermann, J., Schindler, K., and Blumensaat, F.: High-quality observation of surface imperviousness for urban runoff modelling using UAV imagery, Hydrol. Earth Syst. Sci., 19, 4215–4228, https://doi.org/10.5194/hess-19-4215-2015, 2015.

United States Department of Agriculture (USDA): Time of Concentration, in: National Engineering Handbook Hydrology, chap. 15, Part 630 Hydrology, eDirectives site, available at: https://www.nrcs.usda.gov/wps/portal/nrcs/detailfull/national/water/manage/hydrology/?cid=stelprdb1043063 (last access: 26 July 2017), 2010.

van de Beek, C. Z., Leijnse, H., Stricker, J. N. M., Uijlenhoet, R., and Russchenberg, H. W. J.: Performance of high-resolution X-band radar for rainfall measurement in The Netherlands, Hydrol. Earth Syst. Sci., 14, 205–221, https://doi.org/10.5194/hess-14-205-2010, 2010.

van de Giesen, N., Stomph, T., and de Ridder, N.: Surface runoff scale effects in West African watersheds: modeling and management options, Agr. Water Manage., 72, 109–130, 2005.

van de Giesen, N., Stomph, T., Ebenezer Ajayi, A. E., and Bagayoko, F.: Scale effects in Hortonian surface runoff on agricultural slopes in West Africa: Field data and models, Agr. Ecosyst. Environ., 142, 95–101, 2011.

van de Ven, F. H. M.: Water balances of urban areas, Hydrological Processes and Water Management in Urban Areas, Proceedings of the Duisberg Symposium, 24–29 April 1988, Zoetermeer, the Netherlands, IAHS Publ., 198, 1990.

Van Loon, A. F.: Hydrological drought explaines, WIREs Water, 2, 359–392, 2015.

Versini, P. A., Gires, A., Abbes, J. B., Giangola-Murzyn, A., Tchinguirinskaia, I., and Schertzer, D.: Simulation of Green Roof Impact at Basin Scale by Using a Distributed Rainfall-Runoff Model, 13th International Conference on Urban Drainage (ICUD), 7–11 September 2014, Sarawak, Malaysia, 1–9, 2014.

Villarini, G., Mandapaka, P. V., Krajewski, W. F., and Moore, R. J.: Rainfall and sampling uncertainties: A rain gauge perspective, J. Geophys. Res.-Atmos., 113, D11102, https://doi.org/10.1029/2007jd009214, 2008.

Villarini, G., Seo, B. C., Serinaldi, F., and Krajewski, W. F.: Spatial and temporal modeling of radar rainfall uncertainties, Atmos. Res., 135–136, 91–101, https://doi.org/10.1016/j.atmosres.2013.09.007, 2014.

Wang, L. P., Ochoa-Rodriguez, S., Simoes, N., Onof, C., and Maksimovic, C.: Radar-raingauge data combination techniques: a revision and analysis of their suitability for urban hydrology, Water Sci. Technol., 68, 737–747, 2013.

Wang, L. P., Ochoa-Rodriguez, S., Onof, C., and Willems, P.: Singularity-sensitive gauge- based radar rainfall adjustment methods for urban hydrological applications, J. Hydrol., 531, 408–426, 2015a.

Wang, L. P., Ochoa-Rodriguez, S., van Assel, J., Pina, R. D., Pessemier, M., Kroll, S., Willems, P., and Onof, C.: Enhancement of radar rainfall estimates for urban hydrology through optical flow temporal interpolation and Bayesian gauge-based adjustement, J. Hydrol., 531, 408–426, https://doi.org/10.1016/j.jhydrol.2015.05.049, 2015b.

Wilby, R. and Wigley, T.: Downscaling general circulation model output: a review of methods and limitation, Prog. Phys. Geog., 21, 530–48, 1997.

Wilks, D. S. and Wilby, R. L.: The weather generation game: a review of stochastic weather models, Prog. Phys. Geog., 23, 329–357, 1999.

Winchell, M., Gupta, H. V., and Sorooshian, S.: On the simulation of infiltration and saturation excess runoff using radar based rainfall estimates: Effects of algorithm uncertainty and pixel aggregation, Water Resour. Res., 34, 2655–2670, 1998.

WMO, W. M. O.: Guide to Meteorological Instruments and Methods of Observation, Seventh edition, WMO-No. 8, Geneva, Switzerland, 2008.

Wood, S. J., Jones, D. A., and Moore, R. J.: Static and dynamic calibration of radar data for hydrological use, Hydrol. Earth Syst. Sci., 4, 545–554, https://doi.org/10.5194/hess-4-545-2000, 2000.

Wright, D., Smith, J., Villarini, G., and Baeck, M.: Estimating the frequency of extreme rainfall using weather radar and stochastic storm transposition, J. Hydrol., 488, 150–165, 2013.

Wright, D., Smith, J., and Baeck, M.: Flood frequency analysis using radar rainfall fields and stochastic storm transposition, Water Resour. Res., 50, 1592–1615, 2014.

Xu, C.: From GCMs to river flow: a review of downscaling methods and hydrologic modelling approaches, Prog. Phys. Geog., 23, 229–249, 1999.

Yang, L., Smith, J. A., Baeck, M. L., and Zhang, Y.: Flash flooding in small urban watersheds: storm event hydrological response, Water Resour. Res., 52, 4571–4589, https://doi.org/10.1002/2015WR018326, 2016.

Yao, L., Wei, W., and Chen, L.: How does imperviousness impact teh urban rainfall-runoff process under various storm cases?, Ecol. Indic., 60, 893–905, 2016.

Zoppou, C.: Review of urban storm water models, Environ. Model. Softw, 16, 195–231, 2000.

11

The WACMOS-ET project – Evaluation of global terrestrial evaporation data sets

D. G. Miralles[1,2], C. Jiménez[3], M. Jung[4], D. Michel[5], A. Ershadi[6], M. F. McCabe[6], M. Hirschi[5], B. Martens[2], A. J. Dolman[1], J. B. Fisher[7], Q. Mu[8], S. I. Seneviratne[5], E. F. Wood[9], and D. Fernández-Prieto[10]

[1]Department of Earth Sciences, VU University Amsterdam, Amsterdam, the Netherlands

[2]Laboratory of Hydrology and Water Management, Ghent University, Ghent, Belgium

[3]Estellus, Paris, France

[4]Max Planck Institute for Biogeochemistry, Jena, Germany

[5]Institute for Atmospheric and Climate Science, ETH Zurich, Zurich, Switzerland

[6]Division of Biological and Environmental Sciences and Engineering, King Abdullah University of Science and Technology, Thuwal, Saudi Arabia

[7]Jet Propulsion Laboratory, California Institute of Technology, Pasadena, California, USA

[8]Department of Ecosystem and Conservation Sciences, University of Montana, Missoula, Montana, USA

[9]Department of Civil and Environmental Engineering, Princeton University, Princeton, New Jersey, USA

[10]ESRIN, European Space Agency, Frascati, Italy

Correspondence to: D. G. Miralles (diego.miralles@vu.nl)

Abstract. The WAter Cycle Multi-mission Observation Strategy – EvapoTranspiration (WACMOS-ET) project aims to advance the development of land evaporation estimates on global and regional scales. Its main objective is the derivation, validation, and intercomparison of a group of existing evaporation retrieval algorithms driven by a common forcing data set. Three commonly used process-based evaporation methodologies are evaluated: the Penman–Monteith algorithm behind the official Moderate Resolution Imaging Spectroradiometer (MODIS) evaporation product (PM-MOD), the Global Land Evaporation Amsterdam Model (GLEAM), and the Priestley–Taylor Jet Propulsion Laboratory model (PT-JPL). The resulting global spatiotemporal variability of evaporation, the closure of regional water budgets, and the discrete estimation of land evaporation components or sources (i.e. transpiration, interception loss, and direct soil evaporation) are investigated using river discharge data, independent global evaporation data sets and results from previous studies. In a companion article (Part 1), Michel et al. (2016) inspect the performance of these three models at local scales using measurements from eddy-covariance towers and include in the assessment the Surface Energy Bal-

ance System (SEBS) model. In agreement with Part 1, our results indicate that the Priestley and Taylor products (PT-JPL and GLEAM) perform best overall for most ecosystems and climate regimes. While all three evaporation products adequately represent the expected average geographical patterns and seasonality, there is a tendency in PM-MOD to underestimate the flux in the tropics and subtropics. Overall, results from GLEAM and PT-JPL appear more realistic when compared to surface water balances from 837 globally distributed catchments and to separate evaporation estimates from ERA-Interim and the model tree ensemble (MTE). Nonetheless, all products show large dissimilarities during conditions of water stress and drought and deficiencies in the way evaporation is partitioned into its different components. This observed inter-product variability, even when common forcing is used, suggests that caution is necessary in applying a single data set for large-scale studies in isolation. A general finding that different models perform better under different conditions highlights the potential for considering biome- or climate-specific composites of models. Nevertheless, the generation of a multi-product ensemble, with weighting based on validation analyses and uncertainty assessments, is proposed as the best way forward in our long-term goal to develop a robust observational benchmark data set of continental evaporation.

---

1 Introduction

The importance of terrestrial evaporation (or "evapotranspiration") for hydrology, agriculture, and meteorology has long been recognized. In fact, most of our current understanding of the physics of evaporation originated in early experiments during the past 2 centuries (e.g. Dalton, 1802; Horton, 1919; Penman, 1948). However, it has been during the last decade that the interest of the scientific community in land evaporation has increased more dramatically, following the recognition of the key role it plays in climate (Wang and Dickinson, 2012; Dolman et al., 2014). Evaporation is highly sensitive to radiative forcing: changes in atmospheric chemical composition affect the magnitude of the flux, ensuring the propagation of anthropogenic impacts to all the components of the hydrological cycle (Wild and Liepert, 2010) and altering the global availability of water resources (Hagemann et al., 2013). In addition, evaporation regulates climate through a series of feedbacks acting on air temperature, humidity, and precipitation (Koster et al., 2006; Seneviratne et al., 2010), thus affecting climate trends (Douville et al., 2013; Sheffield et al., 2012) and hydro-meteorological extremes (Seneviratne et al., 2006; Teuling et al., 2013; Miralles et al., 2014a). Finally, due to the link between transpiration and photosynthesis, atmospheric carbon concentrations and carbon cycle feedbacks are closely linked to terrestrial evaporation (Reichstein et al., 2013). When these factors are taken together, evaporation represents a crucial nexus of processes and cycles in the climate system.

The rising interest of the climate community has coincided with an unprecedented availability of global field data to scrutinize the response of evaporation to climate impacts and feedbacks. However, due to the limitations in coverage of direct in situ measurements, the scientific community have turned to satellite remote sensing (Kalma et al., 2008; Wang and Dickinson, 2012; Dolman et al., 2014). Consequently, different international activities now focus on the joint advancement of remote sensing technology and evaporation science; these activities include the National Aeronautics and Space Administration (NASA) Energy and Water cycle Study (NEWS, http://nasa-news.org), the European Union WATer and global CHange (WATCH, http://www.eu-watch.org) project, and the Global Energy and Watercycle Experiment (GEWEX) LandFlux initiative (https://hydrology.kaust.edu.sa/Pages/GEWEX_Landflux.aspx). Despite continuing progress in satellite and computing science, to date, the evaporative flux cannot be directly sensed from space; technology thus lags behind our physical knowledge of evaporation. Nonetheless, taking advantage of this existing knowledge, different models have been proposed to combine the physical variables that are linked to the evaporation process and can be observed from space (e.g. radiation, temperature, soil moisture, or vegetation dynamics). Such efforts have yielded a number of global evaporation products in recent years (Mu et al., 2007; Zhang et al., 2010; Fisher et al., 2008; Miralles et al., 2011b; Jung et al., 2010). These data sets are not to be interpreted as the direct result of satellite observations but rather as model outputs generated based on satellite forcing data. The reader is directed to Su et al. (2011) or McCabe et al. (2013) for recent reviews of the state of the art.

Despite the recent initiatives dedicated to exploring these evaporation data sets – LandFlux-EVAL in particular (Jiménez et al. (2011); Mueller et al. (2011, 2013) – the relative merits of each model on the global scale remain largely unexplored. To date, the lack of inter-model consistency in the choice of forcing data has hampered the attribution of the observed skill of each evaporation data set to differences in the models. Only recently, some efforts have been directed towards homogenizing the forcing of these models to allow the assessment of algorithm quality (Vinukollu et al., 2011a; Ershadi et al., 2014; Chen et al., 2014; McCabe et al., 2016). In 2012, the European Space Agency (ESA) WAter Cycle Multi-mission Observation Strategy – EvapoTranspiration (WACMOS-ET) project (http://WACMOSET.estellus.eu) started in response to the need for a thorough and consistent model intercomparison at different spatial and temporal scales. At the same time, WACMOS-ET is a direct contribution to GEWEX LandFlux, sharing the long-term goal of achieving global closure of surface water and energy budgets. The project objectives strive to (a) develop a reference input data set consisting of satellite observations, reanalysis data and in situ measured meteorology, (b) run a group of selected evaporation models forced by the reference input data set, and (c) perform a cross comparison, evaluation, and validation exercise of the evaporation data sets that result from running this group of models. Four algorithms that are commonly used by the research community have been tested: the Surface Energy Balance Model (SEBS; Su, 2002); the Penman–Monteith approach that sets the basis for the official Moderate Resolution Imaging Spectroradiometer (MODIS) evaporation product, hereafter referred to as PM-MOD (Mu et al., 2007, 2011, 2013); the Global Land Evaporation Amsterdam Model, GLEAM (Miralles et al., 2011b); and the Priestley and Taylor model from the Jet Propulsion Laboratory, PT-JPL (Fisher et al., 2008).

In a companion article – henceforth referred to as Part 1 – Michel et al. (2016) describe the results of the local validation activities of WACMOS-ET based on in situ evaporation measurements from eddy-covariance towers. Here, we present the global-scale inter-product evaluation. After forcing the models with the reference input data set (see Sect. 2.2 for the description of the forcing data), the resulting evaporation data sets are evaluated by means of (a) a general exploration of the global magnitude and spatiotemporal variability

Table 1. Inputs from the reference input data set used in each of the models. The specific products chosen for each variable are also noted.

Input	Product	PM-MOD	GLEAM	PT-JPL
Radiation	SRB 3.1	✓	✓	✓
Air temperature	ERA-Interim	✓	✓	✓
Precipitation	CFSR-Land	–	✓	–
Soil moisture	CCI WACMOS	–	✓	–
Air humidity	ERA-Interim	✓	–	✓
Snow cover	GlobSnow, NSIDC	–	✓	–
Vegetation characteristics	Internally produced (except for the vegetation optical depth from AMSR-E, see Sect. 2.2 and Part 1)	✓	✓	✓

of the estimates (Sects. 3.1 and 3.2), (b) a comparison with other, commonly used, evaporation data sets (Sects. 3.1, 3.2, and 3.3), including the model tree ensemble (MTE) estimates by Jung et al. (2009, 2010) and the European Centre for Medium-range Weather Forecasts (ECMWF) Re-Analysis (ERA)-Interim (Dee et al., 2011), (c) an assessment of the skill to close the surface water balance over a broad range of catchments worldwide (Sect. 3.3), and (d) an analysis of the contribution to total terrestrial evaporation from the discrete components or sources of this flux, i.e. transpiration, interception loss, and direct evaporation from the soil (Sect. 3.4). Due to the difficulties that arise from executing SEBS on the global scale (see Su et al., 2010), the current work concentrates on PM-MOD, GLEAM, and PT-JPL, while the local-scale analysis in Part 1 also includes the SEBS model.

2 Methods and data

2.1 Models or algorithms

Here we present a brief description of the three models that are studied in this article. For more exhaustive descriptions the reader is directed to Part 1 and to the original articles describing the parameterizations and algorithms of PM-MOD (Mu et al., 2007, 2011), GLEAM (Miralles et al., 2011b), and PT-JPL (Fisher et al., 2008). A summary of the forcing requirements of PM-MOD, GLEAM, and PT-JPL can be found in Table 1, together with the specific product for each input variable.

2.1.1 PM-MOD

The Penman–Monteith model by Mu et al. (2007, 2011) is arguably the most widely used remote-sensing-based global evaporation model, and, in its latest version, it is also the algorithm behind the official MODIS (MOD16) product (Mu et al., 2013). PM-MOD is based on the Monteith (1965) adaptation of Penman (1948); thus, it is has relatively high demands in terms of inputs. The parameterizations of aerodynamic and surface resistances for each component of evaporation are based on extending biome-specific conductance parameters to the canopy scale using vegetation phenology

and meteorological data. The model applies the surface resistance scheme by Cleugh et al. (2007) – which uses leaf area index as suggested by Jarvis (1976) – in an extended version that considers the constraints of vapour pressure deficit and minimum temperature on stomatal conductance (Mu et al., 2007). However, in contrast to the majority of Penman–Monteith-type models, PM-MOD does not require soil moisture or wind speed data to parameterize the surface and aerodynamic resistances. The non-consideration of wind speed appears as an advantage when aiming for a fully observation-driven product. Snow sublimation and open-water evaporation are not considered independently of other processes. Unlike GLEAM and PT-JPL, which do not require calibration, the resistance parameters in PM-MOD have been calibrated with data from a set of global eddy-covariance towers (see Mu et al., 2011).

2.1.2 GLEAM

GLEAM (www.gleam.eu) is a simple land surface model fully dedicated to deriving evaporation based on satellite forcing only (Miralles et al., 2011b). It distinguishes between direct soil evaporation, transpiration from short and tall vegetation, snow sublimation, open-water evaporation, and interception loss from tall vegetation. Interception loss is independently calculated based on the Gash (1979) analytical model forced by observations of precipitation (Miralles et al., 2010). The remaining components of evaporation are based upon the formulation by Priestley and Taylor (1972), which does not require the parameterization of stomatal and aerodynamic resistances, in contrast to the Penman–Monteith equation. In the case of transpiration and soil evaporation, the potential evaporation estimates – resulting from the application of the Priestley and Taylor approach – are constrained by a multiplicative stress factor. This dynamic stress factor is calculated based on the content of water in vegetation (microwave vegetation optical depth; Liu et al., 2011) and the root zone (multilayer soil model driven by observations of precipitation and updated through assimilation of microwave surface soil moisture; see Martens et al., 2016). The consideration of vegetation water content accounts for the effects of plant phenology, while the root-zone soil moisture accounts for soil water stress. For regions covered by ice and snow, sublimation is calculated using a Priestley and Taylor equation with specific parameters for ice and supercooled waters (Murphy and Koop, 2005). For the fraction of open water at each pixel, the model assumes potential evaporation. GLEAM has recently been applied to look at trends in the water cycle (Miralles et al., 2014b) and land–atmospheric feedbacks (Guillod et al., 2015; Miralles et al., 2014a).

2.1.3 PT-JPL

The PT-JPL model by Fisher et al. (2008) uses the Priestley and Taylor (1972) approach to estimate potential evaporation. Unlike GLEAM, it applies a series of ecophysiological stress factors based on atmospheric moisture (vapour pressure deficit and relative humidity) and vegetation indices (normalized difference vegetation index, i.e. NDVI, and soil adjusted vegetation index) to constrain the atmospheric demand for water. This implies that the set of forcing requirements of PT-JPL is in fact very comparable to that of PM-MOD (see Table 1). In order to partition land evaporation into soil evaporation, transpiration, and interception loss, PT-JPL first distributes the net radiation to the soil and vegetation components and then calculates the potential evaporation for soil, transpiration, and interception separately. The partitioning between transpiration and interception loss is done using a threshold based on relative humidity. As in PM-MOD, snow sublimation and open-water evaporation are not considered independently of other processes. The model has been employed in a number of studies to estimate terrestrial evaporation on regional and global scales in recent years (see, e.g., Sahoo et al., 2011; Vinukollu et al., 2011a, b).

2.2 Input data

One of the objectives of the WACMOS-ET project has been to correct for a recurring issue in inter-product evaluations of global evaporation: due to inconsistencies in the forcing data behind current evaporation products, it is difficult to attribute the observed inter-product disagreements to algorithm discrepancies (Jiménez et al., 2011; Mueller et al., 2013). Consequently, one of the first steps in WACMOS-ET has been to compile a reference input data set that has been used to run all models in a consistent manner. This consistency applies to both local-scale runs (in Part 1) and regional and global runs (in the present study). On the other hand, since neither the required input variables nor the models' sensitivity to these input variables and their uncertainties are the same for all models (see Table 1), it is not possible to fully attribute observed differences in performance to internal model errors. Nonetheless, our efforts to homogenize forcing data in a global evaporation inter-model comparison are unique, with the exception of Vinukollu et al. (2011a), who used off-the-shelf forcing data sets to run earlier versions of SEBS, PT-JPL, and PM-MOD. For all the details on the produc-

tion of the reference input data set, the reader is directed to the thorough descriptions in Part 1 and the supporting documents available on the project website. Nonetheless, a short summary is also provided here.

Some of the variables considered in the reference input data set have been internally generated during the project, while others were selected from the existing pool of global climatic and environmental data sets. Choices regarding the spatial and temporal resolution, period covered, and study domain were made with the support of a large number of end users surveyed via the internet (see project website). The target grid resolution of WACMOS-ET is 25 km, the domain is global and the study period spans 2005–2007. A 3-hourly temporal resolution maximizes the links with the work undertaken by the GEWEX LandFlux initiative to produce sub-daily evaporation estimates (McCabe et al., 2016). The present Part 2 evaluates the outputs after aggregating them to daily, monthly, and annual scales, while the skill of the models to resolve the diurnal cycle of evaporation is explored in Part 1. Although the internally generated input data sets were originally derived at a relatively fine (< 5 km) spatial resolution, critical inputs not generated within the project were only available at 75–100 km (see below). Consequently, all input data sets have been spatially resampled to match the 25 km target resolution and reprojected onto a common sinusoidal grid before using them to run the evaporation models.

Internally developed products include the fraction of photosynthetically active radiation and leaf area index, which are derived to a large extent from European satellites (see Part 1). Data access, product descriptions, and user guidelines for these data sets are available to interested parties upon request via the project website. Whereas PM-MOD and PT-JPL apply these internally generated data sets to characterize vegetation phenology, GLEAM uses observations of microwave vegetation optical depth as a proxy for vegetation water content; these are taken from the data set of Liu et al. (2011) based on the Advanced Microwave Scanning Radiometer – Earth Observing System (AMSR)-E at 0.25° spatial resolution.

The remaining products comprising the reference input data set have been selected from the pool of available community data sets. Surface net radiation is obtained by integrating the upwelling and downwelling radiative fluxes from the NASA and GEWEX Surface Radiation Budget (SRB, Release 3.1), which contains global 3-hourly averages of these fluxes on a 1° resolution grid. The SRB product is based on a range of satellite data, atmospheric reanalysis, and data assimilation (Stackhouse et al., 2004). The meteorology (i.e. near-surface air temperature, air humidity, and wind speed) comes from the ERA-Interim atmospheric reanalysis, provided at 3-hourly resolution (using the forecast fields) and at a spatial resolution of ~ 75 km. The reason for using atmospheric reanalysis data (based on observations assimilated into a weather forecast model), as opposed to direct satellite observations, is that some of these variables (like air temperature and humidity) are presently difficult to observe over continents, if not impossible (as in the case of wind speed), and are not routinely available at sub-daily time steps and for all weather conditions.

Despite its relevance for plant-available water and interception loss, precipitation is not a direct input for most global satellite-based evaporation models. The same applies to surface soil moisture, which can also be observed from space. From the WACMOS-ET models, only GLEAM uses observations of precipitation and surface soil moisture as input. In the reference input data set, precipitation data come from the Climate Forecast System Reanalysis for Land (CFSR-Land; Coccia et al., 2015), which uses the Climate Prediction Center (CPC, Chen et al., 2008) and the Global Precipitation Climatology Project (GPCP, Huffman et al., 2001) daily data sets and applies a temporal downscaling based on the CFSR (Saha et al., 2010). For soil moisture, we use the satellite product of combined active–passive microwave surface soil moisture by Liu et al. (2012), which combines information from scatterometers and radiometers from different platforms, and was developed as part of the ESA Climate Change Initiative (CCI). In addition, GLEAM also uses information on snow water equivalents that is taken from the ESA GlobSnow product, version 1.0 (Luojus and Pulliainen, 2010), based on AMSR-E and corrected using ground-based measurements. Since GlobSnow covers the Northern Hemisphere only, data from the National Snow and Ice Data Center (NSIDC) are used in snow-covered regions of the Southern Hemisphere (Kelly et al., 2003). Observations of soil moisture and snow water equivalents have a native resolution of 0.25° and are imported into GLEAM at daily time steps.

2.3 Data used for evaluation

2.3.1 Other global land evaporation products

For the purpose of comparing our three WACMOS-ET products to related evaporation data sets, we incorporate two additional data sets into the evaluation: the ERA-Interim reanalysis evaporation (Dee et al., 2011) and the MTE product (Jung et al., 2009, 2010). The latter is derived from satellite data and FLUXNET observations (Baldocchi et al., 2001) using a machine-learning algorithm. In the model, tree ensembles are trained to predict monthly eddy-covariance fluxes based on meteorological, climate, and land cover data. It has a monthly temporal resolution and 0.5° spatial resolution. For full details, the reader is referred to Jung et al. (2009).

2.3.2 Catchment water balance data

The mass balance of a catchment implies that the space and time integration of precipitation (P) minus river run-off (Q) should equal evaporation (integrated over the same space and time). This requires the consideration of a long period, so changes in storage within the catchment and the travel time of precipitation through the landscape can be neglected (see discussion in Sect. 3.3). Given that river run-off and precipitation are more easily and extensively measured than evaporation, estimates of $P-Q$ based on ground measurements of these two fluxes provide a convenient means to evaluate evaporation over large domains and long periods (Liu et al., 2014; Miralles et al., 2011a; Vinukollu et al., 2011b; Sahoo et al., 2011). Here, we use globally distributed multi-annual river discharge data for basins larger than 2500 km². Discharge data and watershed boundaries are obtained from the Global Runoff Data Centre (GRDC). Run-off data have been converted from cubic metres per second to millimetres per year using the area of each catchment as reported by the GRDC; basins where the absolute difference between the GRDC-reported area and the area calculated from basin boundaries exceeded 25 % have been excluded from the analyses.

Precipitation for the target period 2005–2007 is taken from GPCP (Huffman et al., 2001) and the Global Precipitation Climatology Centre (GPCC) v6 (Schneider et al., 2013). Two versions of GPCC v6 are processed by applying relative gauge correction factors according to Fuchs et al. (2001) and Legates and Willmott (1990) to the native GPCC products as recommended by the producers. We further discard basins with (a priori) low-quality precipitation due to the low density of rain gauges (< 0.1 per 0.5° latitude–longitude), frequent snowfall (> 25 days per year based on CloudSat), or where cumulative values of discharge exceed those of precipitation over the 3-year period. Finally, radiation data from the NASA Clouds and Earth's Radiant Energy System (CERES) Synoptic Radiative Fluxes and Clouds 1-degree resolution (SYN1deg) product (Wielicki et al., 2000) are used to exclude basins where $P-Q$ exceeds surface net radiation on average.

This results in a record of 837 basins from which $P-Q$ values are calculated. Figure 1 illustrates the locations of the centroids of these catchments. Basins are then clustered in 30 classes based on log-transformed precipitation, net radiation, and evaporative fraction (i.e. evaporation over net radiation). This is done in order to reduce noise and retain clear patterns for evaluation. The clustering algorithm used is a k means with city block distance, with variables transformed to zero mean and unit variance. For clarity, each of the 30 classes is assigned to one of four groups based on thresholds of net radiation (80 W m⁻²) and evaporative fraction (0.5) as shown in Fig. 1. The results of comparing the evaporation products, integrated over the corresponding basins, to the $P-Q$ estimates are presented in Sect. 3.3.

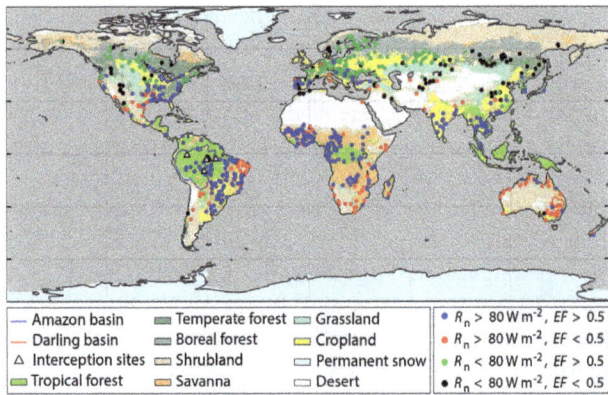

Figure 1. Climatic regimes and biomes considered in the evaluations. The background map illustrates the land use classification scheme of the International Geosphere–Biosphere Programme (IGBP) used in Fig. 8. The Darling Basin in southeastern Australia, as considered in Sect. 3.2, is contoured in red. The Amazon Basin, as considered in Sect. 3.4, is marked in blue, with white triangles indicating the locations of past interception loss campaigns. Dots indicate the centroids of the 837 basins used in the analyses presented in Sect. 3.3.

3   Results and discussion

3.1   Global magnitude of terrestrial evaporation

The global mean annual volume of evaporation has been intensively debated in recent years (see, e.g., Wang and Dickinson, 2012), with the range of reported global averages in current Coupled Model Intercomparison Project Phase 5 (CMIP5) models being large (Wild et al., 2014) and observational benchmark data sets also differing significantly (Mueller et al., 2013). In this section, we aim to give some context to the global magnitude of evaporation that results from the WACMOS-ET analyses by contrasting the results with alternative evaporation data sets and existing literature. Unless otherwise noted, results come from aggregating the outputs of the 3-hourly global runs based on the 25 km spatial resolution of the reference input data set for the period 2005–2007.

Overall, the total annual magnitude of evaporation estimated by the WACMOS-ET models amounts to 54.9×10^3 km^3 for PM-MOD, 72.9×10^3 km^3 for GLEAM, and 72.5×10^3 km^3 for PT-JPL. We further calculated 84.4×10^3 km^3 for ERA-Interim and 68.3×10^3 km^3 for MTE based on the same 2005–2007 period. Unlike the other products, MTE does not include poles and desert regions (as shown in Fig. 2); however, the contribution from these areas to the global volumes is rather marginal ($< 5\%$ based on our analyses). For comparison, values typically found in the literature based on a broad variety of methodologies and forcings are 63.2×10^3 km^3 (Zhang et al., 2016), 65.0×10^3 km^3 (Jung et al., 2010), 65.5×10^3 km^3 (Oki and Kanae, 2006), 65.8×10^3 km^3 (Schlosser and Gao, 2010), 67.9×10^3 km^3

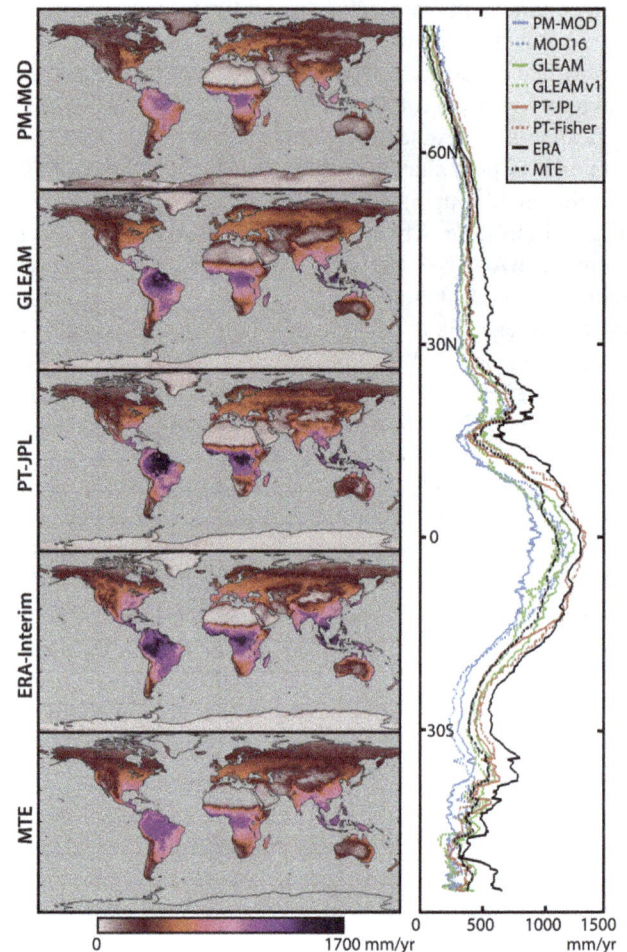

Figure 2. Mean patterns of land evaporation. Average evaporation during 2005–2007 for PM-MOD, GLEAM, and PT-JPL forced by the reference input data set; the ERA-Interim reanalysis and the MTE product are shown for comparison. On the right, the latitudinal profiles of evaporation; the original data sets of PM-MOD, GLEAM, and PT-JPL (i.e. MOD16, GLEAMv1, and PT-Fisher, respectively) are also shown for comparison. We note that the original PT-JPL covers until 2006 only, and therefore its latitudinal profile is based on the 2005–2006 average. Due to the MTE product not reporting values in polar regions and deserts, those areas are excluded from the latitudinal profiles in all models.

(Miralles et al., 2011a), 71×10^3 km^3 (Baumgartner and Reichel, 1975), 73.9×10^3 km^3 (Wang-Erlandsson et al., 2014), and 74.3×10^3 km^3 (Zhang et al., 2015). We note again that some of these studies considered the poles and desert regions, while others did not. Further, the study period considered in WACMOS-ET is 2005–2007, while previously reported annual averages may be based on different periods.

In Fig. 2 the multiannual (2005–2007) mean evaporation is displayed for the different products, including also MTE and ERA-Interim for comparison. All five data sets capture the expected climatic transitions well, although disagreements on the regional scale are still considerable (see below). Latitudinal averages are illustrated in the right panel of Fig. 2. Model estimates are normally contained between the low values from PM-MOD and the high values from ERA-Interim; as an exception, PM-MOD can be comparatively large in Northern Hemisphere high latitudes (see Sect. 3.2). In Fig. 2, the latitudinal profiles from the original and official products of PM-MOD (i.e. MOD16), GLEAM (i.e. GLEAM v1), and PT-JPL (i.e. PT-Fisher) are also displayed for comparison. Note that the main differences between these official products and those developed in WACMOS-ET relate to the choice of forcing – see Mu et al. (2013), Miralles et al. (2011a), and Fisher et al. (2008) for the particular forcing data used to generate these official data sets. In addition, models have been run here on a sub-daily scale (3 hourly) as opposed to their original daily (PM-MOD, GLEAM) or monthly (PT-JPL) temporal resolutions. While for PM-MOD and PT-JPL the choice of temporal resolution and forcing in WACMOS-ET leads to overall lower values (see PM-MOD in tropics), values are slightly higher than in the original version (v1) for GLEAM.

Inter-product differences in mean evaporation become more evident in Fig. 3, which presents the anomalies for each product calculated by subtracting the average of the five-product ensemble. PM-MOD displays lower averages than the multi-product ensemble mean over the entire continental domain, with the exception of high latitudes, as discussed above. GLEAM shows higher than average values in Europe or Amazonia and lower than average values in North America. This pattern is somewhat shared by PT-JPL, although the two models disagree substantially in water-limited regions of Africa and Australia, even if absolute mean values are low in these regions (see Fig. 2). This relates to the different model representation of evaporative stress, with GLEAM being based on observations of rainfall, surface soil moisture, and vegetation optical depth, while PT-JPL is based on air humidity, maximum air temperature, and NDVI. As mentioned in Sect. 2.2, it is important to note that even though we aimed to maximize consistency in forcing data for PM-MOD, GLEAM, and PT-JPL, their disagreement still reflects a combination of algorithm structural errors and input uncertainties, given the use of a distinct range of inputs for each model (Table 1) and the different model sensitivities to each particular driver.

ERA-Interim values are often at the high end of the predictions, consistent with the results by Mueller et al. (2013), more than doubling the evaporation estimated by PM-MOD on some occasions (Fig. 2). MTE values, on the other hand, are lower than the inter-product average in the Himalayas and in tropical forests – which may potentially relate to the lack of a separate computation of interception loss and the open question of whether interception can be measured with eddy-covariance instruments (see van Dijk et al., 2015) – but they agree well with the mean of the multi-product ensemble in other regions (Fig. 3). A quick overview of the range of uncertainty that can be expected may be found in the right

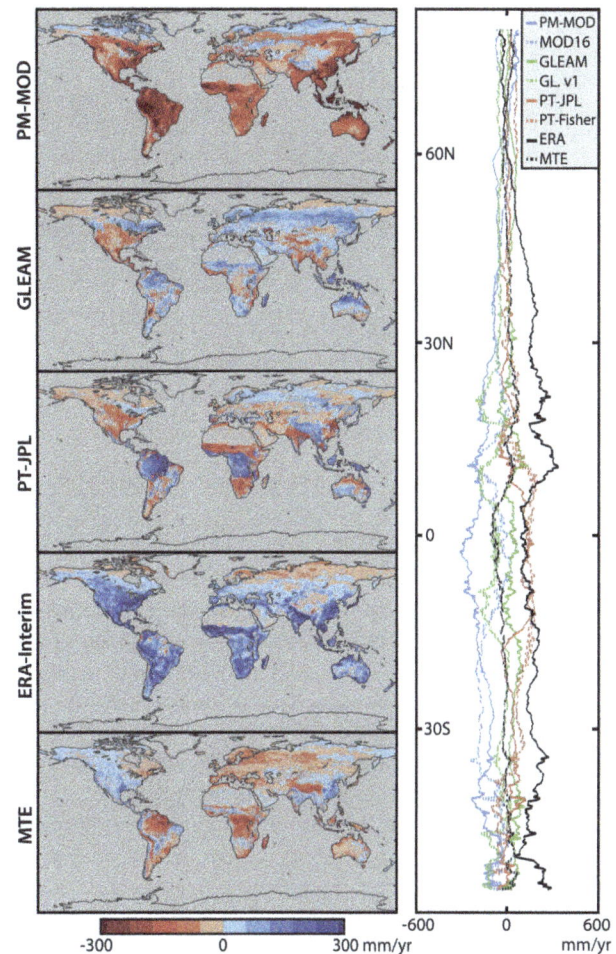

Figure 3. Long-term anomalies of evaporation, as in Fig. 2 but based on the anomalies for each product calculated as the mean of each particular product (i.e. the maps in Fig. 2) minus the inter-product ensemble mean (considering the ensemble of five models). Grey areas over the continents correspond to regions where MTE displays no estimates of evaporation.

panel of Fig. 3, where the latitudinal profiles of anomalies are illustrated. Data sets appear again to be confined between the low values of PM-MOD and the high values of ERA-Interim. If this multi-model range is interpreted as an indication of the uncertainty, it is worth noting that it often amounts to 60–80 % of the mean evaporation, particularly in the subtropics. In the tropics, while the relative uncertainty is lower, the inter-product range still reaches $\sim 500\,\mathrm{mm\,yr^{-1}}$ according to the latitudinal profiles in Fig. 3. To put that volume into context, the mean annual evaporation is below $500\,\mathrm{mm\,yr^{-1}}$ for more than 50 % of continental surfaces, according to the inter-product ensemble mean.

The spatial agreement among models is further explored in Fig. 4, which presents the spatial correlation for each pair of models based on their long-term global means (i.e. the maps in Fig. 2). Each land pixel is an independent point in the scat-

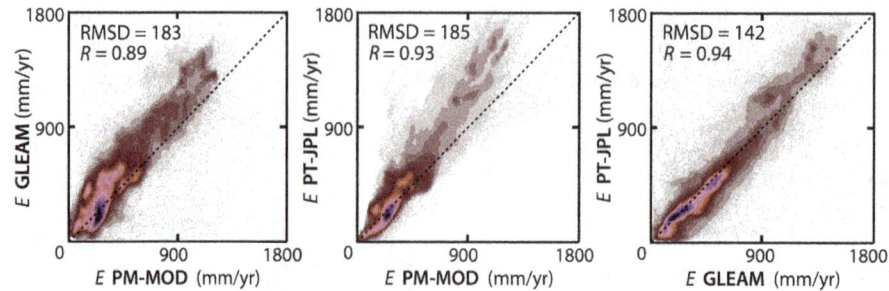

Figure 4. Correspondence in the average spatial patterns for each pair of models. Each point represents a land pixel in Fig. 2. Pearson's correlation coefficients (R) and root mean square differences (RMSDs) are listed.

ter. The lowest spatial correlation occurs between PM-MOD and GLEAM ($R = 0.89$) and the highest between GLEAM and PT-JPL ($R = 0.94$). Although the latter fact may reflect the common choice of a Priestley and Taylor approach to calculate potential evaporation in both models, it occurs despite their large differences in input requirements (Table 1) and in the approach to deriving evaporative stress and interception loss (Sect. 2.1). The agreement in the mean spatial patterns between PM-MOD and PT-JPL is also high in terms of the correlation coefficient ($R = 0.93$), as expected from their shared set of input variables (see Table 1). Nonetheless, their root mean square difference is large (RMSD $= 185 \, \mathrm{mm \, yr^{-1}}$) compared to the difference between PT-JPL and GLEAM (RMSD $= 142 \, \mathrm{mm \, yr^{-1}}$), which mostly reflects the overall lower values of PM-MOD. These low mean values are also accompanied by a low variance, especially in midlatitudes. This is illustrated in Fig. 5, which depicts the standard deviation of the monthly time series at each pixel and as a function of latitude.

3.2 Temporal variability of terrestrial evaporation

In addition to long-term mean differences in evaporation, inter-product discrepancies in temporal dynamics are certainly expected. Temporal correlations based on the (2005–2007) daily time series for each pair of models are illustrated in Fig. 6a. The overall agreement in temporal dynamics is larger in high latitudes, especially between GLEAM and PT-JPL. In semi-arid regions, product-to-product correlations are often below 0.5 and may drop below 0.2 (see, e.g., low correlation between PM-MOD and PT-JPL in southern Africa or Australia). This occurs despite the substantial amplitude of the seasonal cycle in these transitional regimes (see, e.g., Fig. 5), which may, in principle, artificially increase temporal correlations. Overall, Fig. 6a corroborates that, although the agreement between GLEAM and PT-JPL is large, their different approach to estimating water-availability constraints on evaporation and rainfall interception loss leads to significant differences for semi-arid regions and tropical forests.

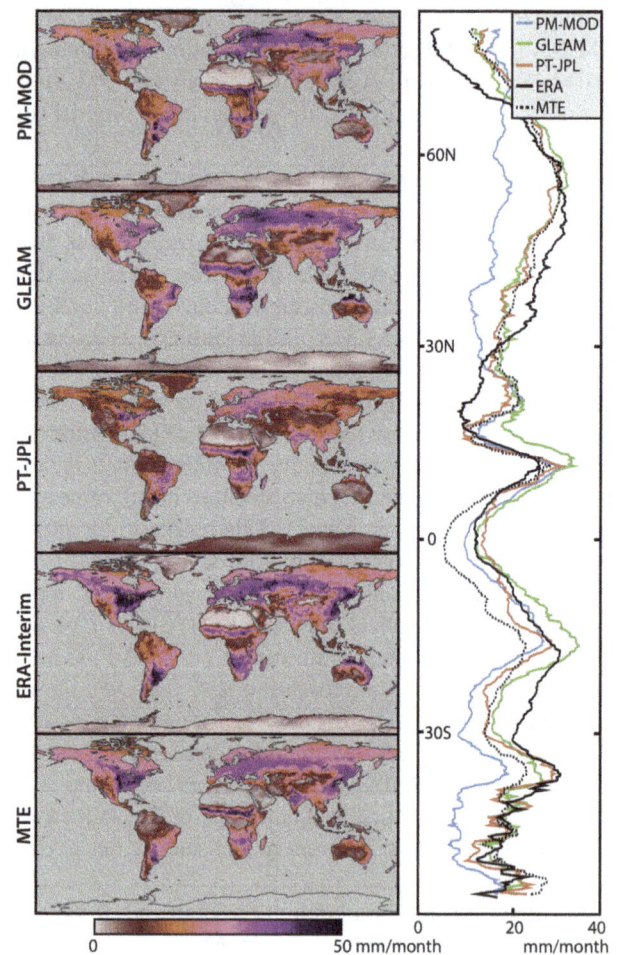

Figure 5. Standard deviation of land evaporation, based on the monthly time series for 2005–2007 at each pixel for PM-MOD, GLEAM, and PT-JPL forced by the reference input data set; the ERA-Interim reanalysis and the MTE product are shown for comparison. The right column illustrates the latitudinal profiles of these standard deviations. Due to the MTE product not reporting values in polar regions and deserts, these areas are excluded from the latitudinal profiles in all models.

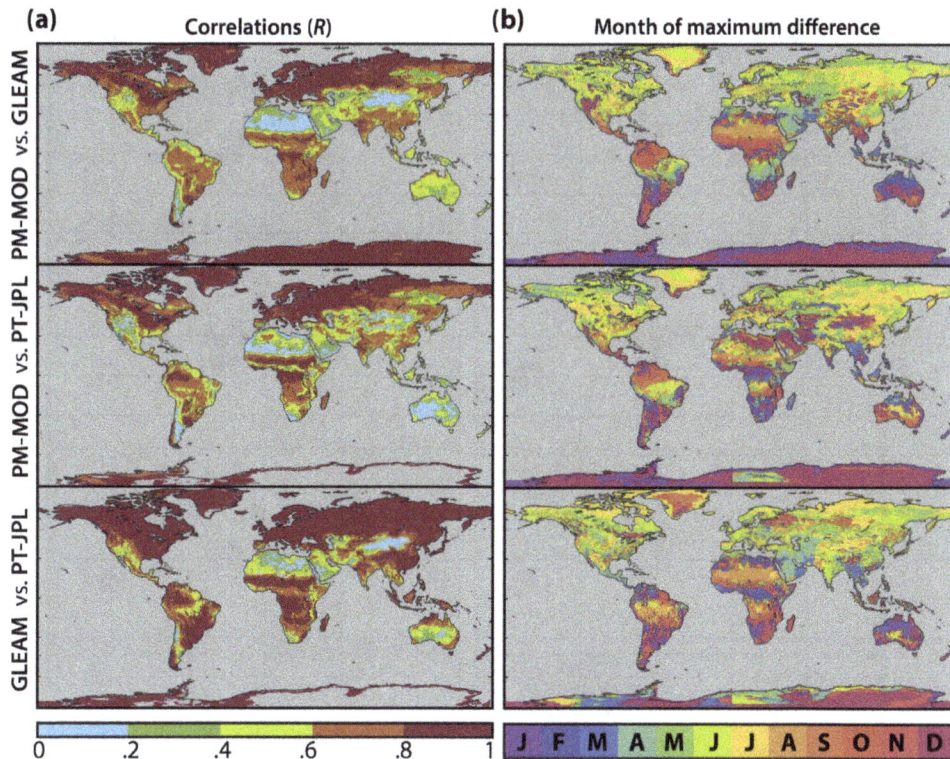

Figure 6. Temporal agreement between the models. Panel (**a**): temporal correlation coefficients between each pair of products based on the daily (2005–2007) time series. Panel (**b**): month of the year in which the maximum (monthly) difference occurs between a particular pair of products based on their monthly climatologies.

Based on the monthly climatology of each model (calculated by averaging the estimates for the same month of the year and considering the multiannual 2005–2007 period), Fig. 6b illustrates the month in which the differences between a given pair of models are the largest. In the Northern Hemisphere, the product-to-product differences are at their maximum during summertime, when the flux of evaporation is high. This is particularly the case in comparisons to PM-MOD, given that the seasonal evaporation peak of PM-MOD is often less pronounced than for the other models (see also Figs. 5, 7, and 8). In the tropics and the Southern Hemisphere, maximum differences between models occur at different times of the year but often coincide with months of higher evaporative demand for water; this is the case for southern Africa, the pampas region or Australia during the Austral summer.

Figure 7 shows the average evaporation for boreal summer (JJA) and winter (DJF) for each model based on the 3-year period of study. MTE and ERA-Interim are again included for comparison. As expected, the seasonal variability of evaporation follows the annual cycle of radiation, except for arid and semi-arid regions that are controlled by the availability of water. The lower values of PM-MOD are again highlighted. The underestimation of PM-MOD, with respect to the other two models, mostly occurs at times and in locations for which both evaporative demand and water availability are high (e.g. midlatitude summer, tropics); thus, evapora-

tion is expected to be high as well. As discussed in Sect. 3.3, this may be associated with an overestimation of evaporative stress in the model. However, PM-MOD is often higher than the other two models in periods and regions where radiation is severely limited, potentially due to the underestimation in Priestley–Taylor-type models (i.e. GLEAM and PT-JPL) when radiation is not the main supply of energy for evaporation (see, e.g., Parlange and Katul, 1992); in these conditions, the Penman–Monteith equation still considers adiabatic sources of energy to drive evaporation. Once more, differences in seasonal means between GLEAM and PT-JPL exist on regional scales, especially in water-limited regimes, with Australia being a clear example (see also Fig. 9).

Nonetheless, Fig. 7 still shows a general agreement amongst the five models in their representation of seasonal dynamics. This agreement also becomes apparent in Fig. 8, which presents the seasonal monthly climatology of evaporation over different biome types. Except for densely vegetated regions (see, e.g., Southern Hemisphere tropical forests), arctic regions, or arid regimes (see, e.g., Northern Hemisphere deserts), all models capture similar monthly dynamics. This occurs despite the systematic differences in the absolute magnitudes of evaporation, which again become apparent, especially between PM-MOD and ERA-Interim, and may indicate limitations in the way models represent the processes governing land evaporation. This highlights the importance

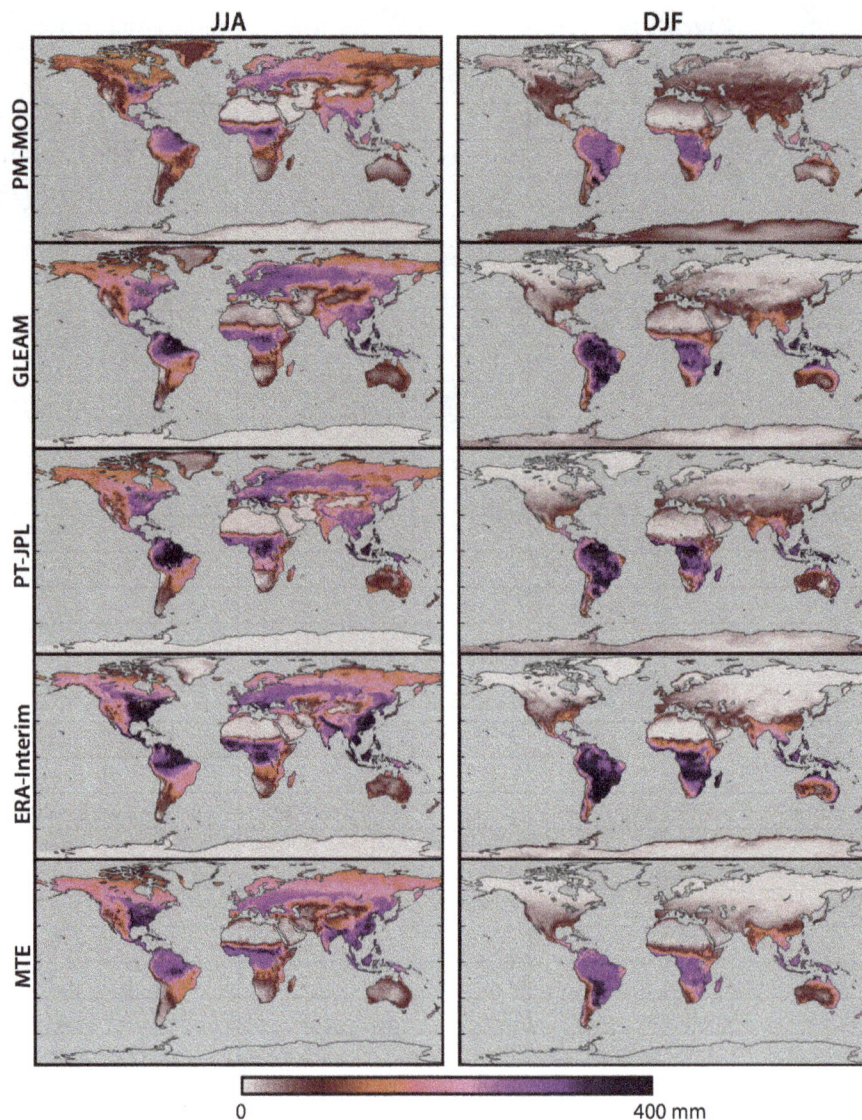

Figure 7. Mean seasonal differences. Average evaporation for PM-MOD, GLEAM, and PT-JPL during boreal summer (June, July, and August) and austral summer (December, January, and February). ERA-Interim reanalysis and MTE are considered for comparison. The 3 years of data (2005–2007) are used in the calculation of these seasonal averages.

of field-based validation activities to improve and select algorithms.

Since the seasonality of evaporation is mostly dominated by the annual cycle of irradiance in nature (especially in energy-limited regions), the skill of these models in correctly capturing these seasonal dynamics relies mostly on adequately representing the sensitivity of evaporation to the (common) net radiation forcing. However, if estimating average seasonal dynamics in evaporation may not appear overly challenging from the modelling perspective, accurately simulating anomalies (i.e. departures) relative to a seasonal expectation is far more problematic. With hydro-meteorological extremes – and particularly droughts – being a target application of these models, correctly reproducing the effect of sur-

face water deficits on evaporation (and vice versa) appears crucial. One of the most remarkable hydro-meteorological extremes that coincide with the WACMOS-ET period is the Australian Millennium Drought, which affected (especially) southeastern Australia and had one of its most severe years of rainfall deficits in 2006 (see van Dijk et al., 2013; Leblanc et al., 2012). Figure 9 (top panel) shows the daily time series of latent heat flux and net radiation for the Darling Basin (area contoured in Fig. 1) from the three WACMOS-ET models during 2005–2007; ERA-Interim is also included for comparison. Figure 9 (bottom panel) presents the monthly aggregates of land evaporation from these models and incorporates the estimates from MTE, precipitation from GPCC v6 (with

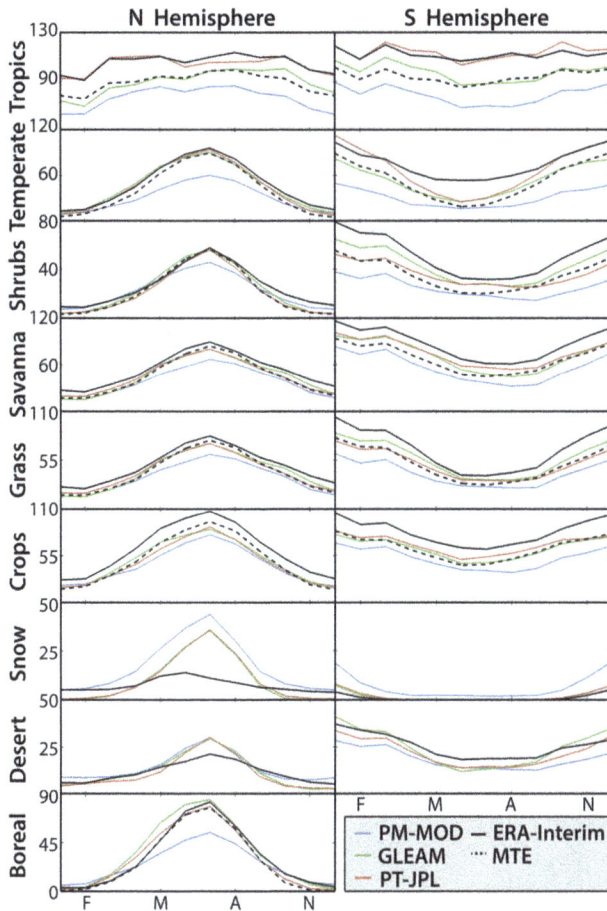

Figure 8. Average seasonal cycle. Monthly climatology of evaporation for each IGBP biome (see Fig. 1 for the global distribution of biomes) based on the 2005–2007 period. Northern Hemisphere (left panels) and Southern Hemisphere (right panels) are presented separately. In addition to the PM-MOD, GLEAM, and PT-JPL results, the evaporation from ERA-Interim and MTE is also shown for completeness. Fluxes are displayed in millimetres per month.

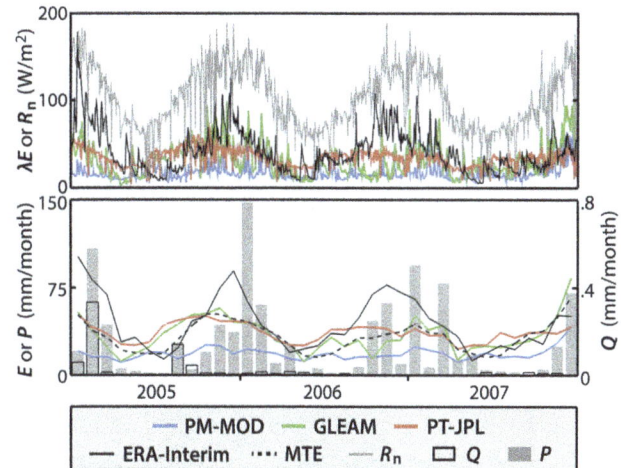

Figure 9. Evaporation during the Australian Millennium Drought. Top panel: daily time series of surface net radiation (SRB 3.1) and latent heat flux from the three WACMOS-ET products for the Darling Basin during 2005–2007. ERA-Interim latent heat flux is also illustrated for comparison. Bottom panel: monthly time series of evaporation, precipitation (GPCC v6 with gauge correction factors from Fuchs et al., 2001), and discharge (GRDC). The contributing area is shown in Fig. 1.

gauge correction factors from Fuchs et al., 2001), and river discharge data from GRDC.

Given the dominant rainfall scarcity, monthly run-off volumes are very low (note the difference of more than 2 orders of magnitude between the left and right axes in the bottom panel of Fig. 9); the river in fact dries out completely for prolonged periods. This indicates that almost the entire volume of incoming rainfall is evaporated. Therefore, cumulative evaporation should approximate cumulative precipitation over the multi-year period. We find, however, that in the case of all models, evaporation exceeds total rainfall, except for PM-MOD, in which evaporation is only 66 % of precipitation. In the case of MTE, the cumulative evaporation is 16 % higher than the precipitation, while it is 21 and 29 % higher for GLEAM and PT-JPL, respectively, and as much as 56 % higher for ERA-Interim. To some extent, this could reflect the progressive soil dry-out as the drought event evolves

(i.e. the negative change in soil storage in time), the use of irrigation, or the accessibility of groundwater for root uptake (see, e.g., Chen and Hu, 2004; Orellana et al., 2012). Nonetheless, there is a general tendency in all models to overestimate evaporation in drier catchments, as shown in the following section (Sect. 3.3). Once more, Fig. 9 shows that the estimates from the different products typically range between the low values of PM-MOD and high values of ERA-Interim and that there is a general agreement on the temporal dynamics between GLEAM, PT-JPL, and MTE. Nevertheless, there are clear differences in the timing of water stress and the rates of evaporation decline (see, e.g., summer 2006), and the inter-product disagreement on short temporal scales (Fig. 9) is considerably larger than the disagreement in mean seasonal cycles (Fig. 8).

3.3 Evaluation of evaporation based on the water balance closure

The skill of the different models to close the water budgets over 837 basins is investigated here. As explained in Sect. 2.3.2, these analyses consist of a comparison of modelled evaporation estimates from PM-MOD, GLEAM, and PT-JPL (forced by the reference input data set over 2005–2007) with estimates of $P-Q$. Such a comparison implies the validity of a series of assumptions (see discussion below), but overall, $P-Q$ estimates remain a valid, recursive means to evaluate long-term evaporation patterns (Liu et al., 2014; Miralles et al., 2011a; Vinukollu et al., 2011b; Sahoo et al., 2011). Here, different criteria have been applied to ensure the quality of the P–Q estimates, and the remaining catchments (Fig. 1) have been clustered into 30 different classes based on average precipitation and evaporative fraction (see Sect. 2.3.2).

The skill of the three WACMOS-ET models to reproduce the general climatic patterns of evaporation becomes apparent from the scatterplots in Fig. 10. All three WACMOS-ET products correlate well with the observations, which implies that their long-term spatial distribution of evaporation (Fig. 2) is, overall, realistic. The general negative bias of PM-MOD becomes discernible again when compared to the P–Q data, a finding which is in agreement with the results by Mu et al. (2013). In addition, there is a tendency in all models to underestimate evaporation in wet regions and overestimate it in dry regions; the latter was already suggested by Fig. 9. While this pattern could potentially be explained by systematic errors in P–Q, the same tendency has been found in Part 1 in comparisons with independent eddy-covariance towers. Once more, it is interesting to see how the independent evaporation data sets, i.e. ERA-Interim and MTE, perform in this comparison; both products correlate well with the P–Q estimates, although the overall higher values of ERA-Interim (and lower of MTE) are again highlighted, together with the tendency to overestimate evaporation in dry catchments and underestimate it in wet ones, which is shared by all five data sets.

As mentioned above, the use of P–Q as a benchmark for evaporation depends on the validity of several assumptions. First, the catchment needs to be watertight (no subsurface leakage to other catchments) and its geographical boundaries must be well defined. Second, the entire volume of river water that is extracted for direct human use must return to the river, and it should do so upstream of the staff gauge location. Third, the lag time between rainfall events and the discharge measured at the station can be neglected when compared to the total period of study. Finally, the changes in soil water storage within the catchment should be insignificant compared to the cumulative volume of the three main hydrological fluxes. Here, by considering long-term averages of P–Q, these assumptions appear to be reasonable for most continental regions. However, for industrialized areas with a dense population, the consumption and export of water and the human regulation of the reservoir storages may compromise these assumptions. Nonetheless, the largest sources of uncertainty regarding the use of P–Q as an estimate of catchment evaporation likely come from (a) the definition of the run-off-contributing area and (b) errors in precipitation and discharge observations. In fact, Fig. 10 shows that the choice of precipitation product can have a significant influence on the results, even despite the existing interdependencies between the gauge-based precipitation data sets tested here (Sect. 2.3.2). On the other hand, uncertainties in observations of river run-off can also be significant and come from errors in the measurements of water height, the discharge data used to calibrate the rating curves, or the interpolation

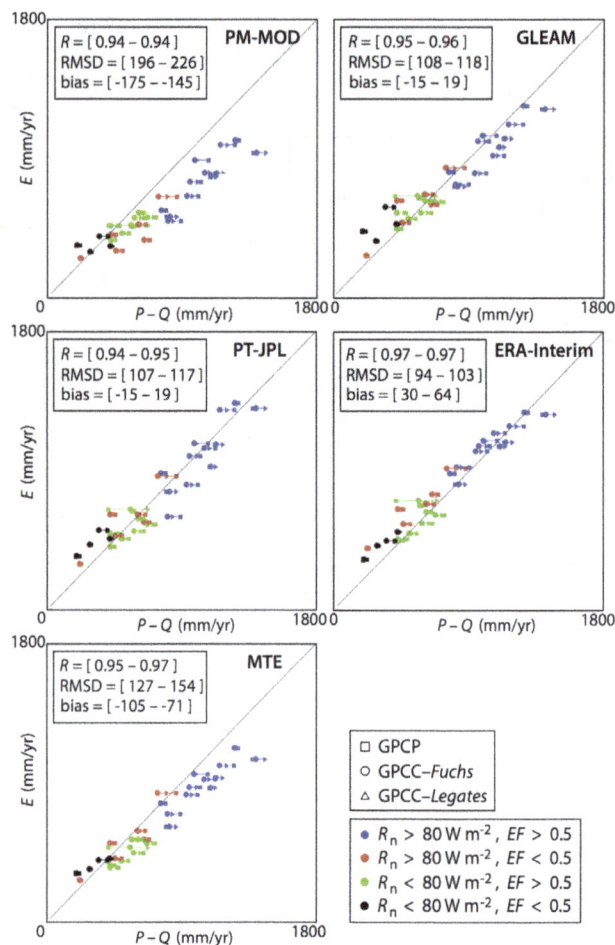

Figure 10. Skill to close catchment water budgets. Correlations between the long-term averages in evaporation from the three WACMOS-ET models and P–Q estimates based on observations from 837 catchments. ERA-Interim and MTE are added for the sake of completeness. Three different precipitation products are considered in the calculation of P–Q: GPCP, GPCC v6 with gauge correction factors from Fuchs et al. (2001), and GPCC v6 with gauge correction factors from Legates and Willmott (1990). The corresponding validation statistics are noted within the scatterplots, and the range displayed for each statistical inference derives from the use of the three different precipitation products.

and extrapolation due to changes in riverbed roughness, hysteresis effects, etc. (see, e.g., Di Baldassarre and Montanari, 2009). Finally, it is important to note that model estimates correspond to the period 2005–2007, while P–Q estimates do not necessarily span the entire period due to limitations in the availability of discharge data.

Additionally, the fit of the models to a Budyko curve (Budyko, 1974) is explored in Fig. 11 as a general diagnostic for the robustness of mean evaporation estimates and their consistency with the input of water and energy. Potential evaporation estimates are taken from the corresponding models, and precipitation is taken from the GPCC v6 prod-

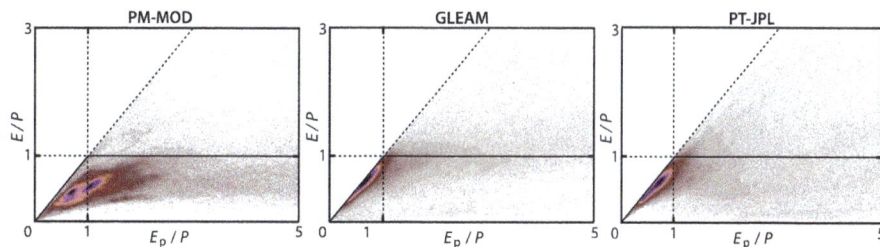

Figure 11. Budyko diagrams for the different models. Budyko curves derived for PM-MOD, GLEAM, and PT-JPL. Each point represents a different land grid cell. The horizontal axis presents the ratio of potential evaporation to precipitation (E_p/P) and the vertical axis presents the ratio of evaporation to precipitation (E/P). Actual and potential evaporation estimates are derived by each of the models, while precipitation comes from GPCC v6 with gauge correction factors from Fuchs et al. (2001). Each land pixel is an independent scatter point.

uct with gauge correction factors from Fuchs et al. (2001), to be consistent with Figs. 9 and 10. Overall, results are in agreement with the water balance scatterplots (Fig. 10). The fraction of precipitation that is evaporated (E/P) is usually lower for PM-MOD; however, this does not happen due to an underestimation of the atmospheric demand for water, as the values of the ratio of potential evaporation over precipitation (E_p/P) are overall comparable to those from GLEAM and PT-JPL. The PM-MOD product, therefore, has a general tendency to overestimate the surface evaporative stress (i.e. underestimate the ratio of E over E_p), which may explain the overall lower estimates of evaporation found across our analyses. GLEAM and PT-JPL show a better fit to the Budyko diagram and a transition from arid to wet climates that is consistent with the average fluxes of precipitation and net radiation. Nevertheless, it is worth noting that all three models estimate average values of evaporation that overcome average precipitation in numerous areas.

3.4 Partitioning of evaporation into separate components

The flux of land evaporation results from the summation of three main components or sources: (a) transpiration (the process that describes the movement of water from the soil, through the plant xylem, to the leaf and finally to the atmosphere), (b) interception loss (the vaporization of the volume of water that is held by the surface of vegetation during rainfall), and (c) soil evaporation (the direct vaporization of water from the topsoil). These processes require separate consideration in models due to their differences in biophysical drivers and rates (Savenije, 2004; Dolman et al., 2014). In addition, two other contributors to evaporation are often considered separately: the direct evaporation (sublimation) from snow- and ice-covered surfaces and the vaporization from continental water bodies (or open-water evaporation).

Transpiration is the component that has received the most attention from the scientific community in recent years, due to its connection to different biogeochemical cycles. The global contribution of transpiration to total average evaporation has been extensively debated recently (Schlesinger and

Jasechko, 2014; Wang et al., 2014). Studies have reported values ranging between 35 and 90 %, based on isotopes (Jasechko et al., 2013; Coenders-Gerrits et al., 2015), sapflow measurements (Moran et al., 2009), satellite data (Miralles et al., 2011a; Mu et al., 2011; Zhang et al., 2016), and modelling (Wang-Erlandsson et al., 2014). Consequently, this large range of uncertainty is also expected in the relative contribution from other evaporation sources. Moreover, reducing this uncertainty appears particularly challenging due to the limited amount of ground data that can be used for validation and the nature of the techniques used to measure latent heat flux: most measuring devices (e.g. lysimeters, eddy-covariance instruments, scintillometers) cannot distinguish between the different sources of evaporation.

All three WACMOS-ET models estimate the components of evaporation separately. In the case of PT-JPL and PM-MOD, the available energy is partitioned into the different land covers to estimate the contribution from each of them. The approach in GLEAM is somewhat different, as the flux of interception loss is calculated using a different algorithm than the one used for transpiration and soil evaporation. Figure 12 illustrates the average contribution of each evaporation component to the total flux as estimated by the WACMOS-ET models. In the case of GLEAM (which calculates sublimation separately), the flux from snow and ice has been added to the bare-soil evaporation in this figure to allow visual comparison to the other two products.

The discrepancy amongst modelled evaporation components shown in Fig. 12 is large and calls for a thorough validation of the way the contribution from different sources is estimated as well as perhaps an in-depth revision to ensure that the conceptual definition of these components is consistent from model to model. Regionally, disagreements are particularly large in transitional regimes; for instance, in the climatic gradient from the Congo rainforest to the savanna, the virtual total of the flux comes from transpiration in the case of GLEAM, while for PM-MOD direct soil evaporation is the dominant component. In tropical forests, the direct soil evaporation can also exceed transpiration in the case of PM-MOD, while for GLEAM and PT-JPL bare-soil evaporation

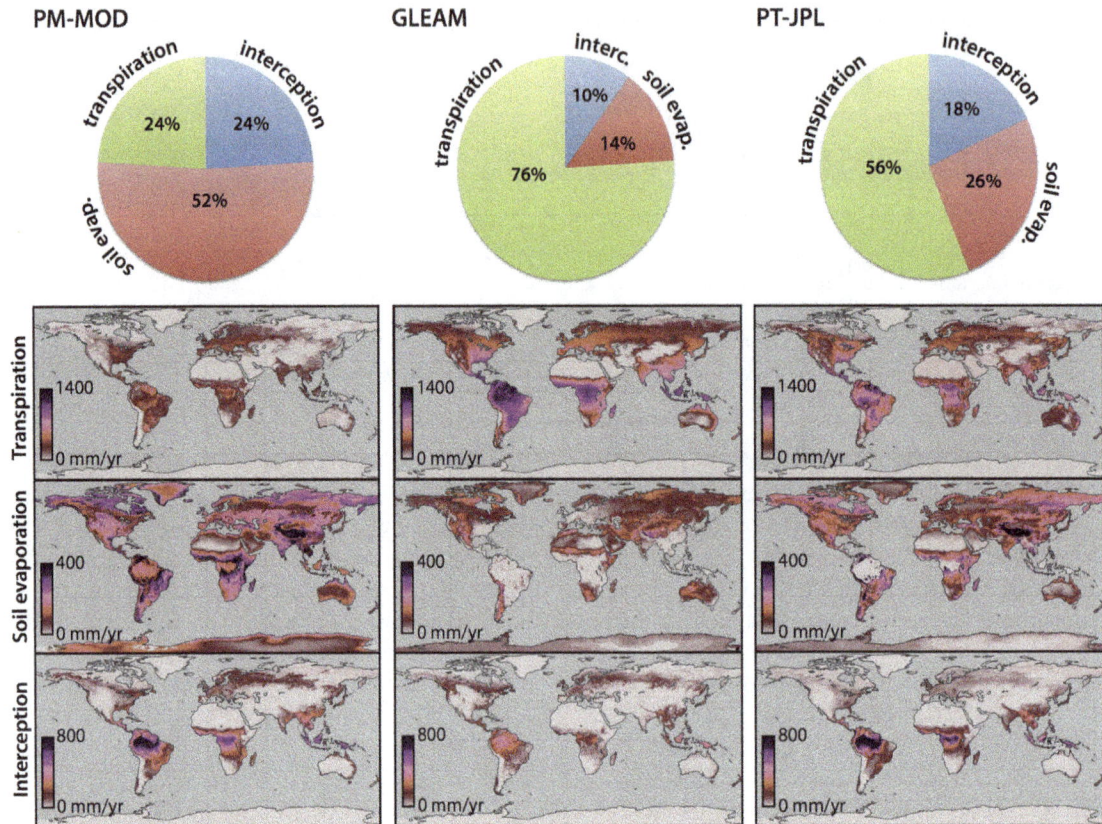

Figure 12. Partitioning evaporation. Maps indicate the average (2005–2007) transpiration, interception loss, and bare-soil evaporation for each of the three WACMOS-ET models. Pie diagrams illustrate the global average contribution to total land evaporation from each component and product.

is almost inexistent. The mean inter-model disagreement is manifest in the pie diagrams in Fig. 12, with GLEAM estimating a large contribution from transpiration (76 %) and a low contribution from soil evaporation (14 %), PM-MOD estimating little transpiration (24 %) and a large contribution from soil evaporation (52 %), and both PM-MOD and PT-JPL yielding a much larger flux of interception loss than GLEAM. Nevertheless, and as discussed above, recent reviews have revealed comparable levels of uncertainty in this partitioning based on a wide range of independent methods (see, e.g., Schlesinger and Jasechko, 2014; Wang et al., 2014).

While the global contribution of transpiration has received much attention in the literature (Jasechko et al., 2013; Coenders-Gerrits et al., 2015), the flux of interception loss has seldom been explored globally (Miralles et al., 2010; Vinukollu et al., 2011b; Wang-Erlandsson et al., 2014). The physical process of interception loss differs from that of transpiration on its sensitivity to environmental and climatic variables: the rates and magnitude of interception are dictated by the aerodynamic properties of the vegetation stand and the occurrence and characteristics of rainfall (Horton, 1919). In fact, while solar radiation is usually the main supply of energy for transpiration and soil evaporation (Wild and Liepert, 2010), the source of energy powering interception loss is still

debated (Holwerda et al., 2011; van Dijk et al., 2015). This limited process understanding, together with the scarcity of ground measurements for validation, makes interception loss particularly challenging to model. Nonetheless, interception has often been reported in units of percentage of incoming rainfall during the restricted number of past in situ measuring campaigns; see, e.g., Miralles et al. (2010) for a nonexhaustive list of these campaigns. This makes interception measurements easy to extrapolate in time and space, and it allows for a relatively straightforward validation of the estimates from our three models. Therefore, Fig. 13 presents the daily time series of interception loss from PM-MOD, GLEAM, and PT-JPL for the average of the Amazon Basin (blue contour in Fig. 1), and it indicates the values reported by past campaigns in Amazonia. According to in situ measurements, there is a more than 2-fold overestimation of the mean flux in the case of PM-MOD and PT-JPL. Temporal dynamics of interception loss from the three products do not correlate well either, as GLEAM tends to follow the occurrence of rainfall, while PM-MOD and PT-JPL are more affected by net radiation variability, as expected from the interception algorithms (i.e. Gash's model for GLEAM, Penman–Monteith for PM-MOD, and Priestley and Taylor for PT-JPL).

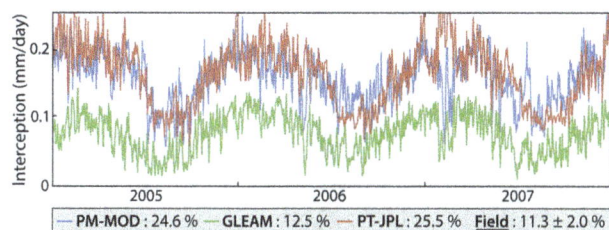

Figure 13. Interception loss in Amazonia. Daily time series of interception ($mm\,day^{-1}$) for 2005–2007 from the three WACMOS-ET products as averaged for the entire Amazon Basin. The average interception (as percentage of rainfall) from the three models is listed, together with the mean ($\pm 1\,SD$ – 1 standard deviation) of past field campaigns by Lloyd et al. (1988) ($\pm 1\,SD$), Czikowsky and Fitzjarrald (2009) (11.6 %), Ubarana (1996) (11.6 %), Cuartas et al. (2007) (13.3 %), Marin et al. (2000) (13.5 %), and Shuttleworth (1988) (9.1 %). See Fig. 1 for the Amazon catchment boundaries and the locations of the field measurements.

Further analyses are needed to explore the skill of these (and other) models to separately derive the different evaporation components or sources. Nevertheless, these preliminary analyses indicate the need for caution when using global estimates of transpiration, soil evaporation, or interception loss from a single model in isolation, as the disagreements can be much larger than for total land evaporation. To date, the lack of in situ networks that measure the components of evaporation independently remains an unsolved problem for the improvement of model estimates.

4 Conclusions

The ESA WACMOS-ET project started in 2012 with the goal of performing a cross comparison and validation exercise of a group of selected global observational evaporation algorithms driven by a consistent set of forcing data. With the project coming to an end, this article has focussed on the global and regional evaluation of the resulting evaporation data sets.

The three main models scrutinized here were the Penman–Monteith approach from the official MODIS evaporation product (Mu et al., 2007, 2011, 2013), GLEAM (Miralles et al., 2011b; Martens et al., 2016), and the Priestley–Taylor JPL (Fisher et al., 2008); the SEBS model (Su, 2002), which was analysed on the local scale in Part 1 (revealing good performance in terms of correlations but a systematic overestimation of evaporation), was not evaluated in this contribution. The spatiotemporal magnitude and variability of the resulting global evaporation products were compared to analogous estimates from reanalysis (ERA-Interim) and eddy-covariance-based global data (MTE). The representation of evaporation dynamics during droughts, the model skill to close the water balance over 837 river basins world-

wide, and the partitioning of evaporation into different components have also been explored.

Despite our efforts to create a homogeneous forcing data set to run the evaporation models, the input requirements of each model are different, which implies that the resulting inter-product disagreements are the result of both internal differences in the models and uncertainties in forcing and ancillary data. This prevents us from making strong claims about the quality of the models. However, these analyses also provide the following take-home messages:

– In agreement with the local-scale validation in Part 1, the PM-MOD product tends to underestimate evaporation (see, e.g., Figs. 3 and 10). This underestimation is systematic, being larger in absolute terms in the tropics (where evaporation is larger) and larger in relative terms in drier subtropical regions (Fig. 3). As an exception, in high latitudes PM-MOD estimates are greater than those from GLEAM and PT-JPL; this may reflect known deficiencies in Priestley–Taylor-based approaches in conditions of low available energy (see, e.g., Parlange and Katul, 1992).

– The global average magnitudes of evaporation from GLEAM and PT-JPL agree well with each other and with the range of literature values (see Figs. 2 and 4). This agreement extends to the average latitudinal patterns, which lie between those of PM-MOD and ERA-Interim (Figs. 2 and 3). In terms of temporal dynamics, there are differences between GLEAM and PT-JPL in dry conditions, as expected from their distinctive approach to representing evaporative stress (see Sect. 2.1). These differences are pronounced in the Southern Hemisphere subtropics (Fig. 6a), reflected more clearly in daily anomalies than in seasonal cycles (Fig. 8), and may be exacerbated during specific drought events (Fig. 9).

– The partitioning of evaporation into different components is a facet of these models that has not received enough attention in previous applications. Each model has a distinct way to estimate these components, and even in cases in which inter-product average evaporation agrees, the separate contribution from these components may fluctuate substantially (Fig. 12). As an example, differences in interception loss amongst models (Fig. 13) may explain a large part of the disagreements in the seasonality of evaporation over tropical forests (Fig. 8). Further exploring the skill of these models at partitioning evaporation into its different sources remains a critical task for the future. This is outside the scope of WACMOS-ET, and it would require innovative means of validation beyond traditional comparisons to eddy-covariance and lysimeter data.

– On a more positive note, the analysis of the skill of different models to close the water balance over particular catchments reveals that the general climatic patterns of evaporation are well captured by all models (Fig. 10). While this comparison has also unveiled the general underestimation by PM-MOD (and overestimation by ERA-Interim), all products correlate well with the cumulative values of $P-Q$. We stress, however, that this agreement does not indicate whether the multi-scale temporal dynamics of evaporation are well captured. For a thorough validation of evaporation temporal variability, we direct the readers to Part 1.

In summary, the activities in WACMOS-ET have demonstrated that some of the existing evaporation models require an in-depth scrutiny to correct for systematic errors in their estimates. This is especially the case over semi-arid regions and tropical forests. In addition, even models that have demonstrated a more robust performance, like GLEAM and PT-JPL, may differ substantially from one another given certain biomes and climates. Overall, our results imply the need for caution in using a single model for any large-scale application in isolation, especially in studies in which transpiration, soil evaporation, or interception loss are investigated separately.

As remote sensing science continues to advance, new long-term records of physical variables to constrain these models are becoming available (e.g. chlorophyll fluorescence, surface soil moisture). While further tools to improve evaporation models become accessible, the possibility of considering biome- or climate-specific composites of flux algorithms is currently being explored, given the general finding that different models may perform better under certain conditions (Ershadi et al., 2014; McCabe et al., 2016). For an inter-product merger to add new skill, the sensitivity of each model to its forcing should be further explored, and a robust propagation of uncertainties appears essential to merge these products efficiently.

The reader is directed to additional supporting documents available form the project website at http://WACMOS-ET.estellus.eu.

Author contributions. D. G. Miralles, C. Jiménez, M. Jung, D. Michel, A. Ershadi, M. F. McCabe, M. Hirschi, and D. Fernández-Prieto designed the content of the manuscript. D. G. Miralles, C. Jiménez, and M. Jung did the analyses. D. G. Miralles wrote the paper. J. B. Fisher and Q. Mu provided the computer codes of the PT-JPL and PM-MOD models, respectively. All authors contributed to carrying out the project and to the discussion and interpretation of results.

Acknowledgements. This work was undertaken as part of the European Space Agency (ESA) project WACMOS-ET (Contract No. 4000106711/12/I-NB). Discharge data were provided by the Global Runoff Data Centre, 56068 Koblenz, Germany. We thank Ulrich Weber and Eric Thomas for processing the catchment data. D. G. Miralles acknowledges the financial support from The Netherlands Organization for Scientific Research through grant 863.14.004, and the Belgian Science Policy Office (BELSPO) in the framework of the STEREO III programme, project SAT-EX (SR/00/306). A. Ershadi and M. F. McCabe acknowledge funding from the King Abdullah University of Science and Technology. J. B. Fisher acknowledges funding under the NASA Terrestrial Hydrology Program.

Edited by: P. Gentine

References

Baldocchi, D., Falge, E., Gu, L., Olson, R., Hollinger, D., Running, S., Anthoni, P., Bernhofer, C., Davis, K., Evans, R., Fuentes, J., Goldstein, A., Katul, G., Law, B., Lee, X., Malhi, Y., Meyers, T., Munger, W., Oechel, W., Paw, K. T., Pilegaard, K., Schmid, H. P., Valentini, R., Verma, S., Vesala, T., Wilson, K., and Wofsy, S.: FLUXNET: A new tool to study the temporal and spatial variability of ecosystem-scale carbon dioxide, water vapor, and energy flux densities, B. Am. Meteorol. Soc., 82, 2415–2434, 2001.

Baumgartner, A. and Reichel, E.: The World Water Balance: Mean Annual Global Continental and Maritime Precipitation, Evaporation and Runoff, Elsevier Scientific Publishing Company, Amsterdam, the Netherlands, Oxford, UK, New York, USA, 1975.

Budyko, M. I.: Climate and life, International Geophysics Series, Academic Press, New York, 1974.

Chen, M., Shi, W., Xie, P., Silva, V. B. S., Kousky, V. E., Wayne Higgins, R., and Janowiak, J. E.: Assessing objective techniques for gauge-based analyses of global daily precipitation, J. Geophys. Res., 113, D04110, doi:10.1029/2007JD009132, 2008.

Chen, X. and Hu, Q.: Groundwater influences on soil moisture and surface evaporation, J. Hydrol., 297, 285–300, doi:10.1016/j.jhydrol.2004.04.019, 2004.

Chen, Y., Xia, J., Liang, S., Feng, J., Fisher, J. B., Li, X., Li, X., Liu, S., Ma, Z., Miyata, A., Mu, Q., Sun, L., Tang, J., Wang, K., Wen, J., Xue, Y., Yu, G., Zha, T., Zhang, L., Zhang, Q., Zhao, T., Zhao, L., Zhou, G., and Yuan, W.: Comparison of satellite-based evapotranspiration models over terrestrial ecosystems in China, Remote Sens. Environ., 140, 279–293, 2014.

Cleugh, H. A., Leuning, R., Mu, Q., and Running, S. W.: Regional evaporation estimates from flux tower and MODIS satellite data, Remote Sens. Environ., 106, 285–304, doi:10.1016/j.rse.2006.07.007, 2007.

Coccia, G., Siemann, A., Pan, M., and Wood, E. F.: Creating consistent datasets by combining remotely-sensed data and land surface model estimates through Bayesian uncertainty post-processing: the case of Land Surface Temperature from HIRS, Remote Sens. Environ., 170, 290–305, doi:10.1016/j.rse.2015.09.010, 2015.

Coenders-Gerrits, A. M. J., van der Ent, R. J., Bogaard, T. A., Wang-Erlandsson, L., Hrachowitz, M., and Savenije, H. H. G.: Uncertainties in transpiration estimates, Nature, 506, E1–E2, doi:10.1038/nature12925, 2015.

Cuartas, L., Tomasella, J., Nobre, A., Hodnett, M., Waterloo, M., and Múnera, J.: Interception water-partitioning dynamics for a pristine rainforest in Central Amazonia: marked differences between normal and dry years, Agr. Forest Meteorol., 145, 69–83, 2007.

Czikowsky, M. and Fitzjarrald, D.: Detecting rainfall interception in an Amazonian rain forest with eddy flux measurements, J. Hydrol., 377, 92–105, 2009.

Dalton, J.: On evaporation, Essay III, in: Experimental essays on the 121 constitution of mixed gases; on the force of steam or vapour from water or other liquids in different temperatures; both in a Torrecellian vacuum and in air; on evaporation; and on the expansion of gases by heat, Mem. Proc. Lit. Phil. Soc. Manchester, 5, 574–594, 1802.

Dee, D. P., Uppala, S. M., Simmons, A. J., Berrisford, P., Poli, P., Kobayashi, S., Andrae, U., Balmaseda, M. A., Balsamo, G., Bauer, P., Bechtold, P., Beljaars, A. C. M., van de Berg, L., Bidlot, J., Bormann, N., Delsol, C., Dragani, R., Fuentes, M., Geer, A. J., Haimberger, L., Healy, S. B., Hersbach, H., Hólm, E. V., Isaksen, L., Kållberg, P., Köhler, M., Matricardi, M., McNally, A. P., Monge-Sanz, B. M., Morcrette, J. J., Park, B. K., Peubey, C., De Rosnay, P., Tavolato, C., Thépaut, J. N., and Vitart, F.: The ERA-Interim reanalysis: configuration and performance of the data assimilation system, Q. J. Roy. Meteorol. Soc., 137, 553–597, 2011.

Di Baldassarre, G. and Montanari, A.: Uncertainty in river discharge observations: a quantitative analysis, Hydrol. Earth Syst. Sci., 13, 913–921, doi:10.5194/hess-13-913-2009, 2009.

Dolman, A. J., Miralles, D. G., and De Jeu, R. A. M.: Fifty years since Monteith's 1965 seminal paper: the emergence of global ecohydrology, Ecohydrology, 7, 897–902, doi:10.1002/eco.1505, 2014.

Douville, H., Ribes, A., Decharme, B., Alkama, R., and Sheffield, J.: Anthropogenic influence on multidecadal changes in reconstructed global evapotranspiration, Nat. Clim. Change, 3, 59–62, 2013.

Ershadi, A., McCabe, M. F., Evans, J. P., Chaney, N. W., and Wood, E. F.: Multi-site evaluation of terrestrial evaporation models using FLUXNET data, Agr. Forest Meteorol., 187, 46–61, doi:10.1016/j.agrformet.2013.11.008, 2014.

Fisher, J. B., Tu, K. P., and Baldocchi, D. D.: Global estimates of the land-atmosphere water flux based on monthly AVHRR and ISLSCP-II data, validated at 16 FLUXNET sites, Remote Sens. Environ., 112, 901–919, 2008.

Fuchs, T., Rapp, J., Rubel, F., and Rudolf, B.: Correction of synoptic precipitation observations due to systematic measuring errors with special regard to precipitation phases, Phys. Chem. Earth, 26, 689–693, 2001.

Gash, J. H.: An analytical model of rainfall interception by forests, Q. J. Roy. Meteorol. Soc., 105, 43–45, 1979.

Guillod, B. P., Orlowsky, B., Miralles, D. G., Teuling, A. J., and Seneviratne, S. I.: Reconciling spatial and temporal soil moisture effects on afternoon rainfall, Nat. Commun., 6, 1–6, doi:10.1038/ncomms7443, 2015.

Hagemann, S., Chen, C., Clark, D. B., Folwell, S., Gosling, S. N., Haddeland, I., Hanasaki, N., Heinke, J., Ludwig, F., Voss, F., and Wiltshire, A. J.: Climate change impact on available water resources obtained using multiple global climate and hydrology models, Earth Syst. Dynam., 4, 129–144, doi:10.5194/esd-4-129-2013, 2013.

Holwerda, F., Bruijnzeel, L. A., Scatena, F. N., Vugts, H. F., and Meesters, A. G. C. A.: Wet canopy evaporation from a Puerto Rican lower montane rain forest: The importance of realistically estimated aerodynamic conductance, J. Hydrol., 414–415, 1–15, doi:10.1016/j.jhydrol.2011.07.033, 2011.

Horton, R. E.: Rainfall interception, Mon. Weather Rev., 47, 603–625, 1919.

Huffman, G. J., Adler, R. F., Morrissey, M., Bolvin, D. T., Curtis, S., Joyce, R., McGavock, B., and Susskind, J.: Global precipitation at one-degree daily resolution from multi-satellite observations, J. Hydrometeorol., 2, 36–50, 2001.

Jarvis, P. G.: The interpretation of the variations in leaf water potential and stomatal conductance found in canopies in the field, Philos. T. Roy. Soc. Lond., 273, 593–610, 1976.

Jasechko, S., Sharp, Z. D., Gibson, J. J., Birks, S. J., Yi, Y., and Fawcett, P. J.: Terrestrial water fluxes dominated by transpiration, Nature, 496, 347–350, 2013.

Jiménez, C., Prigent, C., Mueller, B., Seneviratne, S. I., McCabe, M. F., Wood, E. F., Rossow, W. B., Balsamo, G., Betts, A. K., Dirmeyer, P. A., Fisher, J. B., Jung, M., Kanamitsu, M., Reichle, R. H., Reichstein, M., Rodell, M., Sheffield, J., Tu, K., and Wang, K.: Global intercomparison of 12 land surface heat flux estimates, J. Geophys. Res., 116, D02102, doi:10.1029/2010JD014545, 2011.

Jung, M., Reichstein, M., and Bondeau, A.: Towards global empirical upscaling of FLUXNET eddy covariance observations: validation of a model tree ensemble approach using a biosphere model, Biogeosciences, 6, 2001–2013, doi:10.5194/bg-6-2001-2009, 2009.

Jung, M., Reichstein, M., Ciais, P., Seneviratne, S. I., Sheffield, J., Goulden, M. L., Bonan, G., Cescatti, A., Chen, J., and de Jeu, R.: Recent decline in the global land evapotranspiration trend due to limited moisture supply, Nature, 467, 951–954, 2010.

Kalma, J., McVicar, T., and McCabe, M.: Estimating Land Surface Evaporation: A Review of Methods Using Remotely Sensed Surface Temperature Data, Surv. Geophys., 29, 421–469, 2008.

Kelly, R. E., Chang, A. T., Tsang, L., and Foster, J. L.: A prototype AMSR-E global snow area and snow depth algorithm, IEEE T. Geosci. Remote, 41, 230–242, 2003.

Koster, R. D., Sud, Y., Guo, Z., Dirmeyer, P. A., Bonan, G., Oleson, K. W., Chan, E., Verseghy, D., Cox, P., and Davies, H.: GLACE: the global land-atmosphere coupling experiment. Part I: overview, J. Hydrometeorol., 7, 590–610, 2006.

Leblanc, M., Tweed, S., Van Dijk, A., and Timbal, B.: A review of historic and future hydrological changes in the Murray-Darling Basin, Global Planet. Change, 80–81, 226–246, doi:10.1016/j.gloplacha.2011.10.012, 2012.

Legates, D. R. and Willmott, C. J.: Mean seasonal and spatial variability in gauge-corrected, global precipitation, Int. J. Climatol., 10, 111–127, doi:10.1002/joc.3370100202, 1990.

Liu, Y., Zhuang, Q., Pan, Z., Miralles, D., Tchebakova, N., Kicklighter, D., Chen, J., Sirin, A., He, Y., Zhou, G., and Melillo, J.: Response of evapotranspiration and water availability to the changing climate in Northern Eurasia, Climatic Change, 126, 413–427, doi:10.1007/s10584-014-1234-9, 2014.

Liu, Y. Y., de Jeu, R. A. M., McCabe, M. F., Evans, J. P., and van Dijk, A. I. J. M.: Global long-term passive microwave satellite-based retrievals of vegetation optical depth, Geophys. Res. Lett., 38, L18402, doi:10.1029/2011GL048684, 2011.

Liu, Y. Y., Dorigo, W. A., Parinussa, R. M., De Jeu, R. A. M., Wagner, W., Mccabe, M. F., Evans, J. P., and Van Dijk, A. I. J. M.:

Trend-preserving blending of passive and active microwave soil moisture retrievals, Remote Sens. Environ., 123, 1–18, 2012.

Lloyd, C. R., Gash, J. H. C., Shuttleworth, W. J. and de Marques, F. A. O.: The measurement and modelling of rainfall interception by Amazonian rain forest, Agr. Forest Meteorol., 43, 277–294, 1988.

Luojus, K. and Pulliainen, J.: Global snow monitoring for climate research: Snow Water Equivalent (SWE) product guide, Helsinki, Finland, http://www.globsnow.info/swe/GlobSnow_SWE_product_readme_v1.0a.pdf (last access: February 2016), 2010.

Marin, C., Bouten, W., and Sevink, J.: Gross rainfall and its partitioning into throughfall, stemflow and evaporation of intercepted water in four forest ecosystems in western Amazonia, J. Hydrol., 237, 40–57, 2000.

Martens, B., Miralles, D. G., Verhoest, N. E. C., Lievens, H., and Fernandez-Prieto, D.: Improving terrestrial evaporation estimates over continental Australia through assimilation of SMOS soil moisture, Int. J. Appl. Earth Obs., doi:10.1016/j.jag.2015.09.012, in press, 2016.

McCabe, M. F., Miralles, D. G., Jiménez, C., Ershadi, A., Fisher, J. B., Mu, Q., Liang, M., Mueller, B., Sheffield, J., Seneviratne, S. I., and Wood, E. F.: Global scale estimation of land surface heat fluxes from space: product assessment and intercomparison, in: Remote Sensing of Energy fluxes and Soil Moisture Content, edited by: Petropoulos, G., Taylor and Francis, CRC Press, Boca Raton, FL, 249–282, doi:10.1201/b15610-13, 2013.

McCabe, M. F., Ershadi, A., Jiménez, C., Miralles, D. G., Michel, D., and Wood, E. F.: The GEWEX LandFlux project: evaluation of model evaporation using tower-based and globally-gridded forcing data, Geosci. Model Dev., 9, 283–305, doi:10.5194/gmd-9-283-2016, 2016.

Michel, D., Jiménez, C., Miralles, D. G., Jung, M., Hirschi, M., Ershadi, A., Martens, B., McCabe, M. F., Fisher, J. B., Mu, Q., Seneviratne, S. I., Wood, E. F., and Fernández-Prieto, D.: The WACMOS-ET project – Part 1: Tower-scale evaluation of four remote sensing-based evapotranspiration algorithms, Hydrol. Earth Syst. Sci., 20, 803–822, doi:10.5194/hess-20-803-2016, 2016.

Miralles, D. G., Gash, J. H., Holmes, T. R. H., de Jeu, R. A. M., and Dolman, A.: Global canopy interception from satellite observations, J. Geophys. Res., 115, D16122, doi:10.1029/2009JD013530, 2010.

Miralles, D. G., De Jeu, R. A. M., Gash, J. H., Holmes, T. R. H., and Dolman, A. J.: Magnitude and variability of land evaporation and its components at the global scale, Hydrol. Earth Syst. Sci., 15, 967–981, doi:10.5194/hess-15-967-2011, 2011a.

Miralles, D. G., Holmes, T. R. H., De Jeu, R. A. M., Gash, J. H., Meesters, A. G. C. A., and Dolman, A. J.: Global land-surface evaporation estimated from satellite-based observations, Hydrol. Earth Syst. Sci., 15, 453–469, doi:10.5194/hess-15-453-2011, 2011b.

Miralles, D. G., Teuling, A. J., van Heerwaarden, C. C., and Vilà-Guerau de Arellano, J.: Mega-heatwave temperatures due to combined soil desiccation and atmospheric heat accumulation, Nat. Geosci., 7, 345–349, doi:10.1038/ngeo2141, 2014a.

Miralles, D. G., van den Berg, M. J., Gash, J. H., Parinussa, R. M., De Jeu, R. A. M., Beck, H. E., Holmes, T. R. H., Jiménez, C., Verhoest, N. E. C., Dorigo, W. A., Teuling, A. J., and Dolman, A. J.: El Niño–La Niña cycle and recent trends in continental evaporation, Nat. Clim. Change, 4, 122–126, 2014b.

Monteith, J. L.: Evaporation and environment, Symposia of the Society for Exp. Biol., 19, 205–234, 1965.

Moran, M. S., Scott, R. L., Keefer, T. O., and Emmerich, W. E.: Partitioning evapotranspiration in semiarid grassland and shrubland ecosystems using time series of soil surface temperature, Agr. Forest Meteorol., 149, 59–72, doi:10.1016/j.agrformet.2008.07.004, 2009.

Mu, Q., Heinsch, F. A., Zhao, M., and Running, S. W.: Development of a global evapotranspiration algorithm based on MODIS and global meteorology data, Remote Sens. Environ., 111, 519–536, doi:10.1016/j.rse.2007.04.015, 2007.

Mu, Q., Zhao, M., and Running, S. W.: Improvements to a MODIS global terrestrial evapotranspiration algorithm, Remote Sens. Environ., 115, 1781–1800, 2011.

Mu, Q., Zhao, M., and Running, S. W.: MODIS Global Terrestrial Evapotranspiration (ET) Product (NASA MOD16A2/A3), Algorithm Theoretical Basis Document, Collection 5, NASA HQ, Numerical Terradynamic Simulation Group, University of Montana, Missoula, MT, USA, 20 November 2013.

Mueller, B., Seneviratne, S. I., Jiménez, C., Corti, T., Hirschi, M., Balsamo, G., Ciais, P., Dirmeyer, P., Fisher, J. B., Guo, Z., Jung, M., Maignan, F., McCabe, M. F., Reichle, R., Reichstein, M., Rodell, M., Sheffield, J., Teuling, A. J., Wang, K., Wood, E. F., and Zhang, Y.: Evaluation of global observations-based evapotranspiration datasets and IPCC AR4 simulations, Geophys. Res. Lett., 38, L06402, doi:10.1029/2010GL046230, 2011.

Mueller, B., Hirschi, M., Jiménez, C., Ciais, P., Dirmeyer, P. A., Dolman, A. J., Fisher, J. B., Jung, M., Ludwig, F., Maignan, F., Miralles, D., McCabe, M. F., Reichstein, M., Sheffield, J., Wang, K. C., Wood, E. F., Zhang, Y., and Seneviratne, S. I.: Benchmark products for land evapotranspiration: LandFlux-EVAL multi-dataset synthesis, Hydrol. Earth Syst. Sci., 17, 3707–3720, doi:10.5194/hess-17-3707-2013, 2013.

Murphy, D. and Koop, T.: Review of the vapour pressures of ice and supercooled water for atmospheric applications, Q. J. Roy. Meteorol. Soc., 131, 1539–1565, 2005.

Oki, T. and Kanae, S.: Global hydrological cycles and world water resources, Science, 313, 1068–1072, doi:10.1126/science.1128845, 2006.

Orellana, F., Verma, P., Loheide, I. I. S. P., and Daly, E.: Monitoring and modeling water-vegetation interactions in groundwater-dependent ecosystems, Rev. Geophys., 50, RG3003, doi:10.1029/2011RG000383, 2012.

Parlange, M. B. and Katul, G. G.: An advection-aridity evaporation model, Water Resour. Res., 28, 127–132, doi:10.1029/91wr02482, 1992.

Penman, H. L.: Natural evaporation from open water, bare soil and grass, P. Roy. Soc. Lond. A, 193, 120–145, 1948.

Priestley, C. and Taylor, R.: On the assessment of surface heat flux and evaporation using large-scale parameters, Mon. Weather Rev., 100, 81–92, 1972.

Reichstein, M., Bahn, M., Ciais, P., Frank, D., Mahecha, M. D., Seneviratne, S. I., Zscheischler, J., Beer, C., Buchmann, N., Frank, D. C., Papale, D., Rammig, A., Smith, P., Thonicke, K., van der Velde, M., Vicca, S., Walz, A., and Wattenbach, M.: Climate extremes and the carbon cycle, Nature, 500, 287–295, doi:10.1038/nature12350, 2013.

Saha, S., Moorthi, S., Pan, H.-L., Wu, X., Wang, J., Nadiga, S., Tripp, P., Kistler, R., Woollen, J., Behringer, D., Liu, H., Stokes, D., Grumbine, R., Gayno, G., Wang, J., Hou, Y.-T., Chuang, H.-Y., Juang, H.-M. H., Sela, J., Iredell, M., Treadon, R., Kleist, D., Van Delst, P., Keyser, D., Derber, J., Ek, M., Meng, J., Wei, H., Yang, R., Lord, S., Van Den Dool, H., Kumar, A., Wang, W., Long, C., Chelliah, M., Xue, Y., Huang, B., Schemm, J.-K., Ebisuzaki, W., Lin, R., Xie, P., Chen, M., Zhou, S., Higgins, W., Zou, C.-Z., Liu, Q., Chen, Y., Han, Y., Cucurull, L., Reynolds, R. W., Rutledge, G., and Goldberg, M.: The NCEP Climate Forecast System Reanalysis, B. Am. Meteorol. Soc., 91, 1015–1057, doi:10.1175/2010BAMS3001.2, 2010.

Sahoo, A. K., Pan, M., Troy, T. J., Vinukollu, R. K., Sheffield, J., and Wood, E. F.: Reconciling the global terrestrial water budget using satellite remote sensing, Remote Sens. Environ., 115, 1850–1865, 2011.

Savenije, H. G.: The importance of interception and why we should delete the term evapotranspiration from our vocabulary, Hydrol. Process. 18, 1507–1511, doi:10.1002/hyp.5563, 2004.

Schlesinger, W. H. and Jasechko, S.: Transpiration in the global water cycle, Agr. Forest Meteorol., 189–190, 115–117, doi:10.1016/j.agrformet.2014.01.011, 2014.

Schlosser, C. A. and Gao, X.: Assessing evapotranspiration estimates from the second Global Soil Wetness Project (GSWP-2) simulation, J. Hydrometeorol., 11, 880–897, doi:10.1175/2010JHM1203.1, 2010.

Schneider, U., Becker, A., Finger, P., Meyer-Christoffer, A., Ziese, M., and Rudolf, B.: GPCC's new land surface precipitation climatology based on quality-controlled in situ data and its role in quantifying the global water cycle, Theor. Appl. Climatol., 115, 15–40, doi:10.1007/s00704-013-0860-x, 2013.

Seneviratne, S. I., Lüthi, D., Litschi, M., and Schär, C.: Land–atmosphere coupling and climate change in Europe, Nature, 443, 205–209, 2006.

Seneviratne, S. I., Corti, T., Davin, E. L., Hirschi, M., Jaeger, E. B., Lehner, I., Orlowsky, B., and Teuling, A. J.: Investigating soil moisture–climate interactions in a changing climate: A review, Earth Sci. Rev., 99, 125–161, 2010.

Sheffield, J., Wood, E. F., and Roderick, M. L.: Little change in global drought over the past 60 years, Nature, 491, 435–438, 2012.

Shuttleworth, W. J.: Evaporation from the Amazonian rainforest, Philos. T. Roy. Soc. Lond., 233, 321–346, 1988.

Stackhouse, P. W., Gupta, S. K., Cox, S. J., Mikovitz, J. C., Zhang, T., and Chiacchio, M.: 12-year surface radiation budget dataset, GEWEX News, 14, 10–12, 2004.

Su, Z.: The Surface Energy Balance System (SEBS) for estimation of turbulent heat fluxes, Hydrol. Earth Syst. Sci., 6, 85–100, doi:10.5194/hess-6-85-2002, 2002.

Su, Z., Dorigo, W., Fernández-Prieto, D., Van Helvoirt, M., Hungershoefer, K., de Jeu, R., Parinussa, R., Timmermans, J., Roebeling, R., Schröder, M., Schulz, J., Van der Tol, C., Stammes, P., Wagner, W., Wang, L., Wang, P., and Wolters, E.: Earth observation Water Cycle Multi-Mission Observation Strategy (WACMOS), Hydrol. Earth Syst. Sci. Discuss., 7, 7899–7956, doi:10.5194/hessd-7-7899-2010, 2010.

Su, Z., Roebeling, R., Schulz, J., Holleman, I., Levizzani, V., Timmermans, W., Rott, H., Mognard-Campbell, N., De Jeu, R., and Wagner, W.: Observation of Hydrological Processes Using Remote Sensing, Treatise on Water Science, edited by: Wilderer, P., Academic Press, Oxford, 2, 351–399, 2011.

Teuling, A. J., Van Loon, A. F., Seneviratne, S. I., Lehner, I., Aubinet, M., Heinesch, B., Bernhofer, C., Grünwald, T., Prasse, H., and Spank, U.: Evapotranspiration amplifies European summer drought, Geophys. Res. Lett., 40, 2071–2075, doi:10.1002/grl.50495, 2013.

Ubarana, V.: Observations and modelling of rainfall interception at two experimental sites in Amazonia, in: Amazonian Deforestation and Climate, edited by: Gash, J. H. C., Nobre, C. A., Robert, J. M., and Victoria, R. L., John Wiley, Chichester, UK, 151–162, 1996.

van Dijk, A. I. J. M., Beck, H. E., Crosbie, R. S., De Jeu, R. A. M., Liu, Y. Y., Podger, G. M., Timbal, B., and Viney, N. R.: The Millennium Drought in southeast Australia (2001–2009): Natural and human causes and implications for water resources, ecosystems, economy, and society, Water Resour. Res., 49, 1040–1057, doi:10.1002/wrcr.20123, 2013.

van Dijk, A. I. J. M., Gash, J. H., van Gorsel, E., Blanken, P. D., Cescatti, A., Emmel, C., Gielen, B., Harman, I., Kiely, G., Merbold, L., Montagnani, L., Moors, E., Roland, M., Sottocornola, M., Varlagin, A., Williams, C. A., and Wohlfahrt, G.: Rainfall interception and the coupled surface water and energy balance, Agr. Forest Meteorol., 214–215, 402–415, doi:10.1016/j.agrformet.2015.09.006, 2015.

Vinukollu, R. K., Meynadier, R., Sheffield, J., and Wood, E. F.: Multi-model, multi-sensor estimates of global evapotranspiration: climatology, uncertainties and trends, Hydrol. Process., 25, 3993–4010, doi:10.1002/hyp.8393, 2011a.

Vinukollu, R. K., Wood, E. F., Ferguson, C. R., and Fisher, J. B.: Global estimates of evapotranspiration for climate studies using multi-sensor remote sensing data: Evaluation of three process-based approaches, Remote Sens. Environ., 115, 801–823, doi:10.1016/j.rse.2010.11.006, 2011b.

Wang, K. and Dickinson, R. E.: A review of global terrestrial evapotranspiration: Observation, modeling, climatology, and climatic variability, Rev. Geophys., 50, RG2005, doi:10.1029/2011RG000373, 2012.

Wang, L., Good, S. P., and Caylor, K. K.: Global synthesis of vegetation control on evapotranspiration partitioning, Geophys. Res. Lett., 41, 6753–6757, doi:10.1002/(ISSN)1944-8007, 2014.

Wang-Erlandsson, L., van der Ent, R. J., Gordon, L. J., and Savenije, H. H. G.: Contrasting roles of interception and transpiration in the hydrological cycle – Part 1: Temporal characteristics over land, Earth Syst. Dynam., 5, 441–469, doi:10.5194/esd-5-441-2014, 2014.

Wielicki, B. A., Barkstrom, B. R., Harrison, E. F., Lee Iii, R. B., Louis Smith, G., and Cooper, J. E.: Clouds and the Earth's Radiant Energy System (CERES): An earth observing system experiment, B. Am. Meteorol. Soc., 77, 853–868, 2000.

Wild, M. and Liepert, B.: The Earth radiation balance as driver of the global hydrological cycle, Environ. Res. Lett., 5, 025203, doi:10.1088/1748-9326/5/2/025003, 2010.

Wild, M., Folini, D., Hakuba, M. Z., Schar, C., Seneviratne, S. I., Kato, S., Rutan, D., Ammann, C., Wood, E. F., and König-Langlo, G.: The energy balance over land and oceans: an assessment based on direct observations and CMIP5 climate models, Clim. Dynam., 44, 3393–3429, doi:10.1007/s00382-014-2430-z, 2014.

Zhang, K., Kimball, J. S., Nemani, R. R., and Running, S. W.:
A continuous satellite-derived global record of land surface
evapotranspiration from 1983 to 2006, Water Resour. Res., 46,
W09522, doi:10.1029/2009WR008800, 2010.

Zhang, K., Kimball, J. S., Nemani, R. R., Running, S. W., Hong,
Y., Gourley, J. J., and Yu, Z.: Vegetation greening and climate
change promote multidecadal rises of global land evapotranspiration, Sci. Rep., 5, 15956, doi:10.1038/srep15956, 2015.

Zhang, Y., Peña-Arancibia, J. L., McVicar, T. R., Chiew, F. H. S.,
Vaze, J., Liu, C., Lu, X., Zheng, H., Wang., Y., Liu, Y. Y., Miralles, D. M., and Pan, M.: Multi-decadal trends in global terrestrial evapotranspiration and its components, Sci. Rep., 5, 19124,
doi:10.1038/srep19124, 2016.

PERMISSIONS

LIST OF CONTRIBUTORS

F. Sun, J. Yu and X. Liu
Key Laboratory of Water Cycle and Related Land Surface Processes, Institute of Geographic Sciences and Natural Resources Research, Chinese Academy of Sciences, Beijing 100101, China

C. Du
Key Laboratory of Water Cycle and Related Land Surface Processes, Institute of Geographic Sciences and Natural Resources Research, Chinese Academy of Sciences, Beijing 100101, China
University of Chinese Academy of Sciences, Beijing, 100049, China

Y. Chen
State Key Laboratory of Desert and Oasis Ecology, Xinjiang, Institute of Ecology and Geography, Chinese Academy of Sciences, Urumqi, 830011, China

Iris Manola
Meteorology and Air Quality, Department of Environmental Sciences, Wageningen University, Wageningen, the Netherlands

Hans De Moel and Jeroen C. J. H. Aerts
Institute for Environmental Studies, Vrije Universiteit (VU), Amsterdam, the Netherlands

Bart van den Hurk
Institute for Environmental Studies, Vrije Universiteit (VU), Amsterdam, the Netherlands
The Royal Netherlands Meteorological Institute (KNMI), De Bilt, the Netherlands

Yun Yang, Martha C. Anderson, Feng Gao and William P. Kustas
USDA ARS, Hydrology and Remote Sensing Laboratory, Beltsville, MD, USA

Christopher R. Hain
Marshall Space Flight Center, Earth Science Branch, Huntsville, AL, USA

Kathryn A. Semmens
Nurture Nature Center, Easton, PA, USA

Asko Noormets
Department of Forestry and Environmental Resources, North Carolina State University, Raleigh, NC, USA

Randolph H. Wynne and Valerie A. Thomas
Department of Forest Resources and Environmental Conservation, Virginia Polytechnic Institute and State University, Blacksburg, VA, USA

Ge Sun
Eastern Forest Environmental Threat Assessment Center, Southern Research Station, USDA Forest Service, Raleigh, NC, USA

Natalie C. Ceperley and Marc B. Parlange
Department of Civil Engineering, Faculty of Applied Sciences, University of British Columbia, Vancouver, British Columbia, Canada
Laboratory of Environmental Fluid Mechanics and Hydrology, School of Architecture, Civil and Environmental Engineering, Swiss Federal Institute of Technology, Lausanne, Switzerland

Theophile Mande
Laboratory of Environmental Fluid Mechanics and Hydrology, School of Architecture, Civil and Environmental Engineering, Swiss Federal Institute of Technology, Lausanne, Switzerland

Nick van de Giesen
Department of Civil Engineering and Geosciences, Delft University of Technology, Delft, the Netherlands

Scott Tyler
Department of Geological Sciences & Engineering, University of Nevada, Reno, NV, USA

Hamma Yacouba
Laboratory Hydrology and Resources in Water, International Institute for Water and Environmental Engineering (2iE), Ouagadougou, Burkina Faso

Guillaume Evin, Anne-Catherine Favre and Benoit Hingray
Univ. Grenoble Alpes, CNRS, IRD, Grenoble INP, Grenoble, France

Xing Zhou, Guang-Heng Ni, Chen Shen and Ting Sun
State Key Laboratory of Hydro-Science and Engineering, Department of Hydraulic Engineering, Tsinghua University, Beijing 100084, China

Petra Hulsman, Thom A. Bogaard and Hubert H. G. Savenije
Water Resources Section, Faculty of Civil Engineering and Geosciences, Delft University of Technology, Stevinweg 1, 2628 CN Delft, the Netherlands

Sibylle Kathrin Hassler
Karlsruhe Institute of Technology (KIT), Institute of Water and River Basin Management, Chair of Hydrology, Karlsruhe, Germany
Helmholtz Centre Potsdam, GFZ German Research Centre for Geosciences, Section Hydrology, Potsdam, Germany

Theresa Blume
Helmholtz Centre Potsdam, GFZ German Research Centre for Geosciences, Section Hydrology, Potsdam, Germany

Markus Weiler
Hydrology, Faculty of Environment and Natural Resources, University of Freiburg, Freiburg, Germany

Hong Li
Norwegian Water Resources and Energy Directorate, Oslo, Norway
University of Oslo, Norway

Chong-Yu Xu
University of Oslo, Norway

Jan Erik Haugen
Norwegian Meteorological Institute, Oslo, Norway

Elena Cristiano, Marie-claire ten Veldhuis and Nick van de Giesen
Department of Water Management, Delft University of Technology, 2600 GA, Delft, the Netherlands

A. J. Dolman
Department of Earth Sciences, VU University Amsterdam, Amsterdam, the Netherland

D. G. Miralles
Department of Earth Sciences, VU University Amsterdam, Amsterdam, the Netherlands
Laboratory of Hydrology and Water Management, Ghent University, Ghent, Belgium

B. Martens
Laboratory of Hydrology and Water Management, Ghent University, Ghent, Belgium

C. Jiménez
Estellus, Paris, France

M. Jung
Max Planck Institute for Biogeochemistry, Jena, Germany

D. Michel, S. I. Seneviratne and M. Hirschi
Institute for Atmospheric and Climate Science, ETH Zurich, Zurich, Switzerland

A. Ershadi and M. F. McCabe
Division of Biological and Environmental Sciences and Engineering, King Abdullah University of Science and Technology, Thuwal, Saudi Arabia

J. B. Fisher
Jet Propulsion Laboratory, California Institute of Technology, Pasadena, California, USA

Q. Mu
Department of Ecosystem and Conservation Sciences, University of Montana, Missoula, Montana, USA

E. F. Wood
Department of Civil and Environmental Engineering, Princeton University, Princeton, New Jersey, USA

D. Fernández-Prieto
ESRIN, European Space Agency, Frascati, Italy

Index

A
Aerodynamic Resistance, 31-32
Aridity Index, 1-3, 10-11, 13-14
Averaged Rainfall Intensity, 152

B
Budyko Framework, 1-3, 8, 11, 14, 16

C
Calibrating Models, 103
Calibration-validation Procedure, 89
Convective Available Potential Energy, 21

D
Dew-point Temperature, 17-22, 25
Digital Elevation Model, 5, 87, 122
Directional Radiometric Temperature, 31, 48
Diurnal Cycle, 51, 57-58, 61, 65, 172
Downscaling Methods, 151, 168
Dynamic Stress Factor, 171

E
Enhanced Vegetation Index, 30, 88, 97, 102
Ensemble Prediction System, 19, 27
Equilibrium Line Altitude, 144
Evaporative Fraction, 50-51, 55, 57-58, 60, 62-68, 173, 180
Evapotranspiration, 1-16, 29-30, 41, 45-49, 66-68, 91, 95, 118-119, 132-134, 141, 165, 169-170, 184-188

F
Finer Temporal Scales, 2
Flood-risk Management, 18
Flow Dynamics, 104, 115, 119
Flux Tower, 29-31, 37-39, 184
Fu-type Budyko Equation, 1, 14

G
Geographically Weighted Regression, 88, 101
Global Climate Model, 22

H
Heat Pulse Velocity, 121
Hidden Markov Model, 69-70
Hortonian Overland Flow, 106
Hydraulic Engineering, 2, 7, 87
Hydraulic Radius, 108
Hydrological Cycle, 2, 118, 157-158, 170, 187

Hydrological Models, 69, 102-104, 115, 117, 128-132, 148, 150, 158-159, 163, 166
Hydrological Response Units, 106-108, 114

I
Inverse Distance Weighted, 87

L
Land Pixel, 175-176, 181
Land Surface Temperature, 29, 46, 49, 51, 101, 184
Latent Heat Flux, 32, 51, 53, 55, 58-62, 65-67, 178-179, 181
Leaf Area Index, 30, 47, 88, 171-172
Linear Delta Transformation, 17-18, 22-23, 26

M
Mean Relative Error, 89
Model Tree Ensemble, 169, 171, 185
Moderate Resolution Imaging Spectroradiometer, 29-30, 169-170
Multi-site Precipitation Model, 71, 74, 77
Multi-source Weighted-ensemble Precipitation, 96

N
Net Radiometers, 35, 53, 58, 60
Normalized Difference Vegetation Index, 30, 55, 58, 68, 88, 102, 172
Numerical Weather Prediction, 18, 150, 155

O
Orographic Effects, 93

P
Peaks Over Threshold, 70

R
Rain Disk Cells, 70
Rainfall-runoff Model, 103-104, 106, 116, 166
Rating Curve Extrapolation, 103
Relative Humidity, 19, 27, 35, 53, 172
Remote-sensing-based Precipitation, 87
Residual Standard Error, 123-124
River Flow Velocity, 107
Root Zone Water Storage, 1, 8-10, 12-14

S
Sap Flow, 30, 47, 49, 118-124, 126-133
Sap Velocity, 118-131, 133
Scan-line Corrector, 29, 31

Sensible Heat Flux, 53, 58-62, 67

Soil Heat Flux, 32, 35, 37, 51, 53

Soil Water Storage, 2, 4-5, 7-10, 12-13, 30, 118, 180

Soil Water Stress, 172

Soil-vegetation-atmosphere, 118, 130

Solar Radiation, 5, 32, 53, 68, 97, 120, 122, 182

Spatial Rainfall Variability, 149, 160-161, 164

Spatial Tail Dependence, 69

Spatial Variability, 47, 87, 90, 94-95, 98, 105, 107, 118, 123, 126, 128-130, 132, 135, 137, 141, 149, 152, 154-155, 157-167, 184-185

Spatio-temporal Factor, 162

Spatiotemporal Resolution, 91, 99, 166

Stochastic Precipitation, 69, 86, 164

Stream Flow, 118, 167

Strickler-manning Formula, 103, 108, 110, 112, 115

Subsurface Flow, 106, 132

Surface Energy Balance, 30, 32, 67-68, 169-170, 187

Surface Energy Balance System, 169, 187

T

Thermal Infrared (TIR) Imagery, 30

Thermal View Angle, 31

Turbulent Flux, 50, 62

Two-source Energy Balance Model, 29, 48

V

Vadose Zone, 1, 3, 134

www.ingramcontent.com/pod-product-compliance
Lightning Source LLC
Chambersburg PA
CBHW050455200326
41458CB00014B/5186